Air, Water and Soil Pollution Science and Technology Series

Industrial Pollution Including Oil Spills

AIR, WATER AND SOIL SCIENCE AND TECHNOLOGY

Trends in Air Pollution Research
James, V. Livingston (Editor)
2005. ISBN: 1-59454-326-7

Agriculture and Soil Pollution: New Research
James, V. Livingston (Editor)
2005. ISBN: 1-59454-310-0

Water Pollution: New Research
A.R. Burk (Editor)
2008. ISBN: 1-59454-393-3

Air Pollution: New Research
James, V. Livingston (Editor)
2007. ISBN: 1-59454-569-3

Air Pollution Research Advances
Corin G. Bodine (Editor)
2007. ISBN: 1-60021-806-7

Marine Pollution: New Research
Tobias N. Hofer (Editor)
2008. ISBN: 978-1-60456-242-2

Complementary Approaches for Using Ecotoxicity Data in Soil Pollution Evaluation
M. D. Fernandez and J. V. Tarazona
2008. ISBN: 978-1-60692-105-0

Lake Pollution Research Progress
Franko R. Miranda and Luc M. Bernard (Editors)
2008. ISBN: 978-1-60692-106-7

Lake Pollution Research Progress
(Online Book)
Franko R. Miranda and Luc M. Bernard (Editors)
2008. ISBN: 978-1-60741-905-1

River Pollution Research Progress
Mattia N. Gallo and Marco H. Ferrari (Editors)
2009. ISBN: 978-1-60456-643-7

Heavy Metal Pollution
Samuel E. Brown and William C. Welton (Editors)
2008. ISBN: 978-1-60456-899-8

Cruise Ship Pollution
Oliver G. Krenshaw (Editor)
2009. ISBN: 978-1-60692-655-0

Industrial Pollution including Oil Spills
Harry Newbury and William De Lorne (Editors)
2009. ISBN: 978-1-60456-917-9

Air, Water and Soil Pollution Science and Technology Series

INDUSTRIAL POLLUTION INCLUDING OIL SPILLS

HARRY NEWBURY
AND
WILLIAM DE LORNE
EDITORS

Nova Science Publishers, Inc.
New York

For permission to use material from this book please contact us:
Telephone 631-231-7269; Fax 631-231-8175
Web Site: http://www.novapublishers.com

LIBRARY OF CONGRESS CATALOGING-IN-PUBLICATION DATA

p. cm.
Includes bibliographical references and index.
ISBN 978-1-60456-917-9 (hardcover)
1. Organic compounds--Environmental aspects. 2. Oil spills--Environmental aspects. 3. Factory and trade waste--Environmental aspects. I. Newbury, Harry. II. De Lorne, William.
TD196.O73153 2009
363.73--dc22

2009000971

Published by Nova Science Publishers, Inc. ✝ New York

CONTENTS

PREFACE

Global worries regarding the continuous toxic emissions increase due to the growing problems found in the environment. Some systems have emerged as alternatives to evaluate and avoid risks of exposure of human population to these substances. These systems that are used to evaluate the exposure to these specific substances have been developed and are explored in this book. Ecotoxicological techniques that can lead to risk assessment are also evaluated and shall be available to prevent pollution problems. In addition, early-life-stage tests and biomarkers are important tools for toxicity determination and pollution prevention. This book aims to contribute to biofuels risk assessment modeling by determining the toxic effects caused by the water soluble fractions (WSFs) of three different biofuels. The effect of the addition of different concentrations of crude oil on the physicochemical characteristics of soils was investigated. The adverse health effects of heavy industrial and agricultural pollution, particularly in eastern Slovakia, is explored.

Chapter 1 - Global worries regarding to the continuous toxic emissions increase due to the growing problems found in the environment. Some systems have emerged as alternatives to evaluate and avoid risks of exposure of human population to these substances. Systems to evaluate the exposure to these specific substances have been developed. The systems are named generically Chemical Ranking and Scoring, CRS and are developed permanently in countries like Germany, Canada and the United States.

Chapter 2 - Following the worldwide trend, Brazilian enterprises have been developing research to introduce into the market fuels that could be more eco-compatible. The PNPB (National Program on Biodiesel Production), introduced officially (Law nº 11.097) the biodiesel in the Brazilian Energy Matrix in 2005, making mandatory a partial substitution of diesel fossil by biodiesel, starting with a 2% mixture by 2008 and increasing to a minimum of 5%, target by 2013. Estimates of a national production of 3.8 billion L of biodiesel in 2009, will meet the need for a 3% share, but this acceleration in production may involve environmental threats, depend on how biodiesel is produced, used and traded. Based on the reduction of toxic emissions, biodiesel is considered environmentally less harmful, when compared with fossil diesel. However, ecotoxicological information on products and co-products related to biodiesel production are incomplete, and also are unknown the toxicity of the effluents generated by the productive processes of this fuel. The soluble fractions of the different types of biodiesel, and the effluents generated in the productive processes can contaminate the ground and water. Ecotoxicological techniques that lead to risk assessment shall be available to prevent pollution problems. Early-life-stage tests and biomarkers are important tools for toxicity determination and pollution prevention. This research aimed to contribute to biofuels risk assessment modeling by determining the toxic effects caused by the water soluble fractions (WSFs) of three different biodiesel fuels (B_{100}) originated by transesterifacation of castor oil (*Ricinus communis* L.), palm (dende) oil (*Elaeis guinnensis*) and waste cooking-oil (WCO) through, respectively, homogeneous NaOH catalysis, acid catalysis, and acid and basic catalysis. Different organisms (the microalga *Skeletonema costatum*, the sea-urchin *Echinometra lucunter,* and the fish *Oreochromis niloticus*), and diverse techniques were used, such as microalgae culture growth inhibition, early-life–stage-toxicity tests, and the stability of lysosomes, isolated from hepatocytes of exposed fishes, as a biomarker. The functional integrity of the lysosomal membranes was evaluated via the acid phosphatase activity, expressed in mU/mg of total proteins. The results showed significant changes in the stability of lysosomes from fishes exposed to different biodiesel-WSFs, in relation to control animals. Among the analyzed biodiesel fuels (B_{100}), the *Ricinus communis* (castor oil) biodiesel was the most toxic (5, 0 mU/mg), followed by WCO (3, 9 mU/mg) and *Elaeis guineensis* (palm oil) biodiesel (2,4 mU/mg), when compared to the control (1,6 mU/mg). Similar toxicity trends were obtained for the sea urchin ELS-test, as far as for the *S. costatum* culture growth-inhibition-test, the first providing WSF-EC_{50} values of 2,45%, 8,06% and 22,75%, while the second presented values of WSF-IC_{50} of 1.29%, 8.23% and 5.48%, respectively for biodiesel fuels from castor oil, palm (dende) oil and WCO. Chemical analyses and obtained chromatograms suggest that the relatively high toxicity of the analyzed biodiesel fuels is related to methanol, formed by a hydrolytic reversion of the transesterification reaction.

Chapter 3 - The effect of the addition of different concentrations (0% - control, 0.5%, 1%, 3% and 5%) of crude oil (Brass blend, black colour 'API 41.7 – 43.0') on the physicochemical characteristics of soils was investigated at 24h after pollution and at 18 weeks after pollution. This was to compare the soil characteristics 24h after pollution and several weeks after pollution. The pH values of the control soils and polluted soils at 24 hours after pollution ranged from 6.4 to 7.9. Exchangeable acidity ranged from 0.18 to 0.25. Calcium, magnesium, potassium and sodium values ranged from 1.99 to 6.88, 0.27 to 0.35, 0.09 to 0.24 and 0.03 to 0.06 meq $100g^{-1}$ soil respectively. Available phosphorus, ammonium and nitrate values ranged from 24 to 28 ppm, 15.7 to 19 ppm and 14 to 28ppm respectively. Organic carbon, total nitrogen and carbon/nitrogen (C/N) ratio values ranged from 1.58 to 4.23%, 0.14 to 0.217% and 8.83 to 28.6% respectively. The pH values of the control and oil-polluted soil at 18 weeks after pollution ranged from 6.3 to 7.8. Exchangeable acidity ranged from 0.15 to 0.23. Calcium, magnesium, potassium and sodium values ranged from 1.94 to 6.38 meq $100g^{-1}$ soil, 0.28 to 0.37 meq $100g^{-1}$ soil, 0.11 to 21 meq $100g^{-1}$ soil and 0.02 to 0.14 meq $100g^{-1}$ soil respectively. Available phosphorus, ammonium and nitrate values ranged from 22 to 29 ppm, 15 to 28 ppm, and 24 to 30 ppm respectively. Organic carbon, total nitrogen and carbon/nitrogen ratio values ranged from 1.60 to 3.27%, 0.08 to 0.169% and 10.9 to 23.0 respectively. The percent (%) organic carbon was lowest in control as compared with the polluted soils at 24h. This is as a result of organic carbon being added to the polluted soils through the addition of the crude oil whose major constituents are carbon and hydrogen. While there was a general decrease in % organic carbon in all the polluted soils at the after 18 weeks, there was an increase in the control soil. The % total nitrogen was higher in the control than 0.5% and in 3% polluted soils. Generally, there was a reduction in % total nitrogen in all the soils after 18 weeks. The C: N ratio was lowest in the control. However, there was an increase in the control, 0.5%, and in 1% polluted soils and a reduction in 3% and 5% polluted soils at the end of the investigation.

Chapter 4 - Particulate matter (PM) and bound chemicals like polycyclic aromatic compounds (PAHs) as well as volatile organic compounds (VOC) in air are potential mediators for adverse health effects. The presented study includes a determination of exposure to VOC, determination of PM and thereon bound PAHs as well as investigations of mutagenic, cytotoxic and immunomodulating effects of PM in three regions in La Plata (Argentina), influenced either by industrial or traffic emissions, and in a control area. La Plata was selected as an example for the influence of industrial emission since Argentina's main oil refinery is located nearby.

Following sampling using high volume collectors with cascade impactors to separate six size fractions of PM (from 10 to less than 0.49 μm) PM-absorbed compounds were hexane-extracted. Additionally, 28 different VOC (aromatics, alcanes, terpenes, and halocarbons) were collected by passive sampling and analyzed for typical fingerprints.

For an assessment of toxicity and mutagenicity, extracts of PM were tested for mutagenicity (AMES-test: *Salmonella typhimurium* TA 98/S9+) and cytotoxicity (*Tetrahymena pyriformis*). In addition, cell respiration and ATP production of *T. pyriformis* were determined as parameters for oxidative stress and immunomodulatory effects on human leukocytes were analyzed.

For risk assessment, the lifetime cancer risk (LCR) was calculated for PAHs using the concentration of benzo[a]pyrene (BaP) and BaP equivalents as well as for VOC by concentration of benzene.

Highest total PAH concentrations were found in the industrial area. The size distribution of particles demonstrates the greater relevance of smaller particles since more than 50 % of PAHs were absorbed to particles <0.49m. The highest mutagenic potency and cytotoxicity as well as immunomodulatory effects were also associated with particles <0.49 μm. These effects correlated with PAH concentrations especially of high condensed PAHs with 4 and 5 rings. Investigations of the cell respiration showed an uncoupling of the respiratory chain by extracts of fine particles.

Regarding the VOC concentration a dominance of alkanes and aromatic compounds was observed, even though a different chemical distribution was seen. In the industrial area significant higher concentrations of hexane, heptane and benzene occurred.

According to the higher concentration of benzene and BaP also the highest level for LCR was estimated for the industrial area compared to the others.

Due to the complexity of effects related to the different exposure scenarios, both, exposure analyses and an effect assessment by various bioassays allowed us to estimate the risk of industrial air pollution.

Chapter 5 - Few years after the war three large chemical factories were built in eastern Slovakia which were designed to produce artificial fibers, wood products and various large scale chemical products including polychlorinated biphenyls (PCBs). At the same time the agricultural production in this area also increased which was accompanied by nearly unlimited use of pesticides and fertilizers. Resulting from environmental negligence heavy waterborn and airborn pollution developed, but either any survey or any public or scientific information about the actual status has been strictly banned. From such reason scientific surveys on possible health effects were started as late as after the previous regime has been changed in 1990. These were based on comparing large numbers of adults and children from polluted area (POLL) with these from the area of background pollution located about 60 km upwind and upstream (BCGR). Thus, in POLL very high levels of PCBs, DDE and HCB in environmental as well as in human samples were found. Recently also the data on high level of dioxins, furans as well as selected OH-PCBs., $MeSO_2$-PCBs and $MeSO_2$-DDE in human serum were obtained. Although such levels in BCGR were much lower, the background pollution did not appear quite negligible. Medical examinations showed increased thyroid volume, increased prevalence of thyroid antibodies in blood and thyroid disorders namely of overt and subclinical autoimmune thyroiditis as well as changes in thyroid hormone and thyrotropin level. Moreover, increasing prevalence of impaired fasting glucose level and other dysglycemias was found thus possibly supporting the increasing prevalence of diabetes. In addition, the examinations of cognitive and psychomotoric functions as well as these of hearing apparatus and dental status were conducted in children. All such signs of adverse health status were significantly associated with increasing level of PCBs which were used as a marker of other organochlorines cocktail. Although the manufacturing of PCBs has been terminated in 1984, heavy environmental pollution is still persisting and several investigations are still in progress. Among others, it appears that high organochlorine levels in blood of adult women are influencing their newborns due to transplacental transmission and some adverse effects already appeared in the third generation.

Chapter 6 – This chapter, we compares environmental and technological aspects of some innovative energy recovery systems from industrial waste. Results of a research study are presented conducted on an Italian firm that produces polyethylene terephthalate (PET) supports for waterproof membranes from plastic bottles. For this firm (and all of the firms in the same industrial park), the waste represents only an undesired cost rather than a potential energy source. First, a traditional thermal waste treatment with a molecular dissociator is compared and then with a specific gasifier. All three technologies can be fed with basically every type of waste, and can produce electric and/or thermal energy. In particular, the latter two produce syngas that can be burned after depuration to produce energy. A cost-benefit analysis is then carried out to plan how the whole industrial park can use the industrial waste to produce the energy it needs, with economic and environmental benefits for all.

Chapter 7 - With the growing interest in global warming, climate change, and the influence that human activities have on our environment, there are increasing opportunities for firms to assess their carbon footprint and do something to transform behaviours and invest in the environment with concrete actions.

Moreover, an increasingly common market-based mechanism is the carbon market, which links marketplace values with the sequestration of carbon. In this chapter, some of the carbon footprint assessment uses and the market-based mechanisms (especially voluntary ones) are highlighted that are effective both for the firms and for nature, in that they allow benefits for both.

One research program is currently conducting designs a voluntary approach for firms and individuals to calculate and reduce their carbon footprints. They were analysed and some real cases that have chosen to adhere to an Italian carbon footprint assessment program that entails forestry management and sustainable development behaviour programs.

Chapter 8 - Congenital malformations remain the leading cause of infant mortality, and etiologies for the majority of these conditions are unknown. Several studies on the relation between maternal residence near hazardous waste sites and industrial facilities and risk for birth defects in offspring, show positive associations between these sites and some malformations such as chromosomal anomalies and neural tube defects. This chapter provides a systematic review of published studies on the relation between maternal residential proximity to hazardous waste sites and industrial facilities and risk of birth defects in offspring. Without exception, exposure assessment was a challenge in these studies. Frequently, investigators measured the distance between residence and industrial facilities and waste sites. Residential proximity to such sites does not necessarily equate with chemical exposures at levels that might be considered teratogenic. Newer approaches such as interpolation and dispersion models show promise for refining estimates of chemical exposures in residential areas. Any relation between living in the vicinity of chemical emissions and birth defects likely is confounded by other factors. Maternal race/ethnicity, education, and occupation have been found to be associated with living near such facilities and sites, factors that are related to risk of some birth defects. Variability of risk effect estimates by geographical region and maternal characteristics (e.g., maternal age) may also complicate interpretations of associations between residential proximity to industrial emissions and birth defects .Because of ongoing public concern that industrial pollution might be linked to birth defects, better studies are needed to examine these associations. Such studies should be population-based to assure adequate statistical power, variability of exposure, and generalizability to other populations. More precise methods of exposure assessment coupled with the careful collection and consideration of potential confounding factors will help elucidate whether ambient exposures to air toxics and other types of industrial pollution pose teratogenic risks.

Chapter 9 - The manufacturing and the application of organic dyes involve the production and the handling of many organic compounds hazardous to human health. Many of these substances are considered toxic, even carcinogenic. Over the past couple of decades, manufacturers and users of dyes have faced increasingly stringent legal regulations promulgated to safeguard human health and the environment. So, there is a clear need to treat dye wastewater prior to discharge into the primary effluent. The limitation of traditional wastewater treatment technologies (biological and physical), such as low rate, disability to degrade many of recalcitrant organic dyes and the production of secondary waste which demands further treatment, can be overcome by the utilization of advanced oxidation processes (AOPs). These wastewater treatment methods are considered as low- or even non-waste generation technologies. AOPs are based on the production of very reactive species, such as hydroxyl radicals, able to decolorize and to reduce recalcitrant colored wastewater loads due to the high oxidation power and the lack of selectivity of OH radicals towards a broad range of organic pollutants present in wastewater. Generally, AOPs can be broadly classified concerning the way of OH radicals generation into chemical, photochemical, photocatalytic, mechanical and electrical technologies. This research deals with the application of chemical and photochemical AOPs for the minimization of recalcitrant colored pollutants present in wastewaters. The comparative study of several processes, Fe^0/H_2O_2, $UV/Fe^0/H_2O_2$, UV/O_3 and $UV/O_3/H_2O_2$, was performed through an experimental research of decolorization and mineralization of a reactive azo dye C.I. Reactive Blue 137, as a model wastewater pollutant. Applied processes were optimized according to their process parameters, Fe^0 dosages, initial H_2O_2 dosages and initial pH values of treated solutions. The influence of initial organic dye concentration, as well as the addition of solid particles, synthetic zeolites, on the process effectiveness was also investigated. Studied AOPs were evaluated on the basis of their eco-effectiveness, by the means of color (A_{610}), aromaticity (UV_{280}), TOC and AOX value decrease, and their cost-effectiveness as well.

Short Communication - Contamination of soils by heavy metals from oil spills is a serious problem. The environment of the Arabian Gulf region was exposed to such factors by the 1991 oil spill and is now a focus of study. Sites with apparently different levels of pollution, namely tar mat, soot and clear sites were located at the Greater Al-Burgan oil field. Eight field soil samples were taken from each of the types along with control samples from outside the area in 2001. Ten *Acanthodactylus scutellatus* lizards (5 of each sex) were collected from each field study site, weighed, humanely killed and stored in jars in a freezer at - 20°C until analysis. All samples (both soil and lizard) were analyzed using the Inductively Coupled Plasma/Atomic Emission Spectrophotometry methodology. The mean nickel and vanadium concentrations were highest in the tar mat soils and lowest in the control samples. The mean nickel concentrations in whole body lizard tissue were high in reptiles from both the clear and tar mat sites but lowest at the control sites. Mean concentrations of vanadium were low in lizard tissues and did not significantly differ in animals taken from the different sites. The results confirm that the 1991 oil spill was still resulting in elevated levels of both heavy metals in contaminated areas by 2001. It also confirms that *Acanthodactylus scutellatus* is a useful bioindicator species in this desert environment for nickel but not vanadium.

In: Industrial Pollution Including Oil
Editor: Harry Newbury and William De Lorne
ISBN 978-1-60456-917-9

Chapter 1

TOXICITY AND POLLUTION FACTOR: A PROPOSAL TO EVALUATE THE GASEOUS REFINERY EMISSIONS

Gisela Montero

Global worries regarding to the continuous toxic emissions increase due to the growing problems found in the environment. Some systems have emerged as alternatives to evaluate and avoid risks of exposure of human population to these substances. Systems to evaluate the exposure to these specific substances have been developed. The systems are named generically Chemical Ranking and Scoring, CRS and are developed permanently in countries like Germany, Canada and the United States.

Each of these systems is aimed to measure the amount of pollution that an individual can be potentially exposed to. These systems evaluate or estimate the amount of a chemical substance that a person can be exposed to, based on the national inventory of toxic substance emissions.

Swanson (SETAC, 1997) lists several examples of CRS systems throughout the world:

a) The Wet Milieugevaarlijke Stoffen (WMS) Scoring System, from Germany assigns a classification for environmental exposition in relation to the volume used, the percentage used to the environment, degradation on air, soil and/or water, relative occurrence in these and elements and bioconcentration.
b) The George Washington University (GWU) System evaluates the exposition according to the bioconcentration, persistence and emission or volume of production.
c) The Effluent Monitoring Priority Pollutants List (EMPPL) includes environmental persistence, bioaccumulation and detection in the environment.
d) The German Beratergremium für Umweltrelevante Altstoffe (BUA) System includes bioaccumulation, persistence and volume of production.
e) The United States Environmental Protection Agency (USEPA) Toxics Release Inventory (TRI) Environmental Indicators uses trajectory data, transport and exposition models to estimate a substitute dose or the quantity of a chemical compound to which a person can be exposed to. The Chemical Hazard Evaluation for

Management Strategies (CHEMS-1) Model, uses yearly released pounds from a chemical compound as reported to TRI as a substitute measure of the potential of humans and the environment receptors.

f) The Accelerated Reduction/Elimination of Toxics (ARET) Substance Selection Process, from Canada, evaluates the bioaccumulation and persistency with the following data: environmental life span, bioconcentration factor and water octanol partition coefficient.
g) The Existing Chemicals System uses production and importes volumes of chemical compounds.
h) The Pre-biologic Screen uses the water octanol partition coefficient, Henry's constant and environmental life span to estimate the bioaccumulation, persistency and penetration capacity.

A method has been developed as a result of the systems mentioned above, Davies et al. (1994). This method is used to classify and identify chemical compounds of greater risk to human health as well as to environmental impact. As part of their study, a selection of substances was made based on the *USEPA Toxics Release Inventory* (*TRI). This* list contains the toxic compounds taking into account the chemicals that formed 99% of the total inventory of emissions in the United States. The basic algorithm used to calculate the total risk factor of a chemical compound developed by Davis et al considered the following aspects:

a) The effects on human health
b) The environmental effects
c) The exposure potential

Authors relate some aspects with Equation 1

$$TRV = (EHH + EE) \times EP \qquad \text{Eq. 1}$$

where:

TRV =	*Total Risk Value*
EHH =	*Efectts on Human Heath*
EE =	*Environmental Effects*
EP =	*Exposure Potential*

Giving each value to each of the aspects considered, the maximum values are shown in Table 1. It is important to note that the final risk values were normalized in a 0-100 scale.

Considering some elements of Davis' proposal, as well as some of the parameters used in the CSR systems, a model was developed to calculate the toxicity and pollution of substances in gases emission of the processing units in a refinery plant. For this model, it was necessary to determine the substances found in the effluent gases.

Table 1. Maximum values of variables of the model Davis et al. (1994)

VARIABLE	MAXIMUM VALUE
ESSH	20
EA	15
PE	7.5
VRT	262.5

Gaseous Emissions in a Refinery Process

In the refinery process of petroleum, several operations are needed to remove impurities, divide the components of crude and/or transform the molecular structure of the compounds obtained. The selection of such operations allows confirming the general structure of a refinery, integrated by different plants. The structures of refinery vary according to their type of crude that is being processed, as well as to the products expected to obtain.

A common outline in the refinery process is (Self et al., 2000):

- Combined distillation
- Naphtha hydrodesulfuration
- Jet fuel hydrodesulfuration
- Kerosene hydrodesulfuration
- Diesel hydrodesulfuration
- Naphtha reforming
- Isomerization
- Catalytic cracking
- Coking
- Visbraking
- Alkylation
- Hydrotreating
- Hydrocracking

The main products obtained from the process of refinery are listed:

- Asphalt
- Coke
- Diesel
- Fuel gas
- Fuel oil
- Gas oil
- Gasoline
- Jet fuel
- Kerosine

- LPG
- Lube oils
- Sulfur

Besides the products listed, other disposable products are produced in different phases. For example, in the solid phase wasted catalysts are produced; in the liquid phase, sour waters emitted from hydrodesulfuration units; finally, in the gaseous phase the flue gases from direct fired furnaces and steam generators.

When, we refer to a high energy consumption process, as the one described above, special attention should be focused in the flow of several compounds emitted that impact the environment. Because the issued compounds alter their composition and concentration, generating a phenomenon called pollution, such term is defined as "the presence in the atmosphere of one or more contaminants or their combinations, in such amounts and with certain duration that can affect human life, animals and plants, that interferes with life, or the development of activities" (Wark et al.,1998).

From this perspective, a refinery process of petroleum contaminates the environment, not only in the atmosphere but also in the hydrosphere and lithosphere since it contains gaseous, liquid and solid emissions. This paper only covers the gaseous emissions composed mainly by gases coming from combustion equipment which is mostly constituted by carbon dioxide.

Most of gaseous effluents are emissions of flue gases and hot air. However, there is another type of effluents such as incondensable gases stream, emitted by the combined distillation unit, containing a great diversity of compounds, or the sour gas streams found in exhaust systems from kerosene hydrodesulfuration and jet fuel hydrodesulfuration units, as is shown in Table 2.

Table 2. Example of gaseous effluents of process units of refinery

UNIT	GASEOUS EMISSIONS
Alkylation	Flue gases
Fludized catalytic cracking	Hot air, flue gases
Combined distillation unit	Hot air, flue gases, incondensable gases
Kerosene hydrodesulfuration	Flue gases, sour gas
Naphtha hydrodesulfuration	Flue gases
Jet fuel hydrodesulfuration	Flue gases, sour gas
Isomerization	Hot air
Naphtha reforming	Hot air, flue gases

Flue gases are constituted mainly by water, air carbon dioxide, carbon monoxide, nitrogen oxygen, sulfur oxides and nitrogen oxides. The combined distillation unit is the first plant in the sequence of the process of petroleum refining, in this unit the first process, separation of crude takes place.

This unit emitts flue gas and hot air streams; it also delivers an incondensable stream composed of a large variety of components mostly organic as seen in table 3.

Gaseous emissions from vents of jet fuel hydrodesulfuration and kerosene hydrodesulfuration units are known to contain a variety of compounds as shown in Table 4.

Table 3. Compounds of the mixture of incondensable emitted by the combined distillation unit

	SUBSTANCE	FORMULA		SUBSTANCE	FORMULA
1	Hydrogen sulfide	H_2S	27	Trans-1-pentadiene	C_5H_8
2	Water	H_2O	28	Trans 2-pentene	C_5H_{10}
3	Air		29	1,1-Dimethyl cyclopentane	C_7H_{14}
4	Sulphur	S	30	1-Heptene	C_7H_{14}
5	Benzene	C_6H_6	31	1-Hexene	C_6H_{12}
6	Butane	C_4H_{10}	32	2,2,3-Trimethyl butane	C_7H_{16}
7	Cyclohexane	C_6H_{12}	33	2,2-Dimethyl butane	C_6H_{14}
8	Ciclopentane	C_5H_8	34	2,2-Dimethyl pentane	C_7H_{16}
9	Cyclopentane	C_5H_{10}	35	2,2 Dimethyl propane	C_5H_{12}
10	Cis-1,3-pentadiene	C_5H_8	36	2,3 Dimethyl butane	C_6H_{14}
11	Cis-2-Butene	C_4H_8	37	2,3-Dimethyl pentane	C_7H_{16}
12	Cis-2-Pentene	C_5H_{10}	38	2,4-Dimethyl pentane	C_7H_{16}
13	Dimethyl cyclopentane	C_7H_{14}	39	2-Methyl heptane	C_8H_{18}
14	Ethane	C_2H_6	40	2-Methyl pentane	C_6H_{14}
15	Ethyl cyclopentane	C_7H_{14}	41	2-Methyl-1,3-butadiene	C_5H_8
16	Hexane	C_6H_{14}	42	2-Methyl-1-butene	C_5H_{10}
17	Isobutane	C_4H_{10}	43	2-Methyl-1-hexene	C_7H_{14}
18	Methane	CH_4	44	2-Methyl-1-pentene	C_6H_{12}
19	Methyl cyclohexane	C_7H_{14}	45	2-Methyl-2-butene	C_5H_{10}
20	Methyl cyclopentane	C_6H_{12}	46	2-Methyl-hexane	C_7H_{16}
21	N-heptane	C_7H_{16}	47	2-Methyl-propane	C_5H_{12}
22	Pentane	C_5H_{12}	48	3,3-Dimethyl pentane	C_7H_{16}
23	Pentene	C_5H_{10}	49	3-Ethyl pentane	C_7H_{16}
24	Propylene	C_3H_8	50	3-Methyl heptane	C_8H_{18}
25	Toluene	C_7H_8	51	3-Methyl hexane	C_7H_{16}
26	Trans-1,3-Methyl	C_7H_{14}	52	3-Methyl pentane	C_6H_{14}

Besides the substances found in effluents from combustion equipment, ejectors and vent systems listed in Tables 2 to 4, a large amount of substances is produced. These substances have been detected in the flue gases in a level of traces according to the report published by the Western States Petroleum Association, 1998 and the American Petroleum Institute.

Both institutions financed a project to develop emission factors of combustion equipment used in petroleum industry. For this the California Air Resources Board (1998) reported sources of combustion in the petroleum industry. The project included, among others, the following equipment: process furnaces, gas turbines and steam generators that used as fuel: fuel oil, natural gas, refinery gas, a combination of these two gases and others. The substances found in the effluent were: metal traces, polidibencene-p-dioxins y dibenzofurans polycyclic aromatic hydrocarbons (*PAH*); volatile organic compounds (VOC) and semivolatile organic compounds (SVOC).

Table 4. Compounds of sour gas emitted by the jet fuel hydrodesulfuration and kerosene hydrodesulfuration units

	SUBSTANCE	FORMULA		SUBSTANCE	FORMULA
1	Water	H_2S	18	Trans-1,3- Dimethyl	C_7H_{14}
2	Butane	H_2O	19	1,1-Dimethyl cyclopentane	C_7H_{14}
3	Cyclohexane	C_4H_{10}	20	2,2-Dimethyl pentane	C_7H_{16}
4	Cyclopentane	C_6H_{12}	21	2,3-Dimethyl butane	C_6H_{14}
5	Cis-1,3-Dimethyl	C_5H_{10}	22	2,3-Dimethyl pentane	C_7H_{16}
6	Ethane	C_7H_{14}	23	2,3-Trimethyl butane	C_7H_{16}
7	Ethyl cyclopentane	C_2H_6	24	2,4-Dimethyl pentane	C_7H_{16}
8	Heptane	C_7H_{14}	25	2-Methyl butane	C_5H_{12}
9	Hexane	C_7H_{16}	26	2-Methyl heptane	C_8H_{18}
10	Hydrogen	C_6H_{14}	27	2-Methyl hexane	C_7H_{16}
11	Isobutane	H_2	28	2-Methyl pentane	C_6H_{14}
12	Methane	C_4H_{10}	29	3,3-Dimethyl pentane	C_7H_{16}
13	Methyl Cyclohexane	CH_4	30	3-Ethyl pentane	C_7H_{16}
14	Methyl-ciclopentane	C_7H_{14}	31	3-Methyl heptane	C_8H_{18}
15	Pentane	C_6H_{12}	32	3-Methyl hexane	C_7H_{16}
16	Propylene	C_5H_{12}	33	3-Methyl pentane	C_6H_{14}
17	Water	C_3H_8			

The emission factors calculated are shown in Table 5 (Folwarkow, et al., 1998). This chart shows an adaptation of the original and represents an extract of the same. Also, a conversion to obtain emission factors in Kg/MJ was made. This chart only show the fuels commonly used in the refinery process such as fuel oil, refinery gas and natural gas, it also illustrates the most commonly used equipments as furnaces and steam generators.

1. Selection of Indicators that Integrate the Factor of Toxicity and Pollution

For the development of the model to calculate the toxicity factor some concepts were selected as indicators, these concepts are used by toxicologists in the exposition of human beings and animals to toxic substances as well as to magnitudes related to climate change. Each of these concepts involved is described in detail in section 2. Toxicologists state that the exposition of human beings to toxic chemical substances can produce cancer and non cancer related illnesses.

Non carcinogenic substances show a limit of concentration named risk concentration. Under this concept there are no adverse effects observable, taking into account the average responses collected in a survey on a specific population.

As to human beings, these risk limits are called a) *Chronic Inhalation Reference Concentration, RfC and b Chronic Oral Reference Concentration, RfD.* For animals, different risk values are used such as *Lethal Concentration, LC50* or *Chronic Value, ChV*. The cancer related illnesses do not show a threshold value, instead, they are classified according to the effects caused on human beings as well as on animals.

Besides the fact that some chemical compounds may cause adverse reactions on human beings and animals, some chemical compounds are concentrated in living tissue, in other words they bioconcentrate. Up to now, fish are one of the most studied species related to this behavior. Worries associated to these bioconcentration processes in low trophic levels found in fish are a result of the presence of biomagnifications phenomena as it is directed to the level of human beings. Therefore human beings found at the end of the food chain receive all accumulated toxics gradually from one trophic level into other, as an effect of the bioconcentration found in some species.

The toxic effect in substances emitted in gaseous effluents from the refinery process also affects wildlife where some of these substances reach the ground. When these substances are placed in crops, they constitute a means to be ingested by cattle where they biconcentrate, to later be transferred to humans. An example that illustrates this fact is shown in the case of dioxins and furans of which 90% of the exposition to human beings occurs by a diet based on animal products. One of the main causes of contamination is a result of the transference of the compounds in crops and bodies of water, followed by bioaccumulation of terrestrial and aquatic food chains (WHO, 1998).

Besides the impact produced in wildlife, in other words the biotic resources, the toxic compounds found in gaseous effluents can participating in chemical reactions with natural components from the atmosphere as well, affecting the abiotic resources as the destruction of the ozone layer, causing global warming effects and acid rain, or polluting the soil where some compounds remain for long periods of time.

In fact, the impact produced by some chemical substances emitted to the environment is more serious than it has been described above. It is important to mention that some areas of environmental problems were not considered in this study since the information available was not sufficient to support concrete conclusions, considering the large amount and variety of gaseous compounds emitted in a refinery process of petroleum.

Based on the information given, a toxicity and pollution factor, that represents the impact on the environment of each of the compounds found in the gaseous effluents, was developed. The toxicity factor includes the impact on biotic resources related with living species as well as on abiotic resources those non living species. The biotic resources are formed by fauna and flora, and microorganisms, while abiotic resources are formed by lithosphere, hydrosphere and atmosphere.

Table 5. Emission factors of combustion equipment, which uses petroleum derivates (adapted from Folwarkow, et al., 1998)

EMISSION FACTORS (kg/MJ) (1)						
	EQUIPMENT	STEAM GEN.	STEAM GEN.	FURNACE	FURNACE	FURNACE
	FUEL	FUEL OIL	FUEL GAS	NATURAL GAS	NAT GAS/FUEL GAS	FUEL GAS
CATEGORY	SUBSTANCE					
DIOXINS	2,3,7,8-Tetrachlorodibenzo-p-dioxin	$1.85 x 10^{-15}$				
	1,2,3,7,8-pentachlorodibenzo-p-dioxin	$1.07 x 10^{-15}$				
	1,2,3,4,7,8-hexachlorodibenzo-p-dioxin	$1.07 x 10^{-15}$				
	1,2,3,6,7,8-hexachlorodibenzo-p-dioxin	$1.07 x 10^{-15}$				
	1,2,3,7,8,9-hexachlorodibenzo-p-dioxin	$1.07 x 10^{-15}$				
	1,2,3,4,6,7,8-heptachlorodibenzo-p-dioxin	$9.11 x 10^{-15}$				
	Octachlorodibenzo-p-dioxin	$2.19 x 10^{-15}$				
FURANS	2,3,7,8 Tetrachlorodibenzofuran	$2.38 x 10^{-15}$				
	1,2,3,7,8 Pentachlorodibenzofuran	$1.34 x 10^{-15}$				
	2,3,4,7,8 Pentachlorodibenzofuran	$1.34 x 10^{-15}$				
	1,2,3,4,7,8 Hexachlorodibenzofuran	$1.07 x 10^{-15}$				
	1,2,3,6,7,8 Hexachlorodibenzofuran	$8.00 x 10^{-15}$				
	1,2,3,7,8,9 Hexachlorodibenzofuran	$1.07 x 10^{-15}$				
	2,3,4,6,7,8 Hexachlorodibenzofuran	$1.61 x 10^{-15}$				
	1,2,3,4,6,7,8 Heptachlorodibenzofuran	$4.21 x 10^{-15}$				

Table 5. (Continued)

EMISSION FACTORS (kg/MJ) (1)						
	EQUIPMENT	STEAM GEN.	STEAM GEN.	FURNACE	FURNACE	FURNACE
	FUEL	FUEL OIL	FUEL GAS	NATURAL GAS	NAT GAS/FUEL GAS	FUEL GAS
	1,2,3,4,7,8,9 Heptachlorodibenzofuran	1.35×10^{-15}				
	Octachlorodibenzo-p-furan	2.09×10^{-14}				
METALS	Antimony					2.22×10^{-10}
	Arsenic	4.20×10^{-9}	2.53×10^{-10}			3.65×10^{-10}
	Barium					2.48×10^{-10}
	Beryllium	1.84×10^{-10}	5.63×10^{-11}			1.10×10^{-10}
	Cadmium	2.50×10^{-9}	8.60×10^{-10}			4.25×10^{-10}
	Chromium (Total)	1.41×10^{-8}	4.47×10^{-9}			4.60×10^{-10}
	Copper	1.13×10^{-8}	2.25×10^{-9}			1.81×10^{-9}
	Phosphorus	0.00	0.00			2.76×10^{-9}
	Manganese	1.70×10^{-8}	8.68×10^{-10}			2.93×10^{-9}
	Mercury	3.02×10^{-11}	1.17×10^{-10}			7.74×10^{-11}
	Nickel	9.76×10^{-7}	2.03×10^{-9}			4.05×10^{-9}
	Silver					6.92×10^{-10}
	Lead	4.56×10^{-9}	8.81×10^{-10}			2.10×10^{-9}
	Selenium	8.43×10^{-9}	7.44×10^{-10}			8.43×10^{-12}
	Thallium					2.48×10^{-9}

Table 5. (Continued)

EMISSION FACTORS (kg/MJ) (1)						
	EQUIPMENT	STEAM GEN.	STEAM GEN.	FURNACE	FURNACE	FURNACE
	FUEL	FUEL OIL	FUEL GAS	NATURAL GAS	NAT GAS/FUEL GAS	FUEL GAS
	Zinc	$4.69 x10^{-8}$	$1.22 x10^{-6}$			$8.94 x10^{-9}$
PAHs	Acenaphtene	$2.58 x10^{-12}$	$2.11 x10^{-12}$	$5.85 x10^{-13}$	$1.74 x10^{-12}$	$1.01 x10^{-12}$
	Acenaphtilene	$8.99 x10^{-13}$	$9.16 x10^{-13}$	$5.07 x10^{-12}$	$1.36 x10^{-11}$	$6.66 x10^{-13}$
	Anthracene	$8.90 x10^{-13}$	$8.13 x10^{-12}$	$6.75 x10^{-13}$	$2.40 x10^{-12}$	$1.23 x10^{-12}$
	Benzo(a)anthracene	$5.67 x10^{-13}$	$6.58 x10^{-12}$	$5.93 x10^{-13}$	$2.23 x10^{-12}$	$1.38 x10^{-11}$
	Benzo(a)pyrene	$5.85 x10^{-13}$	$1.23 x10^{-12}$	$4.77 x10^{-13}$	$1.39 x10^{-12}$	$3.85 x10^{-11}$
	Benzo(b)fluoranthene	$3.49 x10^{-12}$	$2.43 x10^{-12}$	$4.77 x10^{-13}$	$6.19 x10^{-13}$	$1.74 x10^{-11}$
	Benzo(e)pyrene	$2.49 x10^{-12}$				
	Benzo(g,h,i)perylene	$2.89 x10^{-12}$	$1.38 x10^{-12}$	$5.25 x10^{-13}$	$9.54 x10^{-14}$	$5.03 x10^{-13}$
	Benzo(k)fluoranthene	$5.76 x10^{-13}$	$9.16 x10^{-13}$	$4.77 x10^{-13}$	$3.28 x10^{-13}$	$1.04 x10^{-11}$
	Chrysene	$1.13 x10^{-11}$	$1.23 x10^{-12}$	$5.85 x10^{-13}$	$1.84 x10^{-13}$	$7.01 x10^{-13}$
	Dibenzo(a,h)anthracene	$2.03 x10^{-12}$	$9.16 x10^{-13}$	$4.77 x10^{-13}$	$3.20 x10^{-14}$	$4.39 x10^{-12}$
	Phenanthrene	$3.18 x10^{-11}$	$2.02 x10^{-11}$	$1.41 x10^{-11}$	$4.77 x10^{-11}$	$6.28 x10^{-12}$
	Fluoranthene	$2.26 x10^{-11}$	$1.53 x10^{-11}$	$4.99 x10^{-12}$	$4.18 x10^{-12}$	$1.32 x10^{-12}$
	Fluorene	$1.35 x10^{-11}$	$3.52 x10^{-12}$	$1.93 x10^{-12}$	$1.50 x10^{-10}$	$4.64 x10^{-12}$
	Indeno(1,2,3-cd)pyrene	$1.31 x10^{-12}$	$9.16 x10^{-13}$	$4.77 x10^{-13}$	$1.05 x10^{-13}$	$4.43 x10^{-11}$
	Naphtalene	$1.74 x10^{-10}$	$7.39 x10^{-11}$	$9.93 x10^{-11}$	$5.33 x10^{-10}$	$1.35 x10^{-10}$

Table 5. (Continued)

EMISSION FACTORS (kg/MJ) (1)						
	EQUIPMENT	STEAM GEN.	STEAM GEN.	FURNACE	FURNACE	FURNACE
	FUEL	FUEL OIL	FUEL GAS	NATURAL GAS	NAT GAS/FUEL GAS	FUEL GAS
	Pyrene	1.17×10^{-11}	2.15×10^{-11}	2.35×10^{-12}	2.90×10^{-12}	1.22×10^{-12}
SVOC	2-Cloronaphtalene	6.62×10^{-14}				
	2-Methylnaphtalene	3.18×10^{-11}				
	Ethyl benzene			9.46×10^{-10}		1.30×10^{-8}
	Phenol		7.87×10^{-10}		3.96×10^{-10}	2.42×10^{-9}
	Perylene	3.20×10^{-13}				
VOC	1,3 Butadiene	1.80×10^{-8}				
	Acetaldehyde	3.01×10^{-9}	1.29×10^{-9}	1.93×10^{-9}	3.02×10^{-9}	6.58×10^{-9}
	Acrolein	6.36×10^{-9}		1.89×10^{-9}	4.64×10^{-10}	
	Benzene	1.35×10^{-8}	7.48×10^{-8}	9.80×10^{-10}	4.34×10^{-9}	2.78×10^{-8}
	Chloroform	1.45×10^{-8}				
	Formaldehyde	1.94×10^{-8}	5.67×10^{-9}	1.99×10^{-9}	9.42×10^{-9}	4.77×10^{-8}
	Propylene	6.41×10^{-8}		1.95×10^{-7}	2.47×10^{-9}	9.33×10^{-10}
	Toluene	1.68×10^{-8}	9.50×10^{-8}	1.35×10^{-8}	5.42×10^{-9}	4.60×10^{-8}
	Xylene (Total)	3.23×10^{-8}		7.82×10^{-9}	6.23×10^{-9}	1.60×10^{-8}

Up to now, world research regarding the impact produced by contaminants on biotic resources has been widely developed and has been focused mainly on fauna. Some of the species widely studied are human beings and fish. The information available for both species is consistent and is constantly updated and extended. Also, some studies have been done where the results of most toxic compounds are similar to the toxic compounds considered in this study.

Based on this information used to evaluate the impact on biotic resources, the following aspects were considered:

a) Risk of cancer in human beings
b) Risk of non carcinogen illnesses in human beings
c) Bioconcentration of chemical compounds in fish
d) Risk of acute toxicity risk in fish
e) Risk of chronic toxicity in fish

Nowadays, the impact produced by human activities on biotic species is commonly studied. For this reason studies aimed to identify causes of temperature increase in atmosphere, thaw of polar caps meteorological phenomena that cause disasters and illnesses associated with environmental contamination among other problems.

In the atmosphere, some tools have been developed to conduct these kind of studies based on indicators that compare the behavior of a substance with another one, related to the effect they cause on the environment. Therefore, the impact associated to the destruction of the ozone layer is evaluated through the *Ozone Potential Depletion (OPD).* To evaluate the impact associated with global warming that a substance generates, the *Warming Global Potential (WGP)* is used. In case the chemical compound remains on the ground (earth, soil) a water soil coefficient partition Kco. In the same manner, acid rain can be evaluated using the negative algorithm of the dissociation constant, pKa.

Based in the consistency found in the information given by coefficient partition values (water-soil) (Kco), the negative logarithm of dissociation constant (pKa), the indicators of Global Warming Potential (GWP), and considering, the importance of each effects produced by chemical compounds issued to environmental, to evaluate the impact over the biotic resources, the following aspects were considered:

a) adsorption in soil,
b) global warming, and
c) acid rain

The development of each of these aspects originated the following toxicity functions noted according to the corresponding risk:

a) toxicity function by cancer on human beings F1
b) toxicity function by inhalation on human beings F2
c) toxicity function by ingestion on human beings F3
d) toxicity function by bioconcentration on fish, F4
e) toxicity function by acute exposition on fish, F5
f) toxicity function by chronic exposition on fish, F6

g) toxicity function by adsorption in soil, F7
h) climatic impact by global warming, F8
i) climatic impact by acid rain, F9

These functions were the based to calculate the adimensional Toxicity and Pollution Factor, TPF, to each one of the chemical substances emitted in the effluent gases, considering the sum of each one of the mentioned functions. The general expression of TPF is summarized according to the equation 2.

$$FTC_i = a1*F1_i + a2*F2_i + a3*F3_i + a4*F4_i + a5*F5_i + a6*F6_i + a7*F7_i + a8*F8_i + a9*F9_i \quad \text{Eq. 2}$$

Next, each of these functions and its integration is developed, considering weight coefficient, ai to each of the functions.

2. Development of the Functions that Integrate the Toxicity and Pollution Factor

In the development of each of the functions considered to integrate the toxicity factor, the following aspects of each indicator were considered:

a) Antecedents
b) Information sources
c) Calculus of regression equation or representative function
d) Calculus of function values to each chemicals substance

To evaluate the impact on human beings, toxicity risks were considered for each compound, for ingestion as well as for inhalation. Evaluating each of these risks according to establish factors as Chronic Reference Dose (RfD) and Chronic Reference Concentration (RfC), each one of them defined by the toxicity effects on human beings. Because only a few chemical substances have been widely studied to obtain one or both risk factors, when the risk factors are unknown the following concepts:

a) Federal Drinking Water Standard, (LAB) (μg/L), established by the Environmental Protection Agency drinking water. It was used to evaluate the risk of toxicity by ingestion, in absence of the value of Oral Reference Dosage (RfD and
b) Permissible Exposure Limit (PEL), (mg/m^3), established by the Occupational Safety and Health Administration (OSHA) from the United States. This element was used to evaluate the risk of toxicity by inhalation, in absence of the value of the Oral Referential Dosage, RfD. PEL value is a type of Threshold Weighted Average (TWA), in 8 hours.

The development of each one of the functions that integrate the toxicity factor and contamination are described in 2.1.

2.1. Toxicity Funtions by Cancer Risk in Human Beings (F1)

Cancer can be defined, according to the American Cancer Society, ACS, as a group of illnesses characterized by the growth and uncontrollable spreading of abnormal cells. This constitutes the second cause of death, 1 of each 4 deaths in the United States.

According to the ASC, ¨Approximately one out of every two American men and one out of every three American women will have some type of cancer at some point during their lifetime¨. This institution has shown an important number of cancer cases is due to the environment to which a person is exposed, this means everything that interacts with the person, what he eats, drinks, smokes, sexual activity, radiation received, whether is natural from the sun exposure or medical; drugs, as well as socioeconomic factors that influence the susceptibility of a person. These effects play an important role in the presence of substances found in the air, water and soil (OTA, 1981).

The association between environmental factors and illnesses on human beings has been observed gradually, for example, it has been proven that some of the most important long term effects of contamination of the air on human health are morbidity and mortality due to chronic illnesses that obstruct the lungs, such as asthma, bronchitis and emphysema.

Generally, urban areas have major mortality rate by person, this is due to chronic obstructive illnesses and lung cancer than in rural areas. In a recent study conducted on residents of 51 metropolitan areas in the United States, in which contamination information was available, an increase of 10 $\mu g/m^3$ was found on the average concentration of particles with a diameter lower than 2.5 μm (PM2.5) on the air, an increase of 6% on risk mortality by cardiopulmonary and 8% on pulmonary cancer (Pope, 2002).

Tomatis et al. (1997), state that cancer can be prevented in two ways a) avoiding the introduction of cancer agents to the environment and b) eliminating or drastically reducing the exposition of cancer agents already found on the environment. To avoid cancer, it is necessary to identify each of the cancerous substances. This activity is affected by organizations such as The International Agency for Research on Cancer, The Environmental Protection Agency and The National Toxicology Program, among others. There are also other institutions involved with the monitoring of the levels of occupation exposition, being the most important the Occupational Safety and Health Administration *(OSHA)*, from the United States.

Minamoto et al. (1999), considers the idea that environmental factors affect carcinogenesis, this ideas is strongly supported not only on epidemiological studies but also experimental ones. Among some of the references used in his article, Minamoto, the case of repetitive gastric cancer, 6 times greater on Japanese than on occidental population. This number decreases strongly on Japanese that migrate to occidental countries.

Some examples of the increment of environmental factors that have been associated with the growth of the risk of cancer on human beings include chemical and mutagens such as cigarette smoke, ultraviolet rays among others, as well as the infection of certain bacterial or viral pathogens.

Even though it is difficult to identify the agent that causes a specific type of cancer, and thus be attributed to an occupational specific risk. As to the cancer risk on humans caused by chemical compounds found in fluent gaseous of a refinery process of petroleum, there are records of some types of cancer widely identified, these types of cancer are associated to specific compounds found in these effluents. Table 6 shows a summary taken from the information published by the *NTP* (2002).

Table 6. Risks of occupational cancer

COMPOUND	TYPE OF CANCER
Arsenic and some arsenic compounds	Lung, skin, escrotum, lymphatic
Benzene	Leukemia
Beryllium and some beryllium compunds	Lung
Sulfur dioxide	Respiratory system
Cadmium and some cadmium compounds	Lung, prostate
Chloroine	Lymphatic system
Cobalt	Lung
Chromium and some chromium compounds	Lung, nasal conduits
Nickel and some nickel compunds	Nasal cavity, lung, larynx
PAH: Acenaphtene, Acenaphthylene, Anthracene, Benzo(a)anthracene, Benzo(a)pyrene, Benzo(b)fluoranthene, benzo(e)pyrene, Benzo(g,h,i)perylene, Benzo(k)fluoranthene, Chrysene, Dibenzo(a,h)anthracene, 2-chloro-Naphthalene, Phenanthrene, Fluoranthene, Fluorene e indeno(1,2,3-cd)Pyrene.	Skin, lung, gastrointestinal tract
Naphthalene	Intestine, larynx.
Perylene	Esophagus, liver, small intestine, colorectal, larynx, lung, bladder.
Toluene	Esophagus, colorectal, bones.
2,3,7,8-Tetrachlorodibenzo-p-dioxin (*TCDD*)	Gastric, lung, prostate, breast.

Several international organisms work with cancer research. One of their activities is to identify the carcinogens with epidemiological studies and lab tests. Among the organisms that classify chemical compounds, mixture or expositions, by their carcinogenicity are:

a) The International Agency for Research on Cancer (IARC), independent organism from the World Health Organization, WHO, with headquarters in France,
b) The Environmental Protection Agency, EPA, US and,
c) The National Toxicology Program, US.

To conduct an evaluation developed studies are used in animals as well as in humans. Giving priority to the evidence emitted by the epidemiological studies. To perform studies on cancer exposition on humans, it is necessary to take into account the latency period than can take, in some cases, from 20 to 30 years or more, before the first cancer symptoms appear.

However, nowadays bioassays are done on animals from which results are projected where the studied substances would cause on humans. Although, these studies are based on the hypothesis that the chemical substances, that provokes cancer in the laboratory animals, will have similar effects in human beings (*NTP*, 2003.) The classification of carcinogenics developed by *EPA, IARC, and NTP* is revised below.

2.1.1. Classification of Carcinogenics, According to the Environmental Protection Agency of United States

The Environmental Protection Agency of United States *(EPA)*, uses its own classification, which consists on 5 main categories, characterize as below (Table 7).

Table 7. Classification used by *EPA* for substances, by its carcinogenicity in human beings

GROUP	RIESGO
A	Carcinogenic to humans
B	Probably carcinogenic to humans
B1	Limited evidence
B2	Inadequate evidence
C	Possible human carcinogen
D	No classifiable as to human carcinogenicity
E	Evidence of non-carcinogenicity for humans

2.1.2. Classification of Carcinogenic, according to International Agency for Research on Cancer

According to *International Agency for Research on Cancer* (*IARC*), the chemical substances may be classified in four groups, according to its capacity to cause cancer in human beings. Until January 2006, *IARC* classified 935 agents. Revision of agents is continuous, and it is update, with respect to existing evidence. In some cases, some of the substances are recategorized by inherent characteristics to some of the used groups by *IARC*. They are described briefly in Table 8.

Table 8. Summary of IARC classification

GROUP	DESCRIPTION
1	Carcinogenic to humans.
2A	Probably carcinogenic to humans
2B	Possibly carcinogenic to humans
3	Not classifiable as to carcinogenicity to humans
4	Probably not carcinogenic to humans

2.1.3. Classification of Carcinogenics, according to the National Toxicology Program, NTP, in United States

The National Toxicology Program started in 1980 with the publication of reports about carcinogenics, developing a continuous revision and updates of compounds' properties, and modifying classifications, according to the obtained results from these revisions. This classification process is conformed by nominee applications that NTP generates for the list or the elimination process of substances, agents, mixtures and exposition circumstances to agencies that represent the Executive Committee of NTP. Agencies that form part of this Committee are:

a) Agency for Toxic Substances and Disease Registry (ATSDR),
b) Consumer Product Safety Commission (CPSC)
c) Environmental Protection Agency (EPA)
d) Food and Drug Administration (FDA)
e) National Center for Toxicological Research (NCTR)
f) National Institute for Occupational Safety and Health (NIOSH)
g) Occupational Safety and Health Administration (OSHA)
h) Department of Health and Human Services (DHHS)
i) National Institutes of Health (NIH)
j) National Cancer Institute (NCI)
k) National Institute of Environmental Health Sciences (NIEHS)

Other nominees were identified from other sources like, IARC or technical reports of bioessays, subsequently, proposals are evaluated and classified, according to NTP criteria, which are described in Table 9.

Table 9. Criteria for listing an agent, substance, mixture or exposure circumstance as carcinogenic

Known to be Carcinogens (K) There is sufficient evidence of carcinogenicity from studies in humans, which indicates a causal relationship between exposure to the agent, substance, or mixture, and human cancer.
Reasonably Anticipated to be Human Carcinogens I There is limited evidence of carcinogenicity from studies in humans , which indicates that causal interpretation is credible, but that alternative explanations, such as chance, bias, or confounding factors, could not adequately be excluded, or there is sufficient evidence of carcinogenicity from studies in experimental animals, which indicates there is an increased incidence of malignant and/or a combination of malignant and benign tumors (1) in multiple species or at multiple tissue sites, or (2) by multiple routes of exposure, or (3) to an unusual degree with regard to incidence, site, or type of tumor, or age at onset, or there is less than sufficient evidence of carcinogenicity in humans or laboratory animals;

however, the agent, substance, or mixture belongs to a well-defined, structurally related class of substances whose members are listed in a previous Report on Carcinogens as either known to be a human carcinogen or reasonably anticipated to be a human carcinogen, or there is convincing relevant information that the agent acts through mechanisms indicating it would likely cause cancer in humans.

The Eleventh Report on Carcinogens of the National Toxicology Program, *NTP*, was used as, another information source, to classify compounds found in the gaseous effluents. Table 10 illustrates assigned categories, by the three organisms, *IARC, EPA, and NTP*, to each of the chemical substances identified in the gaseous effluents.

Table 10. Classification of substances according to *IARC, EPA, and NTP*

SUBSTANCE	CLASSIFICATION			SUBSTANCE	CLASSIFICATION		
	IARC	*EPA*	*NTP*		*IARC*	*EPA*	*NTP*
2,3,7,8-Tetrachlorodibenzo-p-dioxin	1	B2	K	Anthracene	3	D	
1,2,3,7,8-pentachlorodibenzo-p-	3	0		Benzo(a)anthracene	2A	B2	R
1,2,3,4,7,8-hexachlorodibenzo-p-	3	0		Benzo(a)pyrene	2A	B2	R
1,2,3,6,7,8-hexachlorodibenzo-p-	3	0		Benzo(b)fluoranthene	2B	B2	R
1,2,3,7,8,9-hexachlorodibenzo-p-	3	B2		Benzo(e)Pyrene	0	B2	
1,2,3,4,6,7,8-heptachlorodibenzo-p-	3	0		Benzo(g,h,i)perylene	3	D	
Octachlorodibenzo-p-dioxin	3	0		Benzo(k)fluoranthene	2B	B2	R
2,3,7,8 Tetrachlorodibenzofuran	3	0		Chrysene	3	B2	
1,2,3,7,8 Pentachlorodibenzofuran	3	0		Dibenzo(a,h)anthracene	2A	B2	R
2,3,4,7,8 Pentachlorodibenzofuran	3	0		Phenanthrene	3	D	
1,2,3,4,7,8 Hexachlorodibenzofuran	3	0		Fluoranthene	3	D	
1,2,3,6,7,8 Hexachlorodibenzofuran	3	0		Fluorene	3	D	
1,2,3,7,8,9 Hexachlorodibenzofuran	3	0		Indeno(1,2,3-cd)Pyrene	2B	B2	R
2,3,4,6,7,8 Hexachlorodibenzofuran	3	0		Naphthalene	2B	C	
1,2,3,4,6,7,8	3	0		Pyrene	3	D	
1,2,3,4,7,8,9	3	0		Phenol	3	D	
Octachlorodibenzo-p-furan	3	0		Perylene	3	D	
Antimony	2B	0		1,1,1 Trichloroethane	3	D	
Arsenic	1	A	K	1,3 Butadieno	2A	0	K
Barium	0	D		Acetaldehyde	2B	B2	R
Berylium	1	B1	K	Acrolein	3	C	
Cadmium	1	B1	K	Benzene	1	A	K
Chromium	1	A	K	Chloroform	2B	B2	R
Manganese	0	D		Ethylbenzene	3	0	
Mercury	3	D		Formaldehyde	2A	B1	R
Nickel	1	A	K	Heptane	0	D	

Table 10. (Continued)

Silver	0	D		Propene	3	0	
Lead	2B	B2		Toluene	3	D	
Selenium	3	D		Xylene	3	0	
Zinc	0	D		Hydrochloric acid	3	0	
Acenaphthylene	0	D		Sulfur dioxide	3	0	

As may be observe in Table 10, only 17 of the substances found in the gaseous effluents are classified in NTP. For this reason *NTP* classification was omitted, and *IARC* and *EPA* classification were used to integrate *Function of Toxicity by Risk of Cancer in Human Beings (F1)*, which is calculated above.

2.1.4. Description of Function of Toxicity by Risk of Cancer in Human Beings (F1)

With the purpose to assign a value to Toxicity of chemical substances, due to its carcinogenic characteristics, it was necessary to convert qualitative risk classification into quantitative risk classification, as Davies et al. (1994) did it once.

For this reason, validations showed in Table 5.8 were used. From these validations, a discrete function was established *F1*.

Some of the substances in Table 11 are classified with criteria of both agencies, which represent two grades. Other chemical substances just have one classification. And there are some cases, where substances are not classified, by anyone of the two agencies. For this reason, the criteria to calculate function of toxicity was the following: if the compound presents the two values of equivalent risk, the arithmetic mean of both was calculated; if only one risk value is presented, then this value represents the final value, and to conclude, if the compound does not have both of them, then the final value of *Function of Toxicity by Risk of Cancer in Human Beings* is zero.

Values of *F1* for chemicals are showed in Table 38. Because of the continuous revision of the chemicals substances' behavior, this classification must be updated. Values of *F1* only consider the effects by risk of cancer of chemicals compounds, in an individual way, and the synergic effects between them are not taken in consideration, which have been detected in some of the cases. As the complex mixtures of carcinogenic in tobacco's smoke, or gases produced by combustion of fossil fuels.

In a refinery environment, or around it, there are synergic effects that confuse the impact assessment of carcinogenics. However, OTA (1981) estimates, as a reliable number among 10 and 15% of cancer cases, related with the work place. Although, there is a confusion to associate the number and type of cancer with contaminants, this relation is more obvious. By the other hand, it is important to mention that, impact of environmental factors over cancer generation in individuals, depends on their susceptibility to these factors, and their immunological systems. Although, this doubts, toxicity by cancer was included, as one of the parameters of toxicity and pollution factor.

Table 11. Equivalent risk by carcinogenicity

CLASSIFICATION		EQUIVALENT RISK
IARC	*EPA*	
1	A	1
2A	B1	0.5
	B2	0.3
2B	C	0.1
3	D	0
4	E	0

In the following paragraph, information related with non carcinogenic effects of chemical substances over human beings is revised.

2.2. Function of Toxicity by Inhalation in Human Beings (F2)

The study of relations dose-response is part of the everyday theory and practice in Toxicology and Pharmacology. So far, the information of dose-response studies vary from well controlled and conducted studies of exposition in human beings, epidemiological studies, with a large number of individuals and well characterized expositions, support studies with some species of animals, until the lack of toxicity data in animals or human beings, where only there are available predictions based in relations of structure-activity type, in order to assess (Dourson, 1994).

The scientists establish that, in some cases, a chemical substance does not have an area of distinguishable adverse effects and no adverse effects in the dose-response curve. In such as cases, the adverse effect may, or may not occur in any dose (Dourson, 2001).

In the case of the non carcinogen substances, there are studies addressed to evaluate the risk of the non carcinogen, and non mutagenic health effects caused by the chronical exposition to chemicals. In 1986, *EPA* established the risk assessment guide for the chemical substances, which toxic effects are different than cancer and the mutagenic effects; substances referred as systemic toxics, because of its own effects on the function of some organic systems (Barnes, 1988).

For the non carcinogen substances, it is possible to identify a low limit value, in which there are no observable adverse effects. This characteristic of a threshold value, make the difference among the carcinogen substances, which usually, does not identify a limit value. With the purpose of determine a daily level of exposition to non carcinogen substances, the Environmental Protection Agency (EPA) of United States uses the concepts denominated as Chronic Reference Dose (RfD) and Chronic Reference Concentration (RfC), both are considerate as risk values.

Chronic Reference Dose (RfD) is defined as an estimate (with uncertainty spanning perhaps an order of magnitude) of a daily oral exposure for a chronic duration (up to a lifetime) to the human population (including sensitive subgroups) that is likely to be without an appreciable risk of deleterious effects during a lifetime. Chronic Reference Concentration (RfC) is an estimate (with uncertainty spanning perhaps an order of magnitude) of a continuous inhalation exposure for a chronic duration (up to a lifetime) to the human population (including sensitive subgroups) that is likely to be without an appreciable risk of deleterious effects during a lifetime. The determination of RfC is also based in the assumption that there is a limit value, in which there are some toxic effects. The Chronic Reference Concentration considers both, the toxic effects in the respiratory system, which is an entrance point to the human body, and the peripheral effects.

Usually the available data about toxicity in human beings are limited, because of that it is necessary to extrapolate among species. For that reason, the studies that are selected have species that show an adverse effect to the lowest dose registered, and have a careful control. Also, there are analyzed the cases, in which the toxicity data in some specie are clearly atypical compared among other species to if proceed to exclude them.

The units that express the two risk values are: *RfD* (mg/kg/day) and *RfC* mg/m^3). The meaning of these units, in the case of *RfD,* is the amount express in mg of the chemical substance that is assumed an adult human being, with an average weight of 70 kg may consume in a day, for a long period of time, usually his lifetime, and showing no damage. For the determination of this value it is consider that the individual drink daily 2 liters of water.

The units of *RfC*, means the amount in mg of the chemical substance of interest, for each m^3 of contaminated air that an individual, that weight 70 kg., inhalates. The average volume breathed by an adult person considered by EPA to make RfC determinations is 20 m^3/day.

The values of *RfC* and *RfD* are continuously revised, and are frequently used in risk studies, where there is a suspicion or certainty that some chemical substance, with toxic characteristics is in the environment. By the other hand, they are values, which experimental determination are guided by strict rules, and they are careful revised before they are accepted and published be used.

To evaluate the toxicity by inhalation and by ingestion, based on the *RfC* and *RfD* parameters, two functions of continuous type were calculated for every of the chemical substances found in the gaseous emissions coming from the refining process, using the values of risk *RfC* and *RfD,* as independent variables for every function.

2.2.1. Function of Toxicity by Inhalation in Human Beings, Based in the Chronic Reference Concentration (F2)

To structure every of the functions, the risk values RfC and RfD were used, and they are showed in the Table 12, they were taken from (Integrated Risk Information System) IRIS database from EPA.

Table 12. Valor de riesgo *RfC*

SUBSTANCE	*RfC* (mg/m^3)
Beryllium	0.02
Chromium	0.000008
Manganese	0.00005
Mercury	0.0003
Lead	0.0015

Table 12. (Continued)

Naphthalene	0.003
Acetaldehyde	0.009
Acrolein	0.00002
Benzene	0.03
Hexane	0.2
Toluene	0.4
Xylene	0.6
Hydrochloric acid	0.02
Hydrogen sulfide	0.001
Sulfur dioxide	0.08
Nitrogen dioxide	0.1
Carbon monoxide	10.00
Sulfur trioxide	0.157

From the RfC values, the toxicity function denominated *Function of toxicity by inhalation in human beings* (F2a). To develop it, were used the values of *Reference Concentration, RfC,* (mg/m^3), and the considerations shown in the Table 13, assigning the carbon monoxide, that shows the greatest tolerance by inhalation of the evaluated compounds, the lesser value of function; establishing chrome as the maximum value of function because of its severe toxicity.

Table 13. RfC values used to develop the adjustment of F2a

SUBSTANCE	*RfC* (mg/m^3) X	*F2* Y
Carbon monoxide	10	0.1000
Chromium	$8x10^{-6}$	1.0000

The obtained function is expressed by the equation 5.5, starting from it the values corresponding to evaluated function for each of the substances were calculated.

$$F2a = 0.2476 - 0.1476*\log_{10}(RfC) \qquad \text{Eq. 3}$$

With the purpose of characterize, by its toxicity, the greatest amount of chemical substances in this study, a complementary function of toxicity by inhalation in human beings was proposed. For that reason, it was used another risk of value by toxicity in human beings, which limits are settle in based to the inhalated quantity, such as the RfC parameter.

2.2.2 Function of Toxicity by Inhalation in Human Beings, Based in Permissible Exposure Limit Values (F2b)

According with data published by *OSHA* (2002), the disappeared: Union of Soviet Socialist Republics was the first country that established as mandatory the Occupational Exposure Limits. On 1933 published a list, 14 toxic substances, of acceptable concentrations on air.

In the United States, the state of Massachusetts was the first state that published a similar list in 1937. Before that, other professional organizations were the ones that develop the occupational health standards, as the *American Standards Association, ASA* (now The American Standards Institute) which in 1941 published a list of maximum acceptable concentrations (MACs) that jointed the opinion of *ASA* and the *American Conference of Governmental Industrial Hygienists (ACGIH)*. Some other contributions were made, until the 1970 Act of Health and Occupational Safety was announced. This was the beginning of a policy to protect the employees' health, and originated *The Occupational Safety and Health Administration*), in United States (*OSHA*, 2002).

Nowadays, *OSHA* handle the values denominated as *Permissible Exposition Limit (PEL,* mg/m^3), that are the values of air quality, that must be applicable in the work places, those values are obtained by correlation studies among the quantity of toxic substance absorbed by the body, and the health effects (*OSHA*, 2002).

The *PEL* values are revised continuously, and if it is necessary, are modified according to the results of the corresponding studies of occupational health, which are developed in the work places.

Considerating that it is accurate information, to use it as limit of risk to health for inhalation, the values of *Permissible Exposure Limit, PEL*, were used to assess the risk for inhalation, as an alternative option to the *Chronic Reference Concentration* (RfC), *RfC value*. Therefore, basing on PEL values of the substances, it was calculated as a complementary function, the Function of Toxicity by Inhalation in Human Beings, F2b, to assess the toxicity of the studied chemical substances, in those cases where there are not information about the risk values RfC. Table 14 shows the PEL values, of type *TWA* (Time Weighted Average), 8 hours, for the studied chemical substances.

Table 14. *PEL* Values of the substances found in the gaseous emissions

SUBSTANCE	PEL (TWA) OSHA ppm (mg/m^3)	SUBSTANCE	PEL (TWA) OSHA ppm (mg/m^3)
2,3,7,8-Tetrachlorodibenzo-p-dioxin	(0.0000002)	1,1,1 Trichloroethane	250 (1911)
2,3,7,8 Tetrachlorodibenzofuran	(0.001)	1,3 Butadiene	1 (2.2)
Antimony	(0.5)	Acethaldehíde	100 (180)
Arsenic	(0.01)	Acrolein	0.1 (0.25)
Barium	(0.5)	Benzene	10 (32)
Beryllium	(0.002)	Butane	800 (1900)
Cadmium	(0.005)	Cyclohexane	300 (1032)
Copper	(1)	Chloroform	50 (240)
Chromium	(1)	Ethyl benzene	100 (434)

Table 14. (Continued)

Manganese	(5)	Formaldehyde	0.75 (0.92)
Mercury	(0.1)	Heptane	500 (2050)
Nickel	(1)	Hexane	500 (1800)
Silver	(0.01)	Pentane	600 (1772)
Lead	(0.05)	Propane	1000 (1800)
Selenium	(0.2)	Toluene	200 (250)
Thallium	(0.1)	Xylene (Total)	100 (435)
Anthracene	(0.2)	Hydrochloric acid	5 (7)
Benzo(a)pyrene	(0.2)	Hydrogen sulfide	20 (28)
Benzo(b)fluoranthene	(0.2)	Ammonia	50 (35)
Chrysene	(0.2)	Sulfur dioxide	5 (13)
Dibenzo(a,h)anthracene	(0.2)	Carbon dioxide	5000 (9000)
Phenanthrene	(0.2)	Nitrogen dioxide	5 (9.4)
Fluoranthene	(0.2)	Carbon monoxide	50 (58)
Indeno(1,2,3-cd)pyrene	(0.2)	Nitric oxide	25 (34)
Naphthalene	10 (50)	Nitrous oxide	25 (45)
Pyrene	(0.2)	Sulfur trioxide	0.1 (0.2)
Phenol	5 (19)	Sulfuric acid	(0.1)
Perylene	(0.2)	Nitric acid	2 (5)

To develop the *F2b*, the *Permissible Exposition Limit (PEL), (mg/m^3)*, values were employed. They are showed in Table 15, selected among the chemical substances issued in the gaseous emissions. The maximum and minimum values of the PEL parameter are showed.

Table 15. Valores de *PEL*, utilizados para calcular *F2b*

SUBSTANCE	PEL (TWA, OSHA) (mg/m^3)	F2b
	X	Y
Carbon dioxide	9000	0.1000
2,3,7,8-Tetrachlorodibenzo-p-dioxin	$2x10^{-7}$	1.0000

The equation 4 corresponds to the obtained function by the adjustment of PEL values. By this means, the value of each substance was calculated.

$$F2b = 0.4341 - 0.0845*\log_{10}(PEL) \qquad \text{Eq.4}$$

2.2.3. Calculation of Function of Toxicity by Inhalation in Human Beings (F2)

From the analysis of results (from the two functions of toxicity by inhalation proposed) it is released that for some substances, the value of both toxicity values is known; for some others only one of the values is known, and in the worst case none of them is known. According to these results, to calculate the Function *of Toxicity by Inhalation in Human Beings, F2*, priority was given to the value of the calculated function from the risk value RfC; in the absence of this value, the value of the function was taken, calculated from PEL, and finally, in the case of RfC value, and corresponding to PEL value failed to recognize each other, the value 0 was assigned to *Function of Toxicity by inhalation in human beings.*

Decisions corresponding to F2 value are summarized in diagram showed in Figure 1, which shows the proposals of the used functions in each case. The results of F2 are showed in Table 38.

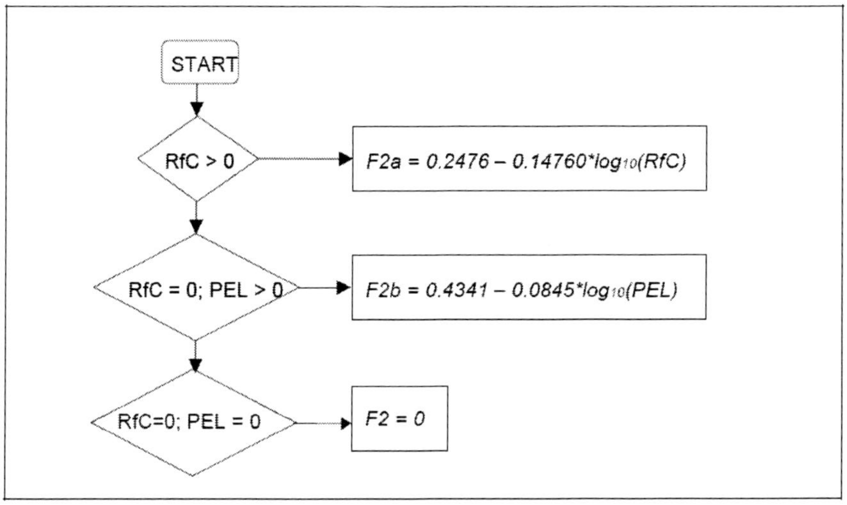

Figure 1. Equations used to calculate the Function of Toxicity by inhalation in Human Beings.

Continuing with the evaluation of the effects provoked by the gas emissions against the biotic resources, mainly human beings, it is described in detail how the functions to assess toxicity by ingestion in human beings were obtained.

2.3. Function of Toxicity by Ingestion in Human Beings (F3)

Using a similar outline to develop F2, it is proposed Function of Toxicity by Ingestion in Human Beings, F3, which is calculated from the values of Chronic Reference Dose, RfD, described in section 5.2.2. Thus, in a complementary manner is proposed the use of the risk values based en the Maximum Contaminant Level (MCL), Drinking Water Standards, which are described in section 2.3.2, in order to structure the mentioned function.

2.3.1. Function of Toxicity by Ingestion in Human Beings, Based in the Values of Oral Reference Dose (F3a)

To structure *F3a*, the risk values RfD showed in Table 16 were used, they were taken from IRIS (*Integrated Risk Information System*) database from *EPA*.

To develop *F3a* Chronic Reference Dose, *RfD,* (mg/kg/day) values were used, they correspond to compounds that observed the greatest, and the least tolerated dose. With this the limit values of function shown in Table 17 were established.

Table 16. *RfD* values reported for substances

SUBSTANCE	*RfD* (mg/kg)/día	SUBSTANCE	*RfD* (mg/kg)/día
2,3,7,8-Tetrachlorodibenzo-p-dioxin	$4x10^{-9}$	Fluoranthene	0.0400
Antimony	0.0004	Fluorene	0.0400
Arsenic	0.0003	Naphthalene	0.0200
Barium	0.0700	Pyrene	0.0300
Beryllium	0.0020	2-Chloronaphthalene	0.0800
Cadmium	0.0005	Phenol	0.3000
Chromium	0.0030	Chloroform	0.0100
Manganese	0.1400	Formaldehyde	0.2000
Nickel	0.0200	Toluene	0.2000
Silver	0.005	Xylene (Total)	2.0000
Selenium	0.0050	Hydrochloric acid	0.0000
Zinc	0.3000	Hydrogen sulfide	0.0030
Acenaphtene	0.0600	Ammonia	0.1000
Anthracene	0.3000		

The obtained function from the semi-log adjustment to RfD values is:

$$F3a = 0.1311 - 0.1035*log_{10}(RfD) \qquad Eq.\ 5$$

Because some of the evaluated substances have no RfD values, an alternative factor of risk to evaluate the toxicity by ingestion was used.

2.3.2. Function of Toxicity by Ingestion in Human Beings, Based in the Maximum Contaminant Level (F3b)

Having the same approach of having an alternative function to evaluate the limit of risk by ingestion in human beings, the Maximum Contaminant Level (MCL), Drinking Water Standards values were used, which is defined as the maximum permissible level of a contaminant in water which is delivered to any user of a public water system. The values are reported in (mg/L), thus in Table 18, are shown in (μg/L). In some cases, the federal MCL values were not available, but state levels, for this reason in Table 18, between parentheses, the state value is indicated.

Table 17. Selected compounds to establish the representative equation of *F3a*

COMPOUND	RfD (mg/kg/dia)	*F3a*
	X	Y
Xylene (Total)	2	0.1000
2,3,7,8-Tetrachlorodibenzo-p-dioxin	$4x10^{-9}$	1.0000

Table 18. Maximum Contaminant Level (*MCL*) values of substances

SUBSTANCE	*MCL (EPA)*[a] (µg/L)	SUBSTANCE	*MCL (EPA)* (µg/L)
2,3,7,8-Tetrachlorodibenzo-p-	0.00003	Benzo(g,h,i)perylene	10 (FL)
1,2,3,7,8,9-hexachlorodibenzo-	0.00025 (FL)[b]	Benzo(k)fluoranthene	4 (FL)
Antimony	6	Chrysene	0.003 (MI)
Arsenic	10	Dibenzo(a,h)anthracene	7.5 (FL)
Barium	2000	Phenanthrene	10 (FL)
Beryllium	4	Fluoranthene	280 (FL)
Cadmium	5	Fluorene	280 (FL)
Copper	1000	Indeno(1,2,3-cd)Pyrene	7.5 (FL)
Chromium	100	Naphthalene	(20)
Manganese	50	Pyrene	200 (MN)
Mercury	2	Phenol	4000
Nickel	100	1,1,1 Trichloroethane	2000
Silver	100	1,3 Butadiene	0.019
Lead	(15)	Acrolein	100 (FL)
Selenium	50	Benzene	5
Thallium	2	Ethylbenzene	700
Zinc	5000	Formaldehyde	1000
Acenaphtene	20 (FL)	Hexane	10 (FL)
Acenaphthylene	10 (FL)	Toluene	1000
Anthracene	2100 (FL)	Xylene (Total)	10000
Benzo(a)anthracene	0.003 (MI)[b]	Nitrogen dioxide	7000
Benzo(a)pyrene	0.2	Nitric oxide	10000 (FL)
Benzo(b)fluoranthene	0.2 (WI)[b]		

[a] The original values are expressed in mg/L.

[b] FL (Florida), MI (Minnesota), WI (Wisconsin).

To develop *F3b* the Maximum Contaminant Level (MCL) (µg/L) values were used. They are showed in Table 19.

Table 19. Maximum Contaminant Level (MCL) values, used to calculate F3b

COMPOUND	*MCL* (µg/L)	*F3b*
	X	Y
Nitric oxide	10000	0.1000
2,3,7,8-Tetrachlorodibenzo-p-dioxin	3x10-5	1.0000

The equation 6 was obtained for *F3b*, semi logarithmic type, by this means, an alternative calculation was developed, and a greater assessment of toxicity for those substances coming from emissions was made.

$$F3b = 0.5224 - 0.1056*\log_{10}(MCL) \quad \text{Eq. 6}$$

2.3.3. Calculation of Function of Toxicity by Ingestion in Human Beings

In the case of Function of Toxicity by Ingestion in Human Beings, F3, the function calculated, from value of risk RfD, had priority. In the lack of this value, the calculated value based in MCL was used.

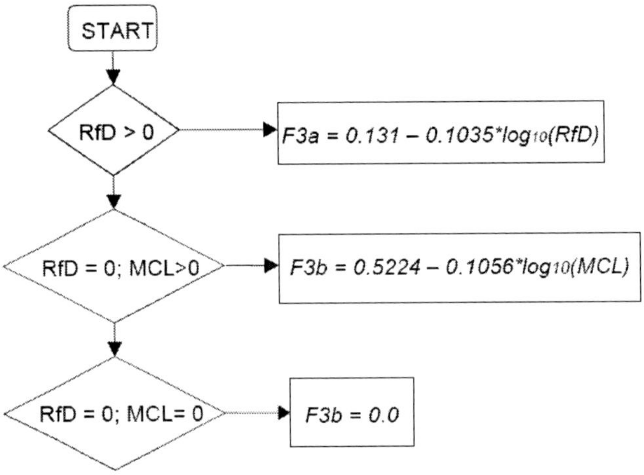

Figure 2. Equations used to calculate Function of Toxicity by Ingestion in Human Beings, F3.

The substances that did not have both of these values were assigned with zero. The diagram shown in Figure 2 summarizes the proposals of the functions mentioned above. The final values are showed in Table 38. As soon as, the functions of toxicity by cancer, by ingestion, and by inhalation in human beings were determinated, these functions assess part of the effect of the substances coming from emissions over the biotic resources. Immediately, it is going to be evaluated the behavior that these substances present on aqueous environment.

2.4. Function of Bioconcentration on Fish (F4)

In the case of compounds denominated as toxic, there are limit values of concentration, reported as safe for organisms. Therefore, if these values are exceeded, the compounds may cause damage to organisms. Because of the continuous increase of toxic emissions in the environment, it is frequent that some aquatic organisms, such as fishes, accumulate contaminants, when birds or human beings eat those fishes the contaminant goes into the next trophic level. The concern in this matter has developed parameters, which are used largely on environmental research. Like the parameter used to express correlation of a contaminant between organism, and aqueous environment denominated as BCF.

The Bioconcentration Factor, is defined as the relation of a balanced concentration of a chemical compound in the organism to the concentration of the compound in water. (Meylan et al, 1996). The BCF values may vary from one to one million (1 to 1 000 000), and usually concentrations are expressed in μg/g.

$$BCF = \frac{\text{Concentration of a chemical in an organism}}{\text{Average concentration in the water}} \qquad \text{Eq. 7}$$

The *BCF* values are obtained in an experimental manner, and if this is not possible, correlations in function of some physical or chemical attribute are used. In this matter, Meylan et al. (1996), recommends the following priority to calculate *BCF* from a correlation:

- coefficient of division octanol-water
- solubility in water
- coefficient of adsorption water-soil

Coefficient of division octanol-water is related directly with the lipophilic character of chemical substances, this characteristic is closely related with their similarity to accumulate in organisms, and it is associated with organic compounds that, generally, show greater values than, inorganic compounds.

Table 20. Aquatic organisms for which BCF values are reported

SUSTANCE	SPECIE	REFERENCE
2,3,7,8-Tetrachlorodibenzo-p-dioxin	Rainbow trout *(Salmo*	EPA-440/5-84-007, 1984.
Chromium (1)		EPA-440/5-84-029, 1984.
Arsenic	Bluegill Sunfish (*Lepomis*	EPA-440/5-84-033, 1984.
Beryllium (2)		EPA-440/5-80-024, 1980.
Cadmium (3)		EPA-822-R-01-001, 2001.
Copper		EPA-440/5-84-031, 1985.
Silver (4)		EPA-440/5-80-071, 1980.
Lead		EPA-440/5-84-027, 1984.
Acrolein		EPA-440/5-80-016, 1980.
Anthracene	*Daphnia Pulex*	Southworth et al., 1978
Benzo(a)-anthracene		Verschueren, 1996.
Mercury	Fathead Minnow (*Pimephales*	EPA-440/5-84-026, 1984.
Selenium		EPA-440/5-87-006, 1987.
Chloroform	Bluegill Sunfish (*Lepomis*	EPA-440/5-80-033, 1980.
Acenaphtene	Bluegill Sunfish (*Lepomis*	EPA-440/5-80-015, 1980.
Phenanthrene	Fathead M innow	Lyman et al., 1990.
Fluorene		
Naphtalene		
1,1,1-Trichloroethane		
Toluene		

Notas:

(1) Reported for K_2PO_4.

(2) Reported for $BeCl_2$ @ 180 mg/L of $CaCO_3$.

(3) Reported for $CdCl_2$ @ 134 g/L of $CaCO_3$.

(4) Reported for $AgNO_3$.

Some regressions are been developed to calculate the value of *Bioconcentration Factor* on fishes, from which equations 8 and 9 are the most recognized and quoted for their accurate predictions. For this reason this equations were used to determinate the missing BFC values. In Table 21 the species of fish used in the experiments, to obtain correlations, are indicated.

$$log(BCF) = 0.76*log_{10}(Kow) - 0.23 \quad Eq.\ 8$$

$$log(BCF) = 2.791 - 0.564*log_{10}(S) \quad Eq.\ 9$$

The information source *Quantitative Structure Property Relationships* (*QSAR*), is a concept based in observations, that usually is presented, as graphics or regression of the attribute of interest; as such as function of appropriated molecular descriptor, obtained from the knowledge of the molecular structure's studied substances (*SETAC,* 1997). In the following Table 22 the BCF values are reported, according to the information source.

From values of Table 22, a representative value was selected for each substance, considerating the following priority order in values, 1° in function of its experimental source; 2° calculated from *Kow*; 3° calculated from solubility; 4°. *QSAR.*

Table 21. Regressions used to determine *BCF* values

EQUATION	SPECIE	REFERENCE
$log(BCF) = 0.76*log_{10}(Kow) - 0.23$	Bluegill bunfish, bathead minnow, rainbow trout	Veith et al., 1980
$log(BCF) = 2.791 - 0.564*log_{10}(S)$	Bluegill sunfish, fathead minnow, rainbow trout, trout of creek	Lyman et al., 1990

Table 22. Reported values and calculations of *BCF*

SUBSTANCE	Source of the information			
	Experimental	Eq. 5.10	Eq. 5.11	QSAR
2,3,7,8-Tetrachlorodibenzo-p-dioxin	9270	54051	281803	
1,2,3,7,8-pentachlorodibenzo-p-dioxin	-	-	-	1.2 -16982
1,2,3,4,7,8-hexachlorodibenzo-p-dioxin	-	418794	455114	5370 -100000
1,2,3,4,6,7,8-heptachlorodibenzo-p-	-	707946	377	4.8 - 316228
Octachlorodibenzo-p-dioxin	-		6497605	-
2,3,7,8 Tetrachlorodibenzofuran	-	54051	37488	-
2,3,4,7,8 Pentachlorodibenzofuran	-	106955	68008	-
Arsenic	4	1	-	-
Beryllium	19	-	-	-
Cadmium	113	-	-	-

SUBSTANCE	Source of the information			
	Experimental	Eq. 5.10	Eq. 5.11	QSAR
Copper	1	-	-	-
Chromium	3	-	-	-
Mercury	4994	-	-	-
Silver	15	-	-	-
Lead	45	-	-	-
Selenium	52	-	-	-
Acenaphtene	387	581	4995	-
Acenaphthylene	-	581	129	-
Anthracene	917	1705	3626	-
Benzo(a)anthracene	10000	9590	8593	-
Benzo(a)pyrene	-	30875	23328	180000
Benzo(b)fluoranthene	-	46174	-	2800 -14500
Benzo(e)pyrene	-	46174	10769	150000
Benzo(g,h,i)perylene	-	139059	65006	64000
Benzo(k)fluoranthene	-	46174	35499	2.7×10^4 -
Chrysene	-	14676	21236	8.2 -100000
Dibenzo(a,h)anthracene	-	159956	44955	652 -773
Phenanthrene	2630	1676	535	-
Fluoranthene	-	4789	1532	2460-6110
Fluorene	1300	-	458	-
Indeno(1,2,3-cd)pyrene	-	131947	2966	10000
Naphthalene	80 - 430	207	91	-
Pyrene	-	30875	1912	-
2-Chloronaphthalene	-	623	154	4266
2-Methylnaphthalene	40 - 23500			190.6
Phenol	-	8	1	1.9 - 200
Perylene	-	48663	50985	-
1,1,1 Trichloroethane	9	46	-	-
1,3 Butadiene	-	19	15	7
2,3 Dimethylbutane	-	234	107	-
2,2 Dimethylpropane	-	136	86	-
2 metil 1,3 butadiene	-	41	16	-
2 methyl 1 butene	-	69	40	25
2 methyl butane	-	33	70	-
2 methyl pentane	-	410	140	-
3 methyl pentane	-	321	147	-
Acetaldehyde	-	1	0.3	3
Acrolein	344	-	-	-
Benzene	-	24	-	1-20

Table 22. (Continued)

Cyclohexane	-	242	-	89
Ciclopentane	-	112	36	41
Cis 2 pentene	-	-	31	19
Chloroform	6	19	-	-
Etilbenzene	-	95	34	-
Formaldehyde	-	1	0.43	3
Heptane	-	2049	337	-
Hexane	-	542	174	-
Hexene	-	222	68	81
Isobutane	-	74	69	-
Methane	-	-	6366	-
Methyl cyclopentane	-	193	75	210
Pentane	-	247	-	-
Propane	-	-	2119	-
Propene	-	13	31	-
Toluene	91.2	70	18	91.2
Trans 2 pentene	-	57	31	19
Xylene (Total)	-	148	-	-
Hydrogen sulfide	-	-	6	-
Ammonia	-	-	0.4	-
Sulfur dioxide	-	-	48	-
Carbon dioxide	-	-	10	-
Nitrogen dioxide	-	-	-	-
Carbon monoxide	-	-	1.6	-
Nitrous oxide	-	-	12.4	-

Franke et al., (1994) noted concerned that chemical substances bioaccumulation in living organisms preceded the apparition of adverse effects in ecosystems. For this reason, they proposed that bioaccumulation was considered, such as a risk characteristic in the chemical substance's general assessment. They structured the classification shown in Table 23, this classification grades chemical compounds according to their capacity for bioconcentrate.

Table 23. Classification of capacity for bioconcentrate

BCF	BIOCONCENTRATION CAPACITY
0 - 30	Low
30 - 100	Moderate
100 -1000	High
>1000	Very high

Having as reference this classification, a continuous function was constructed denominated as *Function of Toxicity for Bioconcentration on fish (F4),* BCF values used are showed in Table 24.

Table 24. Summary of values used, as base, to adjust with *BCF*

SUBSTANCE	BCF X	F4 Y
Copper	1	0.0001
Anthracene	≥917	1.0000

An adjustment of semi-log type was used to find the corresponding toxicity function; the expression was given by the equation 10. Because the substances with FBC values › 1000 have the capacity of high bioconcentration, all the cases, where equation 10 was used, showed values greater than unit, for this reason were taken as equal to one

$$F4 = 0.0001 + 0.3211*\log_{10}(BCF) \qquad \text{Eq. 10}$$

whereby, values of F4 were calculated (they are indicated in Table 38), which addressed to polycyclic aromatic hydrocarbon and some dioxins, as compounds with high tendency to bioconcentrate.

Talking about the aqueous environment, in the following paragraph is described the proposal and calculus of another function of toxicity on fish, considering the risk to acute exposition.

2.5. Function of Toxicity by Acute Exposition on Fish (F5)

Because of the possibility that, one or more animal species may be exposed to an elevated concentration of a substance; it is desirable to know the limits, in which any specific concentration will produce a foreseen effect in the studied specie. For this reason, there are experiments with different concentrations and species in controlled conditions

In studies done in aquatic environment, the most developed six tests; consist on the measurement of a specific concentration series Nabholz, et al., 1993):

a) *A Fish Acute Value* (usually a fish and 96-hour LC50 value)
b) *An Aquatic Invertebrate Acute value* (usually a Daphnid 48-hour LC50 value)
c) *A Green Algal Toxicity value* (usually an Algal 96-hour NEC value for biomass)
d) *A Fish Chronic value* (usually a fish 28- day chronic value (ChV))
e) *An Aquatic Invertebrate Chronic value* (usually a Daphnid 21-day ChV value)
f) *An Algal Chronic value* (usually an Algal 96-hour NEC value for biomass)

In the construction of *Function of Toxicity by Acute Exposition on fish*, the values *LC50 96-h* were used, reported for the substances. It is the fifth function developed to assess the impact of emissions over biotic resources. And this indicator was selected by the close relation among the trophic level of fish, and human beings, who are the receptors in the highest level of the chain.

In this decision, it was important the fact of, availability of substances' information. The fish acute value *LC50*, of a substance is statistically estimate of a concentration of a substance that can be expected to cause death during exposure or within a fixed time after exposure in 50% of animals exposed for a specified time.

The concentration, usually, is expressed in milligrams per liter. In the case of phenolic compounds, for which are not reported LC50 values, experimental determined, the respective LC50 values were calculated by the equation 11, based in *Structure Activity Relationships (SAR)*, according to *EPA (*1996). The *LC50 96-h* concentration for phenols is calculated:

$$Log_{10}LC50 = 0.399 - 0.616*log_{10}kow \quad \text{Eqc. 11}$$

In Table 25 the LC50 96-h values are summarized for some of the identified chemical substances from the gaseous emissions.

Table 25. Reported values of *LC50 96-h* for substances

SUBSTANCE	*LC50 96-h* fish (µg/L)	SPECIES	REFERENCE
TCDD	0.10	rainbow trout, salmo gairdnerii	EPA-440/5-84-007, 1984.
Antimony	6200.00	sheephead minnow	EPA-440/5-80-020, 1980.
Arsenic	13340.00	rainbow trout, salmo gairdnerii	EPA-440/5-84-033, 1984.
Beryllium (1)	150.00	Fathead minnow (pimephales	EPA-440/5-80-024, 1980.
Cadmium (2)	0.71	rainbow trout (salmo gairdnerii)	EPA-822-R-01-001, 2001.
Copper	114.00	fathead minnow (pimephales promelas	EPA-440/5-84-031, 1984.
Chromium	17600.00	fathead minnow (pimephales promelas)	EPA-440/5-84-029, 1984
Mercury	150.00	fathead minnow (pimephales promelas)	EPA-440/5-84-026, 1984.
Nickel	790.00	fathead minnow (pimephales promelas)	EPA-440/5-86-004, 1986.
Silver	8.50	rainbow trout (salmo gairdnerii)	EPA-440/5-80-071, 1980.
Lead	1170.00	rainbow trout (salmo gairdnerii)	EPA-440/5-84-027, 1984.
Selenium	2300.00	fathead minnow (pimephales promelas)	EPA-440/5-87-006, 1987.
Zinc	120.00	bluegill sunfish (lepomis macrochirus)	EPA-440/5-80-079, 1980.
Acenaphtene	1700.00	bluegill sunfish (lepomis macrochirus)	EPA-440/5-80-015, 1980.
Anthracene	11.90	bluegill sunfish (lepomis macrochirus)	Oris, 1984.
Benzo(a)anthrace	10.00	daphnia pulex	Verschueren, 1996.
Benzo(a)pyrene	300.00	poeciliopsis lucida	Verschueren, 1996.
Phenanthrene	234.00	bluegill sunfish (lepomis macrochirus)	Irwin et al., 1997.
Fluoranthene	3980.00	bluegill sunfish (lepomis macrochirus)	EPA-440/5-80-049, 1980.
Fluorene	820.00	rainbow trout (salmo gairdnerii)	Irwin et al., 1997.
Naphthalene	4900.00	fathead minnow (pimephales promelas)	EPA-440/5-80-059, 1980.
2-	1300.00	cancer magister	Caldwell et al., 1977.
Phenol	28904.00	rainbow trout (salmo gairdnerii)	EPA-SAR, 1996.

SUBSTANCE	*LC50 96-h* fish (μg/L)	SPECIES	REFERENCE
1,1,1	70150.00	sheepshead minnow.	EPA-SAR, 1996.
Acetaldehyde	37200.00	fathead minnow (pimephales promelas)	EPA-SAR,1996.
Acrolein	90.00	bluegill sunfish (lepomis macrochirus)	EPA-440/580-016, 1986.
Benzene	5300.00	rainbow trout (salmo gairdnerii)	EPA-440/580-018, 1980.
Cyclohexane	93000.00	fathead minnow (pimephales promelas)	Verschueren, 1983.
Chloroform	43800.00	rainbow trout (salmo gairdnerii)	EPA-440/580-033, 1980.
Ethylbenzene	32000.00	bluegill sunfish (lepomis macrochirus)	EPA-440/580-048, 1980.
Formaldehyde	24100.00	fathead minnow (pimephales promelas)	Geiger et al., 1990.
Toluene	34000.00	fathead minnow (pimephales promelas)	EPA-440/580-075, 1980.
Xylene (Total)	13500.00	rainbow trout (salmo gairdnerii)	Verschueren, 1983.
Hydrochloric acid	24000.00	bluegill sunfish (lepomis macrochirus)	NLM, 2002.
Ammonia	11230.00	rainbow trout (salmo gairdnerii)	EPA-822/R-98-008, 1998.
Nitric acid	90000.00	Crab of sea	EPA, 1973.

Notes:

(1) Reported for BeSO4 @ 20 mg/L of CaCO3.

(2) Reported for CaCl2 @ 30.2 mg/L of CaCO3, pH=7.5 and 8°C.

The fish acute values, (LC50), (μg/L) are showed in Table 26, they were established, as the upper and lower limit values, to develop *F5*.

Table 26. Limit values of *96-h LC50*, used to calculate *F5*

COMPOUND	96 h LC50 (μg/L)	F5
	X	Y
Cyclohexane	93000	0.0100
2,3,7,8-Tetrachlorodibenzo-p-dioxin	0.10	1.0000

Equation 12 corresponds to function *F5*, obtained by the adjustment of *LC50-96h* values, and the results of which are listed in Table 38.

$$F5 = 0.8325 - 0.1675 * \log_{10}(LC50) \quad \text{Eq. 12}$$

In the next paragraph the function, derived from chronic exposition values (ChV) of fish to chemical substances, is described.

2.6. Function of Toxicity by Chronic Exposition on Fish (F6)

The value of chronic exposition (ChV) on fish is one of the commonly determinated values to evaluate the environmental behavior of substances. A great amount of chemical substances, issued in gaseous emissions from refining units stayed for a long time in the environment before they observe any change. For this reason, it is possible to assume a condition of chronic exposition. Therefore, *Function of toxicity by Chronic Exposition on fish (F6)* is proponed, to evaluate the environmental behavior of substances. In Table 27 the ChV reported values are summarized, as well as, the species used in the experimental determination.

Table 27. Reported values of *ChV* for substances

SUBSTANCE	ChV-*Peces* (μg/L)	SPECIES	REFERENCE
TCDD	0.01	Salmon	EPA-440/5-84-007, 1984
Antimony	1600.00	Sheephead minnow	EPA-440/5-80-020, 1980
Arsenic	891.60	Rainbow trout, Salmo gairdnerii	EPA-440/5-84-033, 1984
Beryllium	19.00	Fathead minnow, Pimephales promelas	EPA-440/5-80-024, 1980
Cadmium (1)	2.05	Trout of creek, Salvelinus fontinalis	EPA-822-R-01-001, 2001
Copper (2)	3.87	Trout of creek, Salvelinus fontinalis	EPA-440/5-84-031, 1984
Chromium	265.00	Fathead minnow, Pimephales promelas	EPA-440/5-80-035, 1980
Mercury	0.26	Fathead minnow, Pimephales promelas	EPA-440/5-84-026, 1984
Nickel	160.00	Aquatic life (2)	EPA-SAR, 1996
Silver	0.49	Trout of creek, Salvelinus fontinalis	EPA-440/5-80-071, 1980
Lead	83.00	Trout of creek, Salvelinus fontinalis	EPA-440/5-84-027, 1984
Selenium	565.50	Fathead minnow, Pimephales promelas	EPA-440/5-87-006, 1987
Zinc	110.00	Aquatic life (3)	EPA-SAR, 1996
Acenaphtene	710.00	Bluegill sunfish, Lepomis macrochirus	EPA-440/5-80-015, 1980
Phenanthrene	103.95	Fish (3)	Eq. 14
Fluoranthene	35.46	Fish (3)	Eq. 14
Fluorene	213.84	Fish (3)	Eq. 14
Naphthalene	620.00	Fathead minnow, Pimephales promelas	EPA-440/5-80-059, 1980
2-	327.11	Fish (3)	Eq. 14
Phenol	2560.00	Rainbow trout, Salmo gairdnerii	EPA-440/5-80-066, 1980
1,1,1	4774.42	Fish (3)	EPA-SAR, 1996
Acetaldehyde	202.12	Fish (3)	Eq. 13
Acrolein	21.00	Fathead minnow, Pimephales promelas	EPA-440/5-80-016, 1980
Chloroform	1240.00	Rainbow trout, Salmo gairdnerii	EPA-440/5-80-033, 1980
Ethylbenzene	440.00	Bluegill sunfish, Lepomis macrochirus	EPA-440/5-80-048, 1980
Formaldehyde	89.54	Fish (3)	Eq. 13
Toluene	2038.68	Fish (3)	Eq. 14
Ammonia	11.78	Rainbow trout, Salmo gairdnerii	EPA-822/R-98-008, 1998

Nota:

(1) Reported for $CdCl_2$ @ 44 mg/L of $CaCO_3$.

(2) Reported for $CuSO_4$ @ 37.5 mg/L of $CaCO_3$.

(3) Not specified specie.

Some of the *ChV* values reported in Table 28 were calculated by *Structure Activity Relationships (SAR)*, according with equations 13 for aldehydes, and 14, for neutral organic compounds (*EPA*, 1996).

Prediction of values *ChV (mg/L)*, for aldehydes.

$$log_{10}ChV = -0.81 - 0.68*log_{10}kow \quad \text{Eq. 13}$$

Prediction of values *ChV (mM/L)*, for neutral organic compounds.

$$log_{10}ChV = 0.72 - 0.87*log_{10}kow \quad \text{Eq. 14}$$

To develop *F6*, the values of *Chronic Value, (ChV, mg/m^3)* were used, which correspond to the compounds showed in Table 28.

Equation 15 shows the obtained function by the adjustment of ChV values, whereby the values of the corresponding function were calculated.

$$F6 = 0.6479 - 0.1761*log_{10}(ChV) \quad \text{Eq. 15}$$

This constitutes the last of the proposed functions to evaluate the impact of the studied substances over the biotic resources. Immediately after, the functions related with the impact over the biotic resources, these functions are identified as functions of: *Toxicity by adsorption in soil, Climatic Impact by Global Warming and Environmental Impact by Acid Rain.*

Table 28. Limit values of *ChV*, used to calculate *F6*

COMPOUND	ChV (□g/L)	F6
	X	Y
1,1,1 Trichloroethane	4774.42	0.0001
2,3,7,8-Tetrachlorodibenzo-p-dioxin	0.01	1.0000

2.7. Function of Toxicity by Adsorption in Soil (F7)

One of the process that is developed constantly in the nature, it is the dragging of chemical compounds by rain or snow, which deposits the compounds in soil or sediment. In this place, the phenomenon of adsorption may occur. This process is characterized by the attachment, on surface, of the compound of interest; in this case soil.

About this process, Piwoni (1990) points out some of the more important characteristics that affect the chemicals substances' adsorption on soil:

- Mineralogy
- Permeability/porosity
- Texture
- Homogeneity
- Content of organic carbon

- Superficial load
- Superficial area

Talking about adsorption there are two different types: physical and chemical. The physical adsorption is caused by intermolecular attraction forces between the solid and the compound that it is adsorbed. The chemical adsorbing is given by chemical interaction between a solid and the substance adsorbed, which gives place to chemical bond. In this specific case, a physical adsorption is referred.

In the adsorption process, the simplest relation between a solid phase and the concentration of a fluid phase is the linear isotherm. In the case of adsorption of a compound in soil, it may be consider as adsorption in a heterogeneous surface. For this procedure is recommended to use Freundlich's model of isotherm.

When a balance, in the process of adsorption, is reached; the relation of concentrations in aqueous and solid phases (soil) of the compound of interest may be established by Equation 16 that expresses the isotherm of Freundlich:

$$q = KC^{1/n} \qquad \text{Eq. 16}$$

where:

q = Equilibrium concentration of solid phase (mg adsorbed/g of soil)
C = Equilibrium concentration of solid liquid (mg adsorbed/L of water)
K = Freundlich □dsorption constant (mg adsorbed/g of soil)/(mg adsorbed/L of water)$^{1/n}$
1/n =Freundlich parameter, adimensional

In general, the concentration of contaminant compounds in soils and sediments is low, in this way they may be consider diluted Systems, for which n tends to value 1, simplifying the expression of Freundlich's isotherm.

The soil in its natural state, it is mainly compounded of sand, mud, clay, water, and a variable quantity of organic carbon. Because of the fact, that soil's composition is very variable, determination of each constant of adsorption would be referred to a specific composition. Trying to avoid this limitation, the reported constants are normalized, with regard to concentration of organic matter, which includes: carbon, nitrogen, sulfur, phosphorus, among other organic compounds contained in soil, or normalized to organic carbon concentration.

Some typical contents of organic carbon are (Mihelcic, 1999)

suelo o sedimento:	$Xco = 1 - 5\ \%$	$(Xco = 0.01 - 0.05)$
acuífero arena-grava:	$Xco \leq 0.1\%$	$(Xco \leq 0.001)$

The normalization, with carbon, is made for soils with a concentration of organic carbon greater than 1%, in weight and balance of low concentrations of sorbate, smaller than 10^{-5} Molar or 0.5 times the solubility of sorbate in water (Mihelcic, 1999).

With this consideration the coefficient of partition soil-water is obtained, normalized with respect to the organic carbon, *Kco,* expressed by the Equation 17. In environmental area, this is one of the partition coefficients, studied with great interest by researchers, and the researching development continues, in this matter, for different chemical substances, which may have some toxic characteristic.

$$Kco = \frac{K}{Xco} \qquad \text{Eq. 17}$$

where:

Kco =water-soil partition coefficient, cm^3/g organic carbon
K = □dsorption constant
Xco =weight fraction of carbon in soil

Hamaker (1975) and McCall et al. (1980) demonstrated that a relation exists between the time that a chemical substance takes to move throughout a column, filled with soil, and the Kco value of the substance. From their observations, McCall et al. (1980) obtained data about relation among retention times of some substances and their Kco.

From McCall and colleagues, Swann et al. (1983), they proposed a general classification to characterize the behavior of chemical compounds by mobility in soil, relating intervals of their retention time, and their Kco, with a typical and expected mobility specific characteristic. Table 29 shows this classification.

The chemical compounds, with greater time of permanence in soil, observe more difficulty to be removed, in a natural or artificial way, which disables soil to any productive activity. Thus, chemical compounds have a greater potential to provoke damage in the environment, they can stay as isolated compounds or they can react with other compounds lodged in soil.

Table 29. Classification of mobility proposed by Swann et al. (1983)

RETENTION TIME*	K_{co}	CLASSIFICATION OF MOBILITY
0.0 – 1.0	0 – 50	Very high
1.0 – 1.8	50 – 150	High
1.8 – 2.1	150 – 500	Average
2.1 – 4.5	500 – 2000	Low
4.5 – 7.0	2000 – 5000	Sligth
> 7.0	> 5000	Immobile

Note: * Relative to 2.4-D (retention time of compound/retention time to 2,4-Dichlorophenoxy acetic acid).

Some of the Kco values are experimental results, reported in the corresponding sources; some others were calculated from equations of regression, recommended for the type of compound, as such as the compounds listed in Table 30, where equation 18 was used to calculate the corresponding partition coefficient.

$$log_{10}Kco = 0.544 * log_{10}Kow + 1.377$$ Eq.18

where Kow = partition coefficient water-octanol

Table 30. Compounds for which, *Kco* was calculated with Equation 18

COMPOUND
2,3 Dimethylbutane
2,2 Dimethylpropane
2 methyl 1,3 butadiene
2 methy butane
2 methyl pentane
Heptane
Hexane
Hexene
Methyl cyclopentane

Table 31 lists the values of partition coefficients soil-water, for the substances found in gaseous emissions. In the cases, where intervals are reported, the arithmetic mean of logarithms of both values were obtained.

Table 31. Values of partition coefficient, *Kco*, for substances found in the gaseous emissions

SUBSTANCE	*Kco*	REFERENCE
2,3,7,8-Tetrachlorodibenzo-p-dioxin	2630268	EPA, 1998
1,2,3,4,7,8-hexachlorodibenzo-p-dioxin	37153523	EPA, 1998
1,2,3,4,6,7,8-heptachlorodibenzo-p-dioxin	72443596	EPA, 1998
2,3,4,7,8 Pentachlorodibenzofuran	6309573.4	EPA, 1998
Antimony	2.4	Ambrose, 1999
Arsenic	3.4	Ambrose, 1999
Barium	0.7 – 3.4	Ambrose, 1999
Beryllium	3.1	Ambrose, 1999
Cadmium	2.9	Ambrose, 1999
Copper	2.7	Ambrose, 1999
Chromium	1 – 4.8	Ambrose, 1999
Mercury	3.8	Ambrose, 1999
Nickel	3.1	Ambrose, 1999
Silver	2.6	Ambrose, 1999
Lead	4.2	Ambrose, 1999
Selenium	2.1	Ambrose, 1999
Thallium	3.2	Ambrose, 1999
Zinc	3.1	Ambrose, 1999
Acenaphtene	6310 – 251189	Mackay et al., 1992

SUBSTANCE	*Kco*	REFERENCE
Acenaphthylene	3981 – 6310	Mackay et al., 1992
Anthracene	15849 – 7943282	Mackay et al., 1992
Benzo(a)anthracene	31623 – 5011872	Mackay et al., 1992
Benzo(a)pyrene	1995262 – 3.16×10^8	Mackay et al., 1992
Benzo(b)fluoranthene	1290000 – 5000000	Luers, 1996 – Maruya
Benzo(e)pyrene	15848 932 – 199526232	Mackay et al., 1992
Benzo(g,h,i)perylene	1584893 – 1584893192	Mackay et al., 1992
Benzo(k)fluoranthene	790000	Meylan et al., 1992
Chrysene	398000	EPA, 1998
Dibenzo(a,h)anthracene	630957 – 3.16×10^8	Mackay et al., 1992
Phenanthrene	1289 – 5011872	Mackay et al., 1992
Fluoranthene	39811 – 5011872	Mackay et al., 1992
Fluorene	12590-316228	Mackay et al., 1992
Indeno(1,2,3-cd)pyrene	3981072 – 1.585×10^9	Mackay et al., 1992
Naphthalene	398.11 – 100000	Mackay et al., 1992
Pyrene	67609-83177	Karickhoff, 1981
2-Chloronaphthalene	1130	Lyman et al., 1990
2-Methylnaphthalene	8500	Lyman et al., 1990
Phenol	1 – 3	Jones, 1960
Perylene	35000	Hansch et al., 1995
1,3-Butadiene	288	Hansch et al., 1995
2,3 Dimethyl butane	1700	Lyman et al., 1990
2,2 Dimethyl propane	1200	Lyman et al., 1990
2-methyl 1,3-butadiene	490	Lyman et al., 1990
2-methyl 1-butene	68	Meylan et al., 1992
2 methyl 2 butene	68	Meylan et al., 1992
2-methyl butane	424 – 519	Lyman et al., 1990
2-methyl pentane	1022	Lyman et al., 1990
3-methyl pentane	130	Meylan et al., 1992
Acetaldehyde	1	Meylan et al., 1992
Acrolein	3	Meylan et al., 1992
Benzene	85	Hodson y Williams,
Butane	450 – 900	Lyman et al., 1982
Cyclohexane	160	Meylan et al., 1992W
Cyclopentane	90	Meylan et al., 1992
Cis 2-pentene	81	Meylan et al., 1992
Chloroform	153 – 196	Lyman et al., 1990
Ethane	230 – 460	Lyman et al., 1982
Ethyl benzene	520	Meylan et al., 1992
Formaldehyde	37	Hansch et al., 1995

Table 31. (Continued)

SUBSTANCE	*Kco*	REFERENCE
Heptane	2400 - 8200	Lyman et al., 1990
Hexane	1250 - 4100	Lyman et al., 1990
Hexene	1664	Lyman et al., 1990
Isobutane	35	Meylan et al., 1992
Methane	753	Lyman et al., 1982
Methyl cyclopentane	1600	Lyman et al., 1990
Pentane	580 -1600	Lyman et al., 1982.
Propylene	219 - 237	Lyman et al., 1990
Toluene	166	Kan et al., 1998
Trans 2 pentene	81	Meylan et al., 1992

2.7.1. Description of Toxicity by Adsorption in Soil Function (F7)

A continuous function related with permanence of compounds in soil was calculated from values of partition coefficient Kco, for this reason, the discrete values scale proposed by Swann et al., (1980) was used. The function of toxicity was called Function of Toxicity by attachment to soil (*F7*).

The upper and lower limit values were established based on information showed in Table 32, where Acetaldehyde, with $Kco = 1$, was taken as the lower limit, corresponding to a high mobility characteristic, that means a minimum attachment to soil. In the opposite, Acenaphthylene with $Kco = 5011.90$, was selected, which is closer to the value proposed by Swann et al., for compounds with a maximum attachment to soil.

In the case, of substances with a *Kco* greater than the proposed upper limit, the maximum value of function F7 was established as 1.0, considerating that all substances have the common characteristic of high attachment to soil.

Table 32. Reference values to calculate *F7*

COMPOUND	*Kco*	*F7*
Acetaldehyde	1.00	0.0001
Acenaphthylene	5011.90	1.0000

Based in the reference values, Equation 19 was obtained, which represents the data adjustment of *Kco* with *F8*. The calculated values are showed in Table 38.

$$F8 = 0.001 + 0.27*\log_{10}(Kco) \qquad \text{Eq. 19}$$

Table 33. Mobility of substances

SUSTANCE	MOBILITY	SUSTANCE	MOBILITY
2,3,7,8-Tetrachlorodibenzo-p-dioxin	Immobile	2-Methylnaphthalene	Immobile
1,2,3,7,8-pentachlorodibenzo-p-dioxin	-	Phenol	Immobile
1,2,3,4,7,8-hexachlorodibenzo-p-dioxin	Immobile	Perylene	Immobile
1,2,3,6,7,8-hexachlorodibenzo-p-dioxin	-	1,1,1-Trichloroethane	-
1,2,3,7,8,9-hexachlorodibenzo-p-dioxin	-	1,3-Butadiene	Average
1,2,3,4,6,7,8-heptachlorodibenzo-p-dioxin	Immobile	2,2-Dimethyl butane	-
Octachlorodibenzo-p-dioxin	-	2,3 Dimethyl butane	Low
2,3,7,8 Tetrachlorodibenzofuran	-	2,2 Dimethyl propane	Low
1,2,3,7,8 Pentachlorodibenzofuran	-	2-methyl 1,3-butadiene	Average
2,3,4,7,8 Pentachlorodibenzofuran	Immobile	2-methyl 1-butene	High
1,2,3,4,7,8 Hexachlorodibenzofuran	-	2-methyl 2-butene	High
1,2,3,6,7,8 Hexachlorodibenzofuran	-	2-methyl butane	Average
1,2,3,7,8,9 Hexachlorodibenzofuran	-	2-methyl pentane	Low
2,3,4,6,7,8 Hexachlorodibenzofuran	-	3-methyl pentane	High
1,2,3,4,6,7,8 Heptachlorodibenzofuran	-	Acetaldehyde	Very high
1,2,3,4,7,8,9 Heptachlorodibenzofuran	-	Acrolein	Very high
Octachlorodibenzo-p-furan	-	Benzene	High
Antimony	Very high	Butane	Low
Arsenic	Very high	Cyclohexane	Average
Barium	Very high	Cyclopentane	High
Beryllium	Very high	Cis 2-pentene	High
Cadmium	Very high	Chloroform	Average
Copper	Very high	Ethane	Average
Chromium	Very high	Ethyl benzene	Low
Manganese	-	Formaldehyde	Very high
Mercury	Very high	Heptane	Slight
Nickel	Very high	Hexane	Slight
Silver	Very high	Hexene	Low
Lead	Very high	Isobutane	Very high
Selenium	Very high	Methane	Low
Thallium	Very high	Methyl cyclopentane	Low
Zinc	Very high	Pentane	Low
Acenaphtene	Immobile	Propane	-
Acenaphthylene	Immobile	Propylene	Average
Anthracene	Immobile	Toluene	Average
Benzo(a)anthracene	Immobile	Trans 2-pentene	High
Benzo(a)pyrene	Immobile	Xylene (Total)	-
Benzo(b)fluoranthene	Immobile	Hydrochloric acid	-
Benzo(e)pyrene	Immobile	Hydrogen sulfide	-
Benzo(g,h,i)perylene	Immobile	Ammonia	-
Benzo(k)fluoranthene	Immobile	Sulfur dioxide	-
Chrysene	Immobile	Carbon dioxide	-
Dibenzo(a,h)anthracene	Immobile	Nitrogen dioxide	-
Phenanthrene	Immobile	Carbon monoxide	-
Fluoranthene	Immobile	Nitric oxide	-
Fluorene	Immobile	Nitrous oxide	-
Indeno(1,2,3-cd)pyrene	Immobile	Sulfur trioxide	-
Naphthalene	Immobile	Sulfuric acid	-
Pyrene	Immobile	Nitric acid	-
2-Chloronaphthalene	Low		

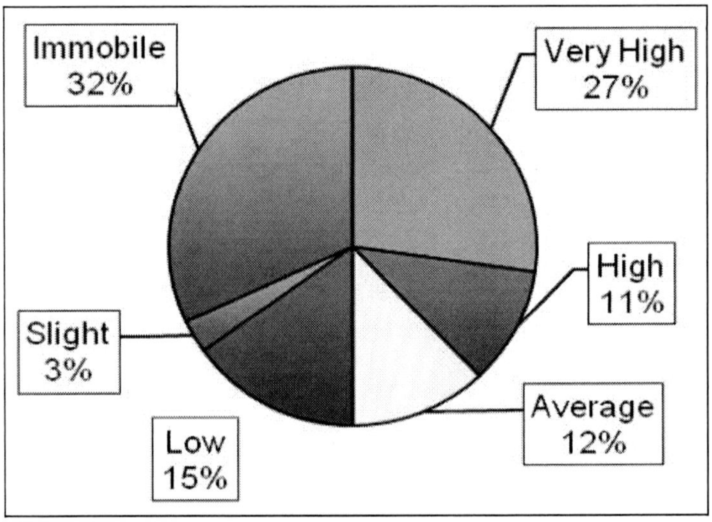

Figure 3. Percentages of the compounds issued in the gaseous emissions, grouped according with prediction of mobility in soil.

Following, it is described the second function proposed to consider the effect of the chemical substances, identified in the gaseous emissions of the refinery process, over the abiotic resources. The proposed function is based in the values of *Global Warming Potential, (GWP),* and constitutes the eighth function of Factor of Toxicity and Pollution, denominated Function of Global warming impact.

2.8. Function of Global Warming Impact (F8)

The changes of weather on earth, in a natural way, due to four main reasons:

a) Movement of tectonic plates
b) Volcano eruptions
c) Solar variations
d) Orbital variations

A balance among input and output energies in the system, constituted by the surface of earth and the atmosphere, determined their temperatures. It has been estimated that earth's atmosphere receives around the 30% of sun's radiant energy, which is reflected by clouds, and by molecules of air. 51% is absorbed by earth's surface, and the 19% is absorbed by atmosphere. The radiant energy that is reflected to space is named albedo (De Nevers, 1998).

Radiation coming from sun is short wave and the absorption combined specter from the atmosphere makes it appear as transparent, mainly in wavelength from 0.3 to 0.7 μm. At the same time as, solar radiation comes to earth's surface, and it starts to heat, because atmosphere retains part of radiation, which provokes that earth reaches a temperature of 300K. This action becomes earth into emissor, with a wavelength that corresponds to infrared. The earth's gases, which intercept radiation, at the same time emit them to earth's surface, which provokes a major warming. The gases that have the characteristic to absorb and emit infrared radiation are denominated as, greenhouse gases, GHG, because of the caused effect. This effect is due to the lack of balance among input and output energies, which warm earth and give place to an adequate temperature for life developing. This effect has been developed for billions of years, because GHG are on earth in a natural way, like water steam, carbon dioxide, ozone, methane and nitrous oxide.

Nowadays, the increase in natural GHG concentration, as the presence of other gases synthetisized by human being is giving an adverse effect over World's weather.

One of the most studied chemicals, in relation with global average temperature, is the carbon dioxide. Its concentration has increased, as is showed in 4; from industry revolution, when the fossil fuels' consumption started intensively, and increased from 283ppm in 1796 to 363 ppm in 1997. Eventhough, a considerable increase took place in the 50's. Measurements developed by Keeling in 1959, in Mauna Loa Observatory, Hawaii, indicated that there was a concentration of 315 ppm approximately. From that period of time, to 1997, CO2 concentration in the atmosphere has increased 48 ppm, which is 2.28 ppm/year in average (Etheridge, et al., 1998; Keeling, et al., 2005).

In nature there are emission sources of CO_2, as such as plants, soil, oceans, their velocities of emission and consumption are in balance. For this reason, the increase in CO_2 is related with industry emissions, with the use of fossil fuels, as such as charcoal and petroleum's derivates. Figure 4 shows variation in CO_2 concentration from 1960 (316.91 ppm) to 2004 (377.43 ppm.2004 (377.43 ppm).

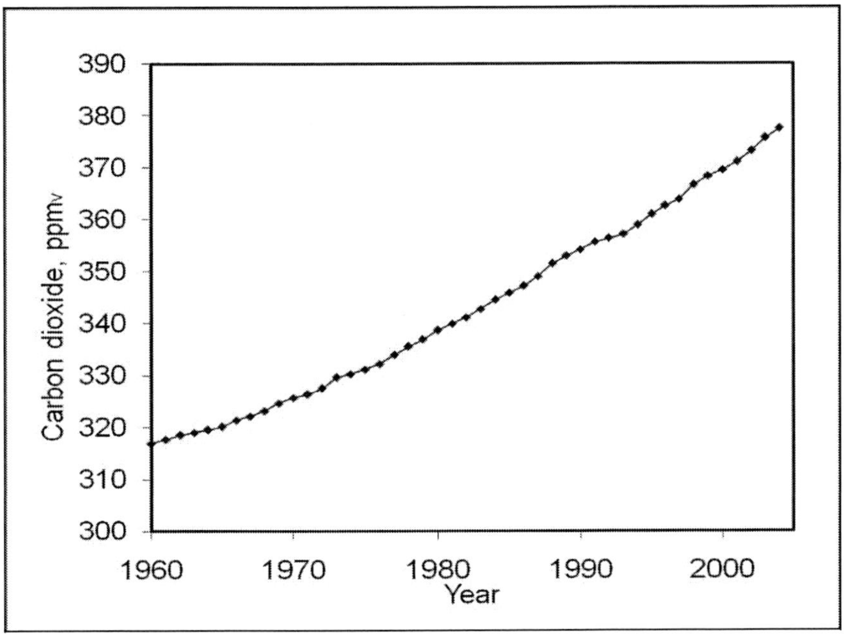

Figure 4. Variation in CO_2 concentration in terrestrial atmosphere.

According to record of average temperature from earth and CO_2 concentration on air, there is a direct relationship between the CO_2 increase and the increase in temperature. The radiant energy that is reflected in clouds, it is because small particles of CO2 form part of clouds, and create mixtures with the name of aerosols. Aerosols can be formed by natural ways; volcano activity generates great quantity of sulfur in stratosphere.

Moreover, the natural aerosols, generated by volcano activity, there are in atmosphere anthropogenic aerosols, coming from combustion of fossil fuels and biomass. Aerosols can reflect and absorb solar radiation. The radiation reflection addresses a cooling effect, while that absorption causes a contrary effect on the weather. Due to their life time average are lower than life time average for green gases, a change in their emissions is associated with a change in atmospherical concentration. In general, aerosols are more associated with a decrease in temperature by avoiding the entrance of solar radiation into atmosphere.

The balance among the input and output energies, it is mainly supported by the great calorific Nevertheless, the fast increase in green gases and aerosols concentrations, due to human activities, is generating changes in a faster velocity than environment can damp, altering as consequence temperature in different areas of World, rainfalls, and atmospheric and ocean circulation.

Because of the evident relationship among greenhouse gases and climatic phenomena, that alters abruptly the weather patterns on earth. The international community concerned by this and other changes, recognize the need to have an organism that may join the opinions and experiences of worldwide experts, in order to handle climatic change. In 1988 *Intergovernmental Panel on Climate Change, IPCC,* was established by United Nations Environment Program, *UNEP*, and World Meteorological Organization, *WMO*, as an initiative to evaluate all the aspects of climatic change, including the way on which human activities may cause these changes.

IPCC published its first report in 1990, and the second one in 1996, in both reports an evaluation was made. In the first report they evaluated human activities that provoke the increase of greenhouse gases and the effect over the climatic change. And the second report, was established that, in order to revert, in a slowly way, warming effect, it was necessary to stabilized greenhouse gases concentration, and avoid, as much as possible, their emission. According to third report of *IPCC (2001)*, among the agents that cause changes in weather the aerosols and greenhouse gases concentrations, and solar variations have the first place. According to Fourth Assessment Report del *IPCC (2007)*, ¨continued GHG emissions at or above current rates would cause further warming and induce many changes in the global climate system during the 21st century that would very likely be larger than those observed during the 20th century¨.

With the purpose of standarize criteria to evaluate global warming effects, each one of the greenhouse gases, and propose measures to control their emissions. *Intergovernmental Panel on Climate Change* proposed the index denominated as *Global Warming Potential, GWP.*

GWP is the global warming relation produced by a mass of greenhouse gas, and the warming produced by a carbon dioxide mass. This index estimates contribution to global warming due to emission of 1 Kg of a greenhouse gas, relative to emission of 1 Kg of CO_2. Election of CO_2 , as reference gas to measure *GWP* is because it is the most abundant specie with a concentration around 98% of total greenhouse gases in atmosphere.

GWP are estimated to different periods, the most frequent reported are 20, 100 and 500 years. Table 34 shows *GWP* values of some of the greenhouse gases, identified en gaseous refinery emissions, for a lifetime horizon of 100 years.

Table 34. *GWP* values for green house gases, for a 100 year time horizon

SUBSTANCE	GLOBAL WARMING POTENTIAL	REFERENCE
1,1,1-Trichloroethane	110	SRC, 2002.
2,3 Dimethyl butane	0	
2-methyl 1,3-butadiene	0	
Benzene	0	
Cyclohexane	0	
Ethane	3	
Heptane	0	
Hexane	0	
Methane	23	IPCC, 2001.
Pentane	0	SRC, 2002.
Propane	3	
Ammonia	APROX 0	
Carbon dioxide	1	IPCC, 2001.
Nitrous oxide	296	IPCC, 2001.

Based in this parameter, a function to evaluate the effect of issued substances in gases from refinery processes over the global warming is proposed.

2.8.1. Description of the Function of Global Warming Impact (F8)

From *Global Warming Impact* values reported in Table 35, *Function of Global Warming Impact, F8* is proposed. Upper and lower limit values were established based in information showed in Table 35, where 2-methyl butane, with $GWP = 0.0004$, was established as lower limit, corresponding to the characteristic of a chemical substance that participates in a low scale in global warming. To set the upper limit, nitrous oxide with $GWP = 296$ was selected, which has the greatest *GWP* values among the studied compounds in this research.

Table 35. Values selected to obtain equation *F8*

COMPOUND	GWP	F8
Nitrous oxide	296.00	1.0000
2-Methyl butane	0.0004	0.0001

The linear function obtained, is expressed by Equation 20.

$$F8 = 0.1704 + 0.579*\log_{10}(GWP) \qquad Eq.\ 20$$

Table 5.39 shows the *F8* values for each one of the compounds listed. In the case, where they do not appear listed, because GWP value is unknown, the value zero was assigned by *F8.* Considerating acid rain, as an environmental phenomenon, which is taking place because of the environmental disasters that has caused in the last years. In the next paragraph, a function that includes some of the issued substances coming from gases in a refinery process is proposed.

2.9. Function of Impact by Acid Rain (F9)

Carbon dioxide emissions, that dissolute themselves on air's humidity produce carbonic acid, which lows pH around values of 5.6. Nevertheless, starting in1970 pH annual media for raining was between 4 and 5 in some areas of United States, such as some European countries. This phenomenon contributes to the increase of sulfur oxides and nitrogen. Information published in 1979 in United States pointed sulfuric acid, as the responsible for the 60 to 70 percent of acidity, in the meanwhile, nitric acid was responsible for the 30 to 40 percent of acidity (Wark y Warner, 1998.)

Concerns about the acid rain phenomenon, is due to several adverse effects. Among the effects both Wark et al., (1998), and Mihelcic, (1999) mentioned that:

- Affects in general aquatic organisms and fish life, because it acids the natural sources of water. Due to many species as too sensitive to low values of pH, in some cases reproduction process stops or some species die. Among the affected organisms are plankton and animal life, which are a main source of food for fish.
- Provokes phenomena of lixiviation of nutrients in soil, living place to lose in productivity and changes in vegetation.
- Damage in buildings and historical monuments.

The problem with acid rain it is because the sulfur oxides and nitrogen react with the atmosphere's water, and form H_2SO_4 and HNO_3, which return to earth's surface as acid waterfall. Therefore, these acids may react with some cation and form sulfates or nitrates that form particles, which provokes visibility problems (De Nevers, 1999.).

Sulfur oxide emissions, produced in combustion gases, are mainly constituted by Sulfur dioxide, and a lower portion of sulfur trioxide. Sulfur trioxide reacts abruptly with water steam that is part of combustion gases, such as part of atmosphere, and produce sulfuric acid (*EPA*, 2002.)

Nitrogen oxides that are formed in combustion processes, are given by three processes (*EPA*, 2002):

- *NOx* thermic is formed by reaction of oxygen and nitrogen o fair, mainly in the high temperature zone, near to burners.
- *NOx* prompt is formed by reaction between nitrogen in air and fuel radicals. The quantity produced by this mechanism is almost null, in comparison with *NOx* thermic.
- *NOx* from fuel, in this *NOx* participates compounds of fuel's nitrogen. This mechanism produces a few quantities of oxides.

The oxides formed by these three mechanisms, are constituted by nitric oxide (*NO*), in quantities of 95%, and by *NO*, and N_2O, having nitrous oxide (N_2O) the lowest proportion. Eventhough, Nitrous oxide is not included within generic term of *NOx,* it is a strong green house gas.

NO and NO_2 reacts to water to form nitric acid, even conversion of NO is slower, because it goes through two steps. In the following equation reactions are expressed (De Nevers, 1999):

$$NO + 0.5O_2 \leftrightarrow NO_2$$

$$NO_2 + H_2O \rightarrow HNO_3$$

Eq. 21

Notwithstanding, NO convertion into Nitric acid can be considerate sufficiently fast in atmosphere to provoke acid rain. By the other hand, in most of the regulations, NOx values referred NO_2 concentration, based in assumptions that all NO converts into dioxide.

Moreover, these acids derived from oxides, in combustion gases are de hydrochloric and sulfhydric acids, which are present in a lower proportion, but they are part of the acid rain.

Based in the behavior background of sulfur oxides and nitrogen, as the importance to consider the impact produced by acid rain, within *Factor of Toxicity and Pollution* of chemical substances a *Function of Impact by Acid Rain, F9* is proposed from negative logarithmic values of dissociation constant, *pKa*.

In spite of, application of this function, most of the gaseous compounds, considered here, were eliminated. Nevertheless, the evaluation is representative, because includes the responsible compounds of acid rain, a phenomenon that keeps causing severe problems worldwide.

Assumptions developed to evaluate this function are the following:

- All sulfur trioxide, SO_3, from flue gases react, instantaneously, with water to form sulfuric acid.
- Both, nitrogen dioxide and nitric oxide turn into nitric acid, but in a lower way.

To structure *F9*, pKa values of hydrochloric and sulfhydric acids were taken, and they are transcribed in the following Table 36.

Table 36. Reference values to establish function *F9*

COMPOUND	*pKa*	F9
Hydrochloric acid	-3.00	1.0000
Hydrogen sulfide	7.04	0.1000

By which, the expressed function was derived by Equation 22, and values calculated are showed in Table 38.

$$F9 = 0.7311 - 0.0896*pKa \qquad \text{Eq. 22}$$

This is the last function, of the proposed functions to evaluate Factor of Toxicity and Pollution from chemical compounds identified in gaseous emissions from a refinery process.

3. Integration of Factor of Toxicty and Pollution of Gaseous Emissions

The values of different functions, that integrate Factor of Toxicity and Pollution, are summarized and expressed in Equation 23:

$$TPF_i = a1*F1_i + a2*F2_i + a3*F3_i + a4*F4_i + a5*F5_i + a6*F6_i + a7*F7_i + + a8*F8_i + a9*F9_i \quad \textit{Eq. 23}$$

where ai coefficients represent weight factors, corresponding to each one of the functions, previously developed. Figure 5 shows a sensibility analysis of factor of toxicity and pollution, in function with variation in weight factor for each of the functions. Function of toxicity by bioconcentration on fish has the greatest effect over TPF value. Meanwhile, TPF stays constant, with respect to variations in weight factors of functions F8 and F9. Due to lack of some of parameters' data, for which functions were defined, in the meanwhile, information more complete is obtained about behavior of substances, weight factors were used to evaluate TPF, as it is showed in Table 37. Based in the assignation, the TPF values, corresponding to each of the substances, were calculated.

Table 37. Factors of weight used to evaluate *TPF*

WEIGHT FACTOR	VALUE	FUNCTION	RISK
a1	1.00	F1	Cancer in human beings
a2	1.00	F1	Toxicity by inhalation in human beings
a3	1.00	F3	Toxicity by ingestion in human beings
a4	1.00	F4	Toxicity by bioconcentration on fish
a5	1.00	F5	Toxicity by acute exposition on fish
a6	1.00	F6	Toxicity by chronic exposition on fish
a7	1.00	F7	Toxicity by adsorption in soil
a8	1.00	F8	Global warming impact
a9	1.00	F9	Acid rain impact

The values obtained, were normalized by a zero to one hundred (0 to 100) scale. And they are showed in Table 38. The group of dioxins and furans is represented by *TCDD*.

Table 38. Summary of values of functions and toxicity and pollution factor

SUSTANCE	F1	F2	F3	F4	F5	F6	F7	F8	F9	TPF
TCDD	0.1000	0.4595	0.4827	0.0000	0.1971	0.3467	0.1028	0.0000	0.0000	100.0000
Antimony	1.0000	0.6030	0.4956	0.2033	0.1414	0.3786	0.1437	0.0000	0.0000	25.3965
Arsenic	0.0000	0.4595	0.2506	0.0000	0.0000	0.0000	0.0510	0.0000	0.0000	44.5962
Barium	0.7500	0.4984	0.4104	0.4317	0.4679	0.5884	0.1329	0.0000	0.0000	11.4452
Beryllium	0.7500	0.6285	0.4727	0.6931	0.8574	0.7099	0.1251	0.0000	0.0000	49.3190
Cadmium	0.0000	0.0000	0.2056	0.0001	0.4879	0.6751	0.1167	0.0000	0.0000	63.7080
Copper	1.0000	1.0000	0.3922	0.1510	0.1212	0.4447	0.0922	0.0000	0.0000	22.3362
Chromium	0.0000	0.8825	0.2195	0.0000	0.0000	0.0000	0.0000	0.0000	0.0000	48.1397
Manganese	0.0000	0.7676	0.4906	1.0000	0.4679	0.8224	0.1568	0.0000	0.0000	16.5715
Mercury	1.0000	0.0000	0.3069	0.0000	0.3470	0.4722	0.1329	0.0000	0.0000	55.7190
Nickel	0.0000	0.6030	0.3692	0.3971	0.6768	0.7878	0.1122	0.0000	0.0000	33.9710
Silver	0.2000	0.6645	0.3982	0.5480	0.3185	0.5080	0.1685	0.0000	0.0000	44.3021
Lead	0.0000	0.4931	0.3692	0.5793	0.2693	0.4034	0.0872	0.0000	0.0000	42.1905
Selenium	0.0000	0.5185	0.5542	0.0000	0.0000	0.0000	0.1366	0.0000	0.0000	33.1051
Thallium	0.0000	0.0000	0.1852	0.0000	0.4841	0.4927	0.1329	0.0000	0.0000	18.1855
Zinc	0.0000	0.0000	0.2576	0.8735	0.2913	0.3910	1.0000	0.0000	0.0000	19.4727
Acenaphtene	0.0000	0.0000	0.4168	0.8228	0.0000	0.0000	1.0000	0.0000	0.0000	42.3062
Acenaphthylene	0.0000	0.4931	0.1852	1.0000	0.6523	0.0000	1.0000	0.0000	0.0000	33.6781
Anthracene	0.4000	0.0000	0.7888	1.0000	0.6649	0.0000	1.0000	0.0000	0.0000	50.0848
Benzo(a)anthracene	0.4000	0.4931	0.5962	1.0000	0.4175	0.0000	1.0000	0.0000	0.0000	57.9510
Benzo(a)pyrene	0.2000	0.4931	0.5962	1.0000	0.0000	0.0000	1.0000	0.0000	0.0000	58.7487
Benzo(b)fluoranthene	0.3000	0.0000	0.0000	1.0000	0.0000	0.0000	1.0000	0.0000	0.0000	49.4634
Benzo(e)pyrene	0.0000	0.0000	0.4168	1.0000	0.0000	0.0000	1.0000	0.0000	0.0000	34.5865
Benzo(g,h,i)perylene	0.2000	0.0000	0.4588	1.0000	0.0000	0.0000	1.0000	0.0000	0.0000	36.3428
Benzo(k)fluoranthene	0.3000	0.4931	0.7888	0.9982	0.0000	0.0000	1.0000	0.0000	0.0000	39.9822
Chrysene	0.4000	0.4931	0.4300	0.9625	0.0000	0.0000	1.0000	0.0000	0.0000	53.8356
Dibenzo(a,h)anthracene	0.0000	0.4931	0.4168	1.0000	0.4356	0.4957	1.0000	0.0000	0.0000	49.4072
Phenanthrene	0.0000	0.4931	0.2758	1.0000	0.2294	0.5544	1.0000	0.0000	0.0000	57.7625
Fluoranthene	0.0000	0.0000	0.2758	1.0000	0.3443	0.4564	1.0000	0.0000	0.0000	53.4233
Fluorene	0.2000	0.4931	0.4300	1.0000	0.0000	0.0000	1.0000	0.0000	0.0000	46.2632

Table 38. (Continued)

SUSTANCE	F1	F2	F3	F4	F5	F6	F7	F8	F9	TPF
Indeno(1,2,3-cd)pyrene	0.1000	0.6200	0.3069	0.7652	0.2143	0.3984	1.0000	0.0000	0.0000	46.9639
Naphthalene	0.0000	0.4931	0.2887	1.0000	0.0000	0.0000	1.0000	0.0000	0.0000	51.1999
Pyrene	0.0000	0.0000	0.2446	1.0000	0.0000	0.0000	0.8252	0.0000	0.0000	41.8318
2-Chloronaphthalene	0.0000	0.0000	0.0000	1.0000	0.3108	0.4332	1.0000	0.0000	0.0000	31.1249
2-Methylnaphthalene	0.0000	0.3260	0.1852	0.4355	0.0851	0.3211	0.0646	0.0000	0.0000	41.2637
Phenol	0.0000	0.4931	0.0000	1.0000	0.0000	0.0000	1.0000	0.0000	0.0000	21.3161
Perylene	0.0000	0.1569	0.1738	0.3206	0.0206	0.2871	0.0000	0.9268	0.0000	37.4904
1,1,1-Trichloroethane	0.5000	0.4051	0.7042	0.2853	0.0000	0.0000	0.6647	0.0000	0.0000	28.3562
1,3-Butadiene	0.0000	0.0000	0.0000	0.9024	0.0000	0.0000	0.0000	0.0000	0.0000	38.4867
2,2-Dimethyl butane	0.0000	0.0000	0.0000	0.7998	0.0000	0.0000	0.8731	0.0000	0.0000	13.5699
2,3 Dimethyl butane	0.0000	0.0000	0.0000	0.7202	0.0000	0.0000	0.8322	0.0000	0.0000	25.1563
2,2 Dimethyl propane	0.1000	0.0000	0.0000	0.4799	0.0000	0.0000	0.7271	0.0000	0.0000	23.3454
2-methyl 1,3-butadiene	0.0000	0.0000	0.0000	0.4720	0.0000	0.0000	0.4953	0.0000	0.0000	19.6549
2-methyl 1-butene	0.0000	0.0000	0.0000	0.6074	0.0000	0.0000	0.4953	0.0000	0.0000	14.5455
2-methyl 2-butene	0.0000	0.0000	0.0000	0.5678	0.0000	0.0000	0.7220	0.0000	0.0000	16.5822
2-methyl butane	0.0000	0.0000	0.0000	0.8032	0.0000	0.0000	0.8134	0.0000	0.0000	19.3949
2-methyl pentane	0.0000	0.0000	0.0000	0.8460	0.0000	0.0000	0.5714	0.0000	0.0000	24.3102
3-methyl pentane	0.2000	0.5496	0.0000	0.1611	0.0668	0.4595	0.0001	0.0000	0.0000	21.3134
Acetaldehyde	0.1000	0.9413	0.3112	0.8563	0.5051	0.5829	0.1290	0.0000	0.0000	21.6106
Acrolein	1.0000	0.4724	0.3750	0.2197	0.2085	0.0000	0.5215	0.0000	0.0000	51.5156
Benzene	0.0000	0.1571	0.0000	0.0921	0.0000	0.0000	0.7578	0.0000	0.0000	42.0616
Butane	0.0000	0.1795	0.0000	0.6581	0.0001	0.0000	0.5957	0.0000	0.0000	15.1415
Cyclohexane	0.0000	0.0000	0.0000	0.5445	0.0000	0.0000	0.5282	0.0000	0.0000	21.5548
Cyclopentane	0.0000	0.0000	0.0000	0.4317	0.0000	0.0000	0.5159	0.0000	0.0000	16.1307
Cis 2-pentene	0.2000	0.2330	0.3381	0.2628	0.0549	0.3606	0.6050	0.0000	0.0000	14.2493
Chloroform	0.0000	0.0000	0.0000	0.4961	0.0000	0.0000	0.6790	0.6603	0.0000	30.8918
Ethane	0.0000	0.2112	0.2220	0.6425	0.0777	0.4171	0.7341	0.0000	0.0000	27.6004
Ethyl benzene	0.5000	0.4371	0.2035	0.1611	0.0984	0.5039	0.4239	0.0000	0.0000	34.6547
Formaldehyde	0.0000	0.1543	0.0000	0.9856	0.0000	0.0000	0.9857	0.0000	0.0000	35.0053
Heptane	0.0000	0.3508	0.4168	0.8396	0.0000	0.0000	0.9067	0.0000	0.0000	31.9627

Table 38. (Continued)

SUSTANCE	F1	F2	F3	F4	F5	F6	F7	F8	F9	TPF
Hexane	0.0000	0.0000	0.0000	0.6443	0.0000	0.0000	0.8706	0.0000	0.0000	37.8038
Hexene	0.0000	0.0000	0.0000	0.6259	0.0000	0.0000	0.4174	0.0000	0.0000	22.7800
Isobutane	0.0000	0.0000	0.0000	0.2021	0.0000	0.0000	0.7775	0.8110	0.0000	15.6883
Methane	0.0000	0.0000	0.0000	0.7839	0.0000	0.0000	0.8660	0.0000	0.0000	26.9261
Methyl cyclopentane	0.0000	0.1596	0.0000	0.4263	0.0000	0.0000	0.8064	0.0000	0.0000	24.8108
Pentane	0.0000	0.1591	0.0000	0.5279	0.0000	0.0000	0.0000	0.6603	0.0000	20.9383
Propane	0.0000	0.0000	0.0000	0.4398	0.0000	0.0000	0.6372	0.0000	0.0000	20.2584
Propylene	0.0000	0.3064	0.2035	0.6617	0.0733	0.3335	0.6001	0.0000	0.0000	16.1957
Toluene	0.0000	0.0000	0.0000	0.4317	0.0000	0.0000	0.5159	0.0000	0.0000	32.7571
Trans 2-pentene	0.0000	0.2804	0.1000	0.7331	0.1405	0.0000	0.0000	0.0000	0.0000	14.2493
Xylene (Total)	0.0000	0.4984	0.0000	0.9583	0.0987	0.0000	0.0000	0.0000	0.6437	18.8567
Hydrochloric acid	0.0000	0.6905	0.3922	0.2569	0.0000	0.0000	0.0000	0.0000	0.0644	33.0676
Hydrogen sulfide	0.0000	0.3036	0.2346	0.0001	0.1539	0.6145	0.0000	0.0000	0.0000	21.1104
Ammonia	0.0000	0.4095	0.0000	0.5675	0.0000	0.0000	0.0000	0.0000	0.0000	19.6495
Sulfur dioxide	0.0000	0.1000	0.0000	0.3403	0.0712	0.0000	0.0000	0.5790	0.0000	14.6925
Carbon dioxide	0.0000	0.3952	0.1164	0.0000	0.0000	0.0000	0.0000	0.0000	0.0000	16.3987
Nitrogen dioxide	0.0000	0.1000	0.0000	0.6418	0.0000	0.0000	0.0000	0.0000	0.0000	7.6931
Carbon monoxide	0.0000	0.3047	0.1000	0.6072	0.0000	0.0000	0.0000	0.0000	0.0000	11.1545
Nitric oxide	0.0000	0.2944	0.0000	0.0148	0.0000	0.0000	0.0000	1.0000	0.0000	15.2158
Nitrous oxide	0.0000	0.3663	0.0000	0.0000	0.0000	0.0000	0.0000	0.0000	0.0000	19.6874
Sulfur trioxide	1.0000	0.5185	0.0000	0.0000	0.0529	0.0000	0.0000	0.0000	1.0000	5.5085
Sulfuric acid	1.0000	0.3750	0.0000	0.0000	0.0025	0.0000	0.0000	0.0000	0.5502	38.6686
Nitric acid	1.0000	0.3750	0.0000	0.0000	0.0025	0.0000	0.0000	0.0000	0.5502	28.9879

The total of values that integrate *Factor of Toxicity and Pollution*, a 42% corresponds to the risk evaluation in human beings; a 25% corresponds to the risk evaluation on fish, and 33% corresponds to the impact over abiotic resources.

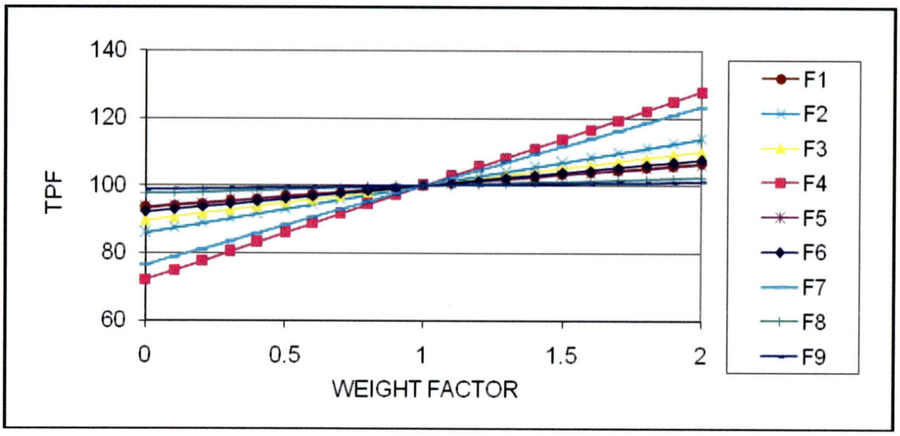

Figure 5. Sensibility of *factor of toxicity and pollution*, and variation of functions that determine it.

4. Application of FTC in Gaseous Emissions from a Petroleum Refinery Process

Next, as an example, the application of Factor of Toxicity and Pollution in gaseous emissions of refinery units: combined distillation, naphtha hydrodesulfuration, jet fuel hydrodesulfuration, kerosene hydrodesulfuration, assuming that combustion equipments use refinery's gas, with compositions showed in Table 39.

Table 39. Composition of refinery gas

SUBSTANCE	FORMULA	MOL FRACTION
Hydrogen sulfide	H_2S	0.001
Hiydrogen	H_2	0.230
Nitrogen	N_2	0.018
Oxygen	O_2	0.018
Carbon monoxide	CO	0.018
Methane	CH_4	0.301
Ethane	C_2H_6	0.099
Ethylene	C_2H_4	0.047
Propane	C_3H_8	0.103
Propylene	C_3H_8	0.134
N-butane	C_4H_{10}	0.031
Low Heat Value: 45.59 MJ/kg		

Table 40 lists the types of effluents that are emitted in each plant, which were taken, as example of FTC application.

Table 40. Types of gaseous effluents from process units in the petroleum refinery

	Combined distillation	Naphtha hydrodesulfuration	Jet fuel hydrodesulfuration	Kerosene hydrodesulfuration
Flue gases	Yes	Yes	Yes	Yes
	Yes			
Sour gas			Yes	Yes

Table 41 summarizes FTC values, multiplied by the mass fraction of the corresponding compound. The obtained values show that naphtha hydrodesulfuration, emmits a draught of flue gases, a lower contaminant than, a draught of ejectors from combined distillation unit. Although, it is observed that FTC, corresponding to sour gas emissions, indicates that these effluents are 8 times more toxic than, the flue gases delivered. This is because; the substances that compound the sour gas emissions have a high toxicity.

Table 41. Valores de *FTC* de las corrientes gaseosas

	Naphtha hydrodesulfuration	Combined distillation	Jet fuel hydrodesulfuration	Kerosene hydrodesulfuration
Flue gases,	2.56	2.56	2.56	2.56
Incondensables,		9.05		
Sour gas,			21.47	20.83

In this paper, the Toxicity and Pollution Factor was calculated, considerating the greatest amount of possible environmental parameters, only limited by information availability for each substance. The purpose was to integrate the toxicity and pollution aspects that define the environmental behavior of substances, when they are issued by gaseous emissions. Having in mind that, in this model there are a great number of excluded parameters, which can be consider in future. The model developed in this study, considered the behavior of each substance independently. Therefore, there is a strong possibility that, among the complex mixtures of substances in the environment, synergic effects may be presented among the complex mixtures of substances in the environment. This can modify total toxicity, resulting in a value that does not express, strictly, the summation of unitary toxicities.

It is important to say that selection of indicators used to calculate *Toxicity and Pollution Factor*, and which integrate the model developed in this paper, may be modified, in order to enhance the proposal. If, physical and chemical properties of substances studied in this research can be determined in future.

References

Ambrose, P.E., *Partition Coefficients for Metals in Surface Water, Soil, and Waste.* Environmental Protection Agency. Office of Solid Waste, Washington, DC 20460, June 22, 1999.

Barnes, G.D.; Dourson, M., Reference Dose (RfD): Description and Use in Health Risk Assessments, *Regulatory Toxicology and Pharmacology*, Vol. 8, 1988, pp. 471-486.

Caldwell, R. S.; Caldarone, E. M.; Mallon, M .H., Effects of a Seawater Soluble Fraction of Cook inlet Crude Oil and its Major Aromatic Components on Larval Stages of the Dungeness Crab, Cancer Magister Dana. In Fate and Effects of Petroleum Hydrocarbons in Marine Ecosystems and Organisms, Wolfe, D. A. (ed.), Pergamon Press, NY, 1977.

Davies, G.A.; Kincaid, L.; Swanson, M.; Schultz, T.; Barmess, J.; Griffith, B.; Jones, S., Chemical Hazard Evaluation for Management Strategies: A Method for Ranking and Scoring Chemicals by Potential Human Health and Environmental Impacts. EPA 600/R-94/177. Risk Reduction Engineering Laboratory, U.S. Environmental Protection Agency, Cincinnati, OH, 1994.

De Nevers, N., Air Pollution Control Engineering, *Ed. Mc Graw Hill*, 1999.

Dourson, M.L., Methods for Establishing Oral Reference Doses, In Risk Assessment of Essential Elements, W. Mertz, C.O., Abernathy and S.S. Olin, (Ed.) ILSI Press, Washington, D.C., 1994, pp. 51-61.

Dourson, M.L.; Andersen, M.E.; Erdreich, L.S.; MacGregor, J.A., Using Human Data to Protect the Public's Health, *Regulatory Toxicology and Pharmacology*, Vol. 33, pp. 234-256.

Environmental Protection Agency, EPA, Appendix, EPA-823-R-00-002. Office of OST Water, 1998.

Environmental Protection Agency, EPA WQC data book, vol. 5, B-5, 1973.

Environmental Protection Agency, EPA, Ambient Quality Criteria for Acenaphtene, EPA-440/5-80-015, Washington, DC, 1980.

Environmental Protection Agency, EPA, Ambient Quality Criteria for Acrolein, EPA-440/5-80-016, Washington, DC, 1980.

Environmental Protection Agency, EPA, Ambient Quality Criteria for Ammonia, EPA-822/R-98-008, Washington, DC, 1998.

Environmental Protection Agency, EPA, Ambient Quality Criteria for Antimony, EPA-440/5-80-020, Washington, DC, 1980.

Environmental Protection Agency, EPA, Ambient Quality Criteria for Arsenic, EPA-440/5-84-033, Washington, DC, 1984.

Environmental Protection Agency, EPA, Ambient Quality Criteria for Barium, EPA-440/9-76-023, Washington, DC, 1976.

Environmental Protection Agency, EPA, Ambient Quality Criteria for Benzene, EPA-440/5-80-018, Washington, DC, 1980.

Environmental Protection Agency, EPA, Ambient Quality Criteria for Beryllium, EPA-440/5-80-024, Washington, DC, 1980.

Environmental Protection Agency, EPA, Ambient Quality Criteria for Cadmium, EPA-822/R-01-001, Washington, DC, 1984.

Environmental Protection Agency, EPA, Ambient Quality Criteria for Chloroform, EPA-440/5-80-033, Washington, DC, 1980.
Environmental Protection Agency, EPA, Ambient Quality Criteria for Chromium, EPA-440/5-80-035, Washington, DC, 1980.
Environmental Protection Agency, EPA, Ambient Quality Criteria for Copper, EPA-440/5-84-031, Washington, DC, 1984.
Environmental Protection Agency, EPA, Ambient Quality Criteria for Ethylbenzene, EPA-440/5-80-048, Washington, DC, 1980.
Environmental Protection Agency, EPA, Ambient Quality Criteria for Fluoranthene, EPA-440/5-80-049, Washington, DC, 1980.
Environmental Protection Agency, EPA, Ambient Quality Criteria for Lead, EPA-440/5-84-027, Washington, DC, 1984.
Environmental Protection Agency, EPA, Ambient Quality Criteria for Mercury, EPA-440/5-84-026, Washington, DC, 1984.
Environmental Protection Agency, EPA, Ambient Quality Criteria for Naphthalene, EPA-440/5-80-059, Washington, DC, 1980.
Environmental Protection Agency, EPA, Ambient Quality Criteria for Nickel, EPA-440/5-86-004, Washington, DC, 1986.
Environmental Protection Agency, EPA, Ambient Quality Criteria for Phenol, EPA-440/5-80-066, Washington, DC, 1980.
Environmental Protection Agency, EPA, Ambient Quality Criteria for 2,3,7,8-Tetrachlorodibenzo-p-dioxin, 440/5-84-007, Washington, DC, 1984.
Environmental Protection Agency, EPA, Ambient Quality Criteria for Selenium, EPA-440/5-87-006, Washington, DC, 1987.
Environmental Protection Agency, EPA, Ambient Quality Criteria for Silver, EPA-440/5-80-071, Washington, DC, 1980.
Environmental Protection Agency, EPA, Ambient Quality Criteria for Toluene, EPA-440/5-80-075, Washington, DC, 1980.
Environmental Protection Agency, EPA, Ambient Quality Criteria for Zinc, EPA-440/5-80-079, Washington, DC, 1980.
Environmental Protection Agency, EPA, SAR, Estimating Toxicity of Industrial Chemicals to Aquatic Organisms Using Structure Activity Relationshiops, Clements, R. (ed.), Washington, DC, 1996.
Environmental Protection Agency, EPA, Chemical Fate Rate Constants for SARA Section 313 Chemicals and Superfund Health Evaluation Manual Chemicals. Prepared by Chemical Hazard Assessment Division, Syracuse Research Corporation, Washington, DC, 1989.
Environmental Protection Agency, EPA, Integrated Risk Information System (IRIS), database, United States, http://www.epa.gov/ncea/iris/help_gloss.htm
Environmental Protection Agency, EPA, Update of Ambient Quality Criteria for Cadmium, EPA-822-R-01-001, Washington, DC, 2001.
Etheridge, D.M.; Steele, L.P.; Langenfelds, R.L.; Francey, R.J.; Barnola, J.M.; Morgan, V.I., Historical CO_2 records from the Law Dome DE08, DE08-2, and DSS ice cores. In Trends: A Compendium of Data on Global Change. Carbon Dioxide Information Analysis Center, Oak Ridge National Laboratory, Oak Ridge, Tenn., U.S.A., 1998.

Folwarkow, S.; Riter, K.; Hansell, D.; England, G., Air Toxic Factors for Combustion Sources using Petroleum Based Fuels. Final Report Volume 2. Development of Emission Factors using CARB Approach, Western States Petroleum Association, American Petroleum Institute and Energy and Environmental Research Corporation, California, 1998.

Franke, C.; Studinger, G.; Berger, G.; Böhling, S.; Bruckmann, U.; Cohors, D.; Jöhncke, U.l., The Assessment of Bioaccumulation. *Chemosphere*, ISSN: 0045-6535, Vol.29, No.7, 1994, pp. 1501-1514.

Geiger, D. L.; Call, D. J.; Brooke, L. T. (ed.), Acute Toxicities of Organic Chemicals to Fathead Minnows (Pimephales Promelas), Vol V. Center for Lake Superior Environmental Studies, University of Wisconsin – Superior, 1990.

Hamaker, J. W. 1975. The interpretation of soil leaching experiments, pp. 115-133. In: R. Haque and V. H. Freed, eds. Environmental dynamics of pesticides. Plenum Press, NY.

Hansch, C.; Leo, A.; Hoekman, D., Exploring QSAR - Hydrophobic, Electronic, and Steric Constants. Washington, DC: American Chemical Society, 1995.

Hodson, J.; Williams, N.A., The estimation of the Adsorption Coefficient (Koc) for Soils by High Performance Liquid Chromatography. *Chemosphere*, ISSN: 0045-6535, Vol. 17, No. 1, 1988, pp. 67-77.

International Agency for Research on Cancer (IARC). Monograph: Overall Evaluations of Carcinogenicity to Humans. 2006. Available at: http://monographs.iarc.fr/ENG/Classification/crthgr01.php. Accessed March 30, 2008.

IPCC, Climate Change 2001: Synthesis Report. A Contribution of Working Groups I, II and III to the Third Assessment Report of the Intergovernmental Panel on Climate Change, Watson, R.T, (ed.), Cambridge University Press, New York, USA, 2001.

IPCC, Climate Change 2001: Synthesis Report. A Contribution of Working Groups I, II and III to the Fourth Assessment Report of the Intergovernmental Panel on Climate Change, Watson, R.T, (ed.), Cambridge University Press, New York, USA, 2007.

Irwin, R. J.; Van Mouwerik, L.; Stevens, L.; Seese, M.D.; Basham, W., Environmental Contaminants Encyclopedia. National Park Service, Water Resources Division, Fort Collins, Colorado, Distribuited as an Electronic Document, 1997.

Jones, A.H., Sublimation Pressure Data for Organic Compounds. *Journal of Chemical and Engineering Data*, ISSN: 0021-9568, Vol. 5, No. 2, 1960, pp. 196-200.

Kan, A.T.; Fu, G.; Hunter, M.; Chen, W.; Ward, C.H.; Tomson, M.B., Irreversible Sorption of Neutral Hydrocarbons to sediments: Experimental Observations and Model Predictions. *Environ. Sci. Technol.*, ISSN: 0013-936X, Vol. 32, No. 7, 1998, pp. 892-902.

Karickhoff, S.W., Semi-empirical Estimation of Sorption of Hydrophobic Pollutants on Natural Sediments and Soils. *Chemosphere*, ISSN: 0045-6535, Vol. 10, No.8, 1981, pp. 833-846.

Keeling, C.D.; Whorf, T.P., Atmospheric CO_2 concentrations (ppmv) derived from in situ air samples collected at Mauna Loa Observatory, Hawaii. The Carbon Dioxide Research Group. Scripps Institution of Oceanography (SIO,) University of California, La Jolla, California USA 92093-0444, 2005.

Lüers, F.; Ten Hulscher, T.E.N., Temperature Effect on the Partitioning of Polycyclic Aromatic Hydrocarbons between Natural Organic Carbon and Water. *Chemosphere*, ISSN: 0045-6535, Vol. 33, No. 4, 1996, pp. 643-657.

Lyman, W.J.; Reehl, W.F.; Rosenblatt, D.H., Handbook of Chemical Property Estimation Methods, Environmental Behavior of Organic Compounds, American Chemical Society, ISBN: 0-8412-1761-0, Washington, DC, 1990.

Mackay, D.; Shiu, W.Y.; Ma, K.C., Illustrated Handbook of Physical Chemical Properties and Environmental Fate for Organic Chemicals. Vol. II: Polynuclear Aromatic Hydrocarbons, Polychlorinated Dibenzodioxins and Dibenzofurans, ISBN: 1566702550, Lewis Publishers, Boca Raton, FL., 1992.

Maruya, K.A.; Risebrough, R.W.; Horne, A.J., Partitioning of Polynuclear Aromatic Hydrocarbons between Sediments from San Francisco Bay and Their Porewaters. *Environ. Sci. Technol.*, ISSN: 0013-936X, Vol. 30, No. 10, 1996, pp. 2942-2947.

McCall, P.J.; Swann, R.L.; Laskowski, D.A.; Unger, S.M.; Vrona, S.A.; Dishburger, H.J., Estimation of Chemical Mobility in Soil from Liquid Chromatographic Retention Times. *Bull. Environ. Contam. Toxicol.* Vol. 24, No. 190, 1980.

Meylan, W.M.; Howard, P.H.; Boethling, R.S., Molecular Topology/Fragment Contribution Method for Predicting Soil Sorption Coefficients. *Environ. Sci. Technol.*, ISSN: 0013-936X, Vol. 26, No. 8, 1992, pp. 1560-1567.

Meylan,WM, Howard,PH, Boethling,RS et al. "Improved Method for Estimating Bioconcentration / Bioaccumulation Factor from Octanol/Water Partition Coefficient", *Environ. Toxicol. Chem.* 18(4): 664-672 (1996).

Mihelcic, J.R.; Auer, M.T.; Hand, D.W.; Honrath, R.E.; Perlinger, J. A.; Urban, N. R.; Penn, M. R., *Fundamentals of Environmental Engineering* , John Wiley and Sons, New York, 1999.

Minamoto, T.; Mai, M.; Ronai, Z., Environmental Factors as Regulators and Effectors of Multistep Carcinogenesis, *Carcinogenesis*, ISSN: 0143-3334, Vol. 40, No. 4, 1999, pp. 519-527.

Nabholz, J.; Clements, R.;; Zeeman, M.; Osborn, K.; Wedge, R., "Validation of Structure Activity Relationships Used by the U.S. EPA's Office of Pollution Prevention and Toxics for Environmental Hazard Assessment of Industrial Chemicals". *Environmental Toxicology and Risk Assessment, 2nd Volume*. ASTM STP 1216. Gorsuch, F.; Dwyer J.; Ingersoll, C.; La Point, T. (Eds.). American Society for Testing and Materials, Philadelphia, 1993.

National Library of Medicine, Hazardous Substances Databank (HSDB), a database of the National Library of Medicine's TOXNET system(http://toxnet.nlm.nih.gov), October, 2002.

National Toxicology Program, Eleventh Report on Carcinogens, U.S. Department of Health and Human Services, Public Health Service, Available at: http://ntp.niehs.nih.gov/ index.cfm.

National Toxicology Program, Listing Criteria. Available at: http://ntp.niehs.nih.gov/ index.cfm.

International Agency for Research on Cancer (IARC). Monograph: Overall Evaluations of Carcinogenicity to Humans. 2006. Available at: http://monographs.iarc.fr/ENG/

Oris, J. T.; Giesy, P. M.; Alfred, P. M.; Grant, D. F.; Landrum, P. F., Photoinduced Toxicity of Anthracene in Aquatic Organisms: An Environmental Perspective. Veziroglu, T. N. (ed.), The Biosphere: Problems and Solutions. Elsevier Sci. Publ., Amsterdam, 1984.

OSHA, Occupational Safety and Health Administration, VI Health Effects Discussion and Determination of Final PEL.US Department of Labor, May 2002.

OTA, Office of Technology Assessment, Congress of the United States, Assessment of Technologies for Determining Cancer Risks from the Environment. PB81-235400, U.S. Government Printing Office, Washington, D.C., 1981.

Piwoni, M.D; Keeley, J.W., Basic Concepts of Contaminant Sorption at Hazardous Waste Sites. EPA/540/4-90/053. Environmental Protection Agency, Office of Research and Development, Ground Water Issue, October 1990.

Pope, A.; Burnett, R.; Thun, M.; Calle, E.; Krewski, D.; Ito, K.; Thurston, G., Lung Cancer, Cardiopulmonary Mortality, and Long-term Exposure to Fine Particulate Air Pollution. *Journal of American Medical Association*, Vol. 287, No. 9, 2002, pp. 1132-1141.

Self F., Ekholm E, Bower K. Refining overview-petroleum, processes and products. American Institute of Chemical Engineers, Fuels and petrochemical divisions, 2000.

SETAC, Society of Environmental Toxicology and Chemistry, Chemical Ranking and Scoring: *Guidelines for Relative Assessments of Chemicals,* Swanson, M., Socha, A. (ed.), SETAC Press, ISBN: 1-880611-12-0, FL, USA, 1997.

Southworth, G.; Beauchamp, J.; Schmieder, P., Bioaccumulation Potential of Polyciclic Aromatic Hydrocarbons in Daphnia pulex, *Water Res*., Vol. 12, 1978, pp., 373-977.

SRC, Compilation of Ozone Depletion Potentials (ODP) and Global Warming Potentials (GWP), SRC Products and Services, http://esc.svrres.com, 2002.

Swann, R.L.; Laskowski, D.A.; McCall, P.J.; Vander Kuy, K.; Dishburger, H.J., A Rapid Method for the Estimation of the Environmental Parameters Octanol/Water Partition Coefficient, Soil Sorption Constant, Water to Air Ratio, and Water Solubility. *Res. Rev*., Vol. 85, No. 23, 1983.

Tomatis, L.; Huff, J.; Hertz, I.; Sandler, D.; Bucker, J.; Boffetta, P.; Axelson, O.; Blair, A.; Taylor, J.; Stayner, L.; Barret, C., Avoided and Avoidable Risks of Cancer. *Carcinogenesis*, ISSN: 0143-3334, Vol. 18, No. 1, 1997, pp. 97-105.

U.S. Department of Health and Human Services, Report on Carcinogens, 10th Ed.; Public Health Service, National Toxicology Program, December 2002.

Veith, G.D., et al. 1980. "An Evaluation of Using Partition Coefficient and Water Solubility to Estimate Bioaccumulation Factors for Organic Chemicals in Fish". In: *Aquatic Toxicology,* J.G. Eaton, et al. (Eds.), American Society for Testing Materials, Philadelphia, PA. ASTM 707

Verschueren, K., *Handbook of Environmental Data on Organic Chemicals*. 2nd Ed. Van Nostrand Reinhold. ISBN: 0442288026 New York, 1983.

Verschueren, K., *Handbook of Environmental Data on Organic Chemicals*, 3rd Ed., Van Nostrand Reinhold Co, ISBN: 0442024215, New York, 1996.

Wark, K.; Warner, C.F.; Davis W.T. Air Pollution: Its Origin and Control, *Addison Wesley*, 1998.

World Health Organization, WHO. European Centre for Environmental and Health International Programme on Chemical Safety. Assessment of the Health Risk of Dioxins: Re-evaluation of the Tolerable Daily Intake (TDI), *Executive Summary*, WHO Consultation, May 25-29 1998, Geneva, Switzerland.

World Health Organization, WHO, Environmental Health Criteria 88: Polychlorinated Dibenzo-para-dioxins and Dibenzofurans, 1989.

In: Industrial Pollution Including Oil
Editor: Harry Newbury and William De Lorne
ISBN 978-1-60456-917-9

Chapter 2

IS BIODIESEL AN ECO-COMPATIBLE FUEL? TOXICITY ESTIMATION TO ORGANISMS OF DIFFERENT TROPHIC LEVELS

Iracema Andrade Nascimento**[*1,2]**, Solange Andrade Pereira**[1]**, Maria Bernadete Neiva Lemos Leite[*1,2]**, Andréa Cristina Santos da Cruz**[1,2]**, Jársia de Melo Santos***[*1,2]**, Débora Andrade Barros**[1]**, Thaís Fernandes Veras**[1]**, Heiddy Márquez Alvarez**[2]**, Maurício Andrade Nascimento**[2]*

[1]Laboratório de Biologia Marinha e Biomonitoramento – IBIO/ UFBA
[2]Faculdade de Tecnologia e Ciência (FTC)

ABSTRACT

Following the worldwide trend, Brazilian enterprises have been developing research to introduce into the market fuels that could be more eco-compatible. The PNPB (National Program on Biodiesel Production), introduced officially (Law n° 11.097) the biodiesel in the Brazilian Energy Matrix in 2005, making mandatory a partial substitution of diesel fossil by biodiesel, starting with a 2% mixture by 2008 and increasing to a minimum of 5%, target by 2013. Estimates of a national production of 3.8 billion L of biodiesel in 2009, will meet the need for a 3% share, but this acceleration in production may involve environmental threats, depend on how biodiesel is produced, used and traded. Based on the reduction of toxic emissions, biodiesel is considered environmentally less harmful, when compared with fossil diesel. However, ecotoxicological information on products and co-products related to biodiesel production are incomplete, and also are unknown the toxicity of the effluents generated by the productive processes of this fuel. The soluble fractions of the different types of biodiesel, and the effluents generated in the productive processes can contaminate the ground and water. Ecotoxicological techniques that lead to risk assessment shall be available to prevent pollution problems. Early-life-stage tests and biomarkers are important tools for

[*] toiracema@oi.com.br

toxicity determination and pollution prevention. This research aimed to contribute to biofuels risk assessment modeling by determining the toxic effects caused by the water soluble fractions (WSFs) of three different biodiesel fuels (B_{100}) originated by transesterifacation of castor oil (*Ricinus communis* L.), palm (dende) oil (*Elaeis guinnensis*) and waste cooking-oil (WCO) through, respectively, homogeneous NaOH catalysis, acid catalysis, and acid and basic catalysis. Different organisms (the microalga *Skeletonema costatum*, the sea-urchin *Echinometra lucunter,* and the fish *Oreochromis niloticus*), and diverse techniques were used, such as microalgae culture growth inhibition, early-life–stage-toxicity tests, and the stability of lysosomes, isolated from hepatocytes of exposed fishes, as a biomarker. The functional integrity of the lysosomal membranes was evaluated via the acid phosphatase activity, expressed in mU/mg of total proteins. The results showed significant changes in the stability of lysosomes from fishes exposed to different biodiesel-WSFs, in relation to control animals. Among the analyzed biodiesel fuels (B_{100}), the *Ricinus communis* (castor oil) biodiesel was the most toxic (5, 0 mU/mg), followed by WCO (3, 9 mU/mg) and *Elaeis guineensis* (palm oil) biodiesel (2,4 mU/mg), when compared to the control (1,6 mU/mg). Similar toxicity trends were obtained for the sea urchin ELS-test, as far as for the *S. costatum* culture growth-inhibition-test, the first providing WSF-EC_{50} values of 2,45%, 8,06% and 22,75%, while the second presented values of WSF-IC_{50} of 1.29%, 8.23% and 5.48%, respectively for biodiesel fuels from castor oil, palm (dende) oil and WCO. Chemical analyses and obtained chromatograms suggest that the relatively high toxicity of the analyzed biodiesel fuels is related to methanol, formed by a hydrolytic reversion of the transesterification reaction.

Keywords: biodiesel toxicity, acid phosphatase biomarker, ELS-test, microalgae growth-inhibition- test.

Introduction

The use of fossil fuels has advanced technology–based civilization to unprecedented levels (Burkhardt et al. 2006). However, this energy choice has lead to pollution and to climate changes, possibly enhanced by the excessive production of greenhouse gases, including CO_2 emissions. Though the amount of CO_2 emissions by mankind is relatively small as compared to CO_2 emissions by the nature, the anthropogenic participation in climate changes is considered as relevant, since the population may control some key issues related to these changes (IPCC, 2007). In view of the environmental damaging tendencies, consolidated at the end of the last century, this industrial civilization is facing the necessity to adopt new paradigm to redesign the political-financial system in order to maintain its supporting environmental system, thus permitting the continuity of the economical advances, which should be based on cleaner energy sources (Nascimento et al., 2008)

Healthy ecosystems are the major supplier of vital resources to humans. The estimated world population growth, from 6 billion to 9 billion by 2050 (WCED, 2007), represents an increased threat of global food and goods shortage, considering the maintenance of the current consumption rates. Presently, 150 major nations of the world already show an ecological deficit (Global Footprint Network, 2003). Taken together, the ecological footprint (resource demand = population X per capita consumption) of all nations in the year 2001, was about 20% higher than the ecological capacity of the Earth, causing losses on biodiversity. A population decline of 1.145 terrestrial marine and freshwater species around the world was recorded by the Millennium Ecosystem Assessment (2005). The energy system, developed historically on fossil fuel resources, is being pointed out as the base of these losses of biodiversity and other environmental damages.

In assessing the possible impact of rising atmospheric concentrations of CO_2 and other greenhouse gases (GHGs) the WMO/UNEP Intergovernmental Panel on Climate Changes (IPCC, 1996) projected a temperature increase in a range of 1 to 3,5 ^{O} C, and a rising of the sea level within a range of 15 to 95cm, by the year 2100 . These estimates have been lately adjusted to more robust model responses, analyzed under several different scenarios (IPCC, 2007); however, they still point out the emission of CO_2 from fossil fuel use and from the effects of land use change on plant and soil carbon, as the primary sources of increased atmospheric CO_2 and the Earth temperature, which may lead to a reduced productivity in the tropics and sub-tropics, and to the displacement of millions of people living in low areas, due to sea level rising. In a warmer world, the standing biomass of forests may not increase, because of extended ranges of pathogens and increasing frequency and intensity of fires, which will contribute to the enhancement of atmospheric CO_2. In spite of the uncertainties of these predictions, the balance of evidences suggests that there is a discernible human influence on global climate, and that future GHGs emissions will be a function of global energy demand and the development and introduction of carbon free and low carbon energy technologies, as referred by the International Energy Outlook (IEO, 2007).

Estimates of the world energy demand point out a 57% increase from 2002 to 2025 (IEO, 2007). Petroleum, charcoal and natural gas, which represent most of the world consumed energy, are supposed to be exhausted in the near future (Schuchardt et al., 1998). World petroleum reserves are calculated as totaling 1.147,80 billions of barrels. Considering the estimated annual consumption of this fossil fuel as 80 millions of barrels/day, it can be concluded that these reserves are supposed to be exhausted in the year 2046. Even though punctual new petroleum sources, such as the recently discovered reserve in Brazil (ANP, 2007), may represent additional availability of petroleum, it is estimated, for a next future, an increase in the international petroleum prices, which enhances the necessity of exploring alternative energy sources. One suggestion, contested by some research groups (Doornbosch and Steenblik, 2007; Searchinger et al., 2008), is to meet part of the present and future energy demand by increasing the production of biofuels.

Biofuels as Alternative to Fossil Fuels

Over the past three decades, the search for alternatives to finite fossil fuels resources has resulted in the development of fuel technologies based on the use of renewable agriculture-generated materials as feedstock (Haas et al., 2006). Two oxygenated biofuels (biodiesel and ethanol) have received intensive attention as potential petroleum substitute, due to their properties and environmental friendliness (Yang et al., 2000). Even though their environmental friendliness is recognized, renewable fuels are not strictly eco-compatible (Righelato and Spracklen, 2007), considering that, if they are obtained through energetic processes of first generation (vegetal oils to biodiesel), its production may be associated with the expansion of monoculture, demanding very large areas of agriculture land, which directly or indirectly will put further pressure on natural forests and grasslands (Doornbosch and Steenblik, 2007).

The global marketing for biodiesel is poised for explosive growth in the next ten years. It is possible that biodiesel could represent as much as 20% of all on-road diesel used in Brazil, Europe, USA, China and India by the year 2020 . The trend is towards lower cost multiple feedstock options and non-food feedstock for sustainable biofuels (Liebreich, 2008). However, a 10% substitution of petrol and diesel fuel is estimated to 43% and 38% requirement of current cropland area in the United States and Europe, respectively (International Energy Authority, 2004), which may lead to lowering of habitats, losses of biodiversity and competition for food production (Shuchardt et al, 1998). Moreover, clearance of land for energy crops, results in oxidation of carbon stores in the vegetation and soil (Searchinger et al., 2008). According to Righelato and Spracklen (2007) the consequent liberation of CO_2 would out-weigh the possible emissions avoided by the use of the biofuel.

Due to its biodiversity, climatic conditions and large areas for cropland (90 millions ha available), the Brazilian potential, as producer of biofuels (both, ethanol and biodiesel), is recognized worldwide (World Bank, 2004; World Bank, 2008; ORNL, 2008). Ethanol is a low cost oxygenated alcohol with high oxygen content (35%), which has been produced in Brazil from sugarcane and used, since the seventies, in cars engines, in total substitution, or mixed (25%) with gasoline. Biodiesel consists of long chain fatty acids esters, derived from renewable sources such as vegetable oils, waste greases or animal fats by conversion of the triglyceride fats to esters via transesterification, pyrolysis and micro-emulsification. Pyrolysis refers to chemical change caused by the application of thermal energy, in the presence of an air or nitrogen sparge; it is strictly defined as the conversion of one substance into another by means of heat or by heat with the aid of a catalyst (Ma and Hanna, 1999). Micro emulsions are isotropic, clear, or translucent and thermodynamically stable dispersions of oil, water, a surfactant and often a small amphiphilic molecule as co-surfactant (Fukuda et al., 2001). However, the most common way to produce biodiesel is by transesterification, which requires a catalyzed chemical reaction involving a vegetable oil and an alcohol to yield fatty acid alkyl esters (biodiesel) and glycerol (by-product). Methanol is the most commonly used alcohol, (Ma and Hanna, 1999; Zhang et al., 2003), especially due to its low cost and its physical and chemical advantages (polar and shortest chain alcohol).

Production of alkyl esters (biodiesel) from vegetable oils can be carried out using different processes, involving inorganic acid, inorganic base or enzymatic catalysis (Haas 1997; Fukuda et al., 2001; Royon et al., 2007). In view of product separation problems, such as glycerol recovery and removal of inorganic salts (Mittelbach, 1990; Basri et al., 1997) when using the chemical catalyzed processes, the enzymatic synthesis of biodiesel has shown significant progress (Shimada et al, 2002; Zhang et al., 2003). The use of lipases in transesterification of oils for biodiesel production addresses these problems and offers an environmentally more attractive option to the conventional processes (Iso et al., 2001; Kose et al., 2002; Shimada et al., 2002; Modi et al., 2007); however, this approach makes process costs comparatively higher, unless the immobilization of the enzyme, which allows for its recovery and reuse, be successful (Xu et al., 2005; Balcao et al.,1996; Modi et al., 2007).

The utilization of biodiesel in Brazil is associated with the substitution of the diesel oil in engines, as pure fuel (B_{100}) blended with fossil diesel, denoted as B_{XX}, where XX represents the percentage of biodiesel in the blend. This substitution, however, still requires increased efforts in research and development of alternative methods for producing biodiesel from renewable resources, at competitive costs and with chemical characteristic, which fit strictly the industry specifications of a fuel-grade biodiesel, which conform to ASTM D76751. Moreover, the biodiesel market still involves uncertainties due to the constraints of feedstock availability, even though it reached impressive production capacity levels in the last three years, due to the security of regulation and the production trading guarantee.

At present, the high cost of biodiesel is the major obstacle to its commercialization. Its cost depends on several factors, the most important being feedstock and production processes. Calculated costs of production (Haas et al., 2006) ranged from US$ 0.30/L ($1.14/gal) for fuel produced from soybeans, to US$ 0.69/L ($2.62/gal), when rapeseed is the feedstock. Usually biodiesel costs more than US$ 0.50/L compared to US$0.35/L for petroleum-based diesel (Lott, 2002; Prokop, 2002; ANP, 2007). In Italy the biodiesel price is 2-3 times higher than that for diesel fuels (Carraretto et al., 2004). In Malaysia, its price is 5 times higher for soybean and 9 times for corn biodiesel (Kalam and Masjuki, 2002). In Brazil, the price is around 50-60% higher than diesel (EMBRAPA, 2005). According to Connemann and Fischer (1998), these high prices are due to the cost of virgin vegetable oil.

There are diverse attempts to minimize the raw material cost of biodiesel production; one of them is exploring the use of cooking oil (about half the price of virgin oil), as effective options of cost reduction and waste oil disposal (Wiltsee, 1998). Another one is exploring the possibility of getting the oil from microalgae biomass, at competitive prices, based on the principle of avoiding competition with food production, in relation to the use of agriculture land (Nascimento et al., 2008).

BIODIESEL AND BLENDS: ENVIRONMENTAL PERFORMANCE AND RISKS

Biodiesel gives generally comparable performance engine to that of conventional diesel fuel, while reducing engine emissions of hydrocarbons and carbon monoxide (Graboski and McCormick, 1998). Information on the production, quality specifications, performance and emissions properties of biodiesel (Pinto et al., 2005) became very sound in support of its increasing production and consumption.

In Brazil, the interest on biodiesel was developed from 1980, when the first patent on biodiesel production process was filed. However, it was on late 2004 that the Brazilian Federal Governor announced the Production and Use of Biodiesel Program. On 2005, the introduction of biodiesel in the Brazilian Energy Matrix was made official through legislation (Law 11.097). The Brazilian Program involved a partial substitution of diesel fossil for biodiesel, starting with a 2% mixture by 2008, and increasing by to 5% by 2013. Recently the National Council of Energetic Politics (CNPE) made mandatory the mixture of 3% in 2009 and there is a clear proposal to reach a higher percentage of diesel substitution, depending on the industrial production capacity and the availability of feedstock.

Based on the reduction of toxic emissions, biodiesel is considered environmentally less harmful compared with fossil diesel (Sharp et al., 2000; Wang et al., 2000; Cardone et al., 2002). However, the national acceleration in production may involve threats depend on how biodiesel is produced, used and traded (OECD, 2006). The actual biodiesel production capacity installed in Brazil is 3.8 billions L/year. However the prospects exist of increasing the use of this fuel in the Brazilian Energetic Matrix (ANP, 2007). The estimated production for 2009 and beyond, may involve environmental risks: first, because the first generation biofuels still includes inefficient conversion technologies, requiring the extension of crops, which is likely to place pressure on natural ecosystems and biodiversity, leading to losses in habitats or to soil and water pollution, due to pesticides and fertilizers run-off, caused by inappropriate agricultural practices (ORNL, 2008); second, because there is no sound knowledge relative to the ecotoxicological characteristics of effluents generated in the biofuels production processes, the toxicity of biofuels water-soluble-fractions (Nascimento et al, 2008) or the biodiesel exhausts, when reach soil or water.

Many authors refer to biodiesel as being biodegradable and non-toxic (Zhang et al. 1998; Fukuda et al., 2001; Zhang et al., 2003; Leung et al. 2006). Studies have clearly indicated that biodiesel can substantially reduce emissions of PM and CO (Sharp et al., 2000; Wang et al., 2000; Cardone et al., 2002), as well as be free of SOx and aromatic compounds (Knothe et al., 1997; Al-Widyan et al., 2002). There is a slight increase in NOx emissions, which can be positively influenced by delaying the injection timing in engines (Agarwal et al., 2006). Compared to petroleum–based diesel, biodiesel has a more favorable combustion emission profile, such as low emission of carbon monoxide, particulate matter and unburned hydrocarbons (Agarwal and Das, 2001). An analysis on life cycle (Sheehan et al., 1998) showed that the overall life cycle emissions of CO_2 from 100% biodiesel are 78.45% lower than those of petroleum diesel, and a blend with 20% biodiesel fuel reduces net CO_2 emissions by 15.66%. Biodiesel has a relatively high flash point (150^{0}C), which makes it less volatile and safer to transport or handle than petroleum diesel.

However, the above studies have not investigated the unregulated emissions from diesel engine using biodiesel-diesel blends, such as carbonyl compounds. Very limited information is available on the unregulated emissions. One of the most important studies on non-regulated pollutants (Turrio-Baldassarri et al., 2004) evaluated the emissions of PAHs, nitro PAHs, carbonyl compounds and light aromatics hydrocarbons of mixtures biodiesel/diesel as far as its mutagenicity is considered. They reported overall small reduction of the aromatic and polyaromatic compounds and their increase by 18% with biodiesel blend. Some carbonyl such as formaldehyde, acetaldehyde, acrolein, and methyl ethyl ketone are toxic, mutagenic and even carcinogenic to human (Carlier et al., 1986). Furthermore carbonyls are important precursors to free radicals (HOx), ozone, and peroxyacylnitrates (Carter, 1995; Gaffney et al., 1997; Pang et al., 2006).

Corrêa and Arbilla (2006) investigated the effect of the biodiesel addition to fossil diesel on the emission profile of mono-and poliaromatics hydrocarbons (MAHs and PAHs) in a six-cylinder heavy-duty engine. The average MAHs reductions were 4.2% (B_2), 8.2% (B_5) and 21.1% (B_{20}), while for PAHs, were 2.7% (B_2), 6.3% (B_5) and 17.2% (B_{20}). However, some PAHs and MAHs emissions increased with the use of biodiesel blends, like phenanthrene, ethyl-benzene and trimethyl-benzene. Based on the overall results they concluded with the beneficial effects of the blends. However, the authors called the attention for the necessity of analysis of carbonyls and other volatile organic non-regulated compounds for a more realistic interpretation of the beneficial effects detected in biodiesel/diesel blends in relation to aromatics.

Comparing the mutagenic and cytotoxic effects of diesel and biodiesel (rapeseed oil methyl esters) as fuels, by analyzing the engine (modern passenger car) exhausts, Bünger et al.,(1998), showed a higher mutagenic potency of diesel, compared to biodiesel. This probably was due to the lower content of PAHs and sulfur content in biodiesel exhausts, although the emitted particulate matter, aldehydes such as acrolein (2-propenal), acetaldehyde and formaldehyde and ketones emission increased in biodiesel in relation to diesel emissions exhausts, determining however, not significant higher cytotoxicity.

Most of the studies on biodiesel toxicity refer only to regulated emissions and very scarce data are available on the toxicity of products and especially co-products related to biodiesel production (Knothe et al. 2006). The soluble fractions of the different types of biodiesel, and the effluents generated in the productive processes can contaminate land and water as a consequence of circumstantial spillage, negligence or incompetence. The scarcity of research dealing with the effects of the effluents from the biodiesel production processes or the soluble fractions of spilled biodiesel on aquatic biota does not correspond to the precautionary principle prescribed by law, in countries where the estimated production and use of biodiesel may represent environmental risks.

PREVENTIVE TOOLS TO ASSESS BIODIESEL-RELATED ENVIRONMENTAL RISKS

While many potential causes for the loss of aquatic species have been identified, chemical contaminants are suspected as major contributors (Nascimento et al. 2000a ; Paixão et al., 2007). The greater danger to coastal and freshwater habitats is determined by chronic toxicity associated with urban run-off, in particular, with the water-soluble-fractions (WSFs) originated from toxic products. The incorporation of sub-lethal quantities of contaminants may reduce the organism´s tolerance to other stress factors (Ralph and Burchett, 1998; Nascimento et al., 1998a; Leite et al., 2005) and may lead to long-term population and ecosystems damage.

Most of the techniques used to assess the toxic effects of chemical substances on aquatic organisms have been based on chemical analysis of pollutant agents, by using samples taken from the ecosystems; or they were assessed through toxicity tests, also carried out in laboratories, generally using one of the species from the threatened environment (Winkaler et al., 2001).

Ecotoxicological techniques that can lead to risk assessment are available to detect pollution problems. However, innovative approaches are required that are easy and cheap to deploy and able to provide risk analysis based on estimates of the harmful effects of the waste on the ecosystem health. Approaches such as bioassays can provide readily accessible and useful information to support the precautionary principle and conservative strategies of source control. Since the first eco-toxicological effects on organisms result from interactions between the pollutant agent and the biota at molecular, cellular and tissue levels (Fent, 2001), early-life-stage tests and biomarkers are important tools for toxicity determination and pollution prevention (Nascimento, 1996; Nascimento et al., 1998b ; Nascimento et al, 1999; Nascimento et al 2000b; Martins et al. 2005; Evangelista et al., 2005;; Paixão et al., 2007; Cruz et al., 2007)

Biomarkers, can be defined as measurements carried out in biological liquids, cells or tissues which indicate, in biological or cellular terms, the presence of contaminants or the magnitude of the response of organisms to their toxic effects (Bodin et al., 2004; Nascimento et al. 2006). These ecotoxicological tools can detect cyto-physiological or histo-pathological alterations that occur in aquatic organisms when exposed to contaminants, previously to disease installation and death (Petrovic et al., 2004).

This research aimed to contribute to biofuels risk assessment modeling by determining the toxic effects caused by the water soluble fractions (WSFs) of 3 different biodiesels (B_{100}) originated through transesterifacation of castor oil (*Ricinus communis* L.), palm (dendê) oil (*Elaeis guinnensis*) and waste-cooking-oil (WCO). The toxic effects were estimated by using different organisms (the microalga *Skeletonema costatum*, the sea-urchin *Echinometra lucunter,* and the fish *Oreochromis niloticus*), and techniques such as microalgae culture-growth-inhibition test, early-life–stage-toxicity test, and the stability of lysosomes, isolated from hepatocytes of exposed fishes, as biomarker.

Materials and Methods

Biodiesel samples were taken from the experimental Energy Laboratory at the Federal University of Bahia, as B_{100} (neat biodiesel), obtained by transesterification of commercial oils from *Ricinus communis* (castor oil), *Elaeis guinnensis* (palm oil) and waste-cooking-oil (WCO), by using, respectively, homogeneous NaOH catalysis, acid catalysis, and acid and basic catalysis. All the steps of the production process were carefully controlled in order to obtain biodiesel meeting the ANP specifications, in terms of their physical and chemical properties. At the Biomonitoring Laboratory (Institute of Biology, Federal University of Bahia), the biodiesel samples were treated according to U.S EPA (1984) method, to obtain the water-soluble-fraction (WSF) from each oil. The samples have been stirred in one homogenizer (1500rpm) and diluted (1:9 v/v) in filtered fresh or seawater, dependent on the requirements of the test-species, and stirred at a constant velocity in a covered Mariotti flask (150 rpm, vortex of 25% of the Mariotti flask height), for 20 hours. After decantation and a clear separation of oil/water phases, the WSF was collected from the tap located at the bottom of the Mariotti flask, and distributed in test containers, according to the specified concentrations, required for each test protocol (Nascimento et al., 2002).

Toxicity Bioassays

The bioassays have been carried out on three organisms: a fish, the Tilapia, *Oreochromis niloticus*, the sea urchin *Echinometra lucunter* and the microalga *Skeletonema costatum.* Three different test protocols (Nascimento et al. 2002) have been applied, providing results based on different end-points: culture growth inhibition for microalgae, net risk based on percent abnormal, for the embryonic development test using sea-urchin, and the stability of lysosomes, isolated from hepatocytes of exposed fishes, as a biomarker.

I. The Use of the Lysosomal Stability as Biomarker

I.1. Test conditions

Each juvenile Tilapia was exposed to 10%-WSF/freshwater (v/v), inside 4L-aquarium, for 24h, at a constant temperature (22°) and without any food. Each experiment was controlled by another, in which fish were held in similar aquaria containing only freshwater. All the experiments were carried out in triplicate. Physical and chemical parameters such as the amount of dissolved oxygen (Orion OD meter), pH (Procyon pH meter), ammonium (Procyon ion meter) and temperature, were evaluated at the beginning and the end of each experiment.

I.2. Separation of the Lysosomal Fraction in Liver Tissue

Fishes were dissected alive. Liver was rapidly removed and cleaned of accessory connective and adipose tissues in an ice-cold sucrose solution 25.10^{-2} M, pH 7.4, refrigerated at 4°C. The fragments of each liver were washed in a proportion of 1gr of tissue to 10 mL of a special buffered solution (manitol 25.10^{-2} M, TRIS 2.10^{-3} M, EDTA 4.10^{-4} M, dissodium phosphate 10^{-2} M and monopotassium phosphate 10^{-2} M, adjusted to pH 7.4). After suspended in the same buffer free of phosphate, they were homogenized in a Potter-Elvejhem apparatus, at 250 rpm for a maximum of 2 minutes, in a bath of melting ice. The various components of the hepatocytes were separated by fractioned centrifugation (900 to 22000g) at 4 °C, according to the diagram showed below.

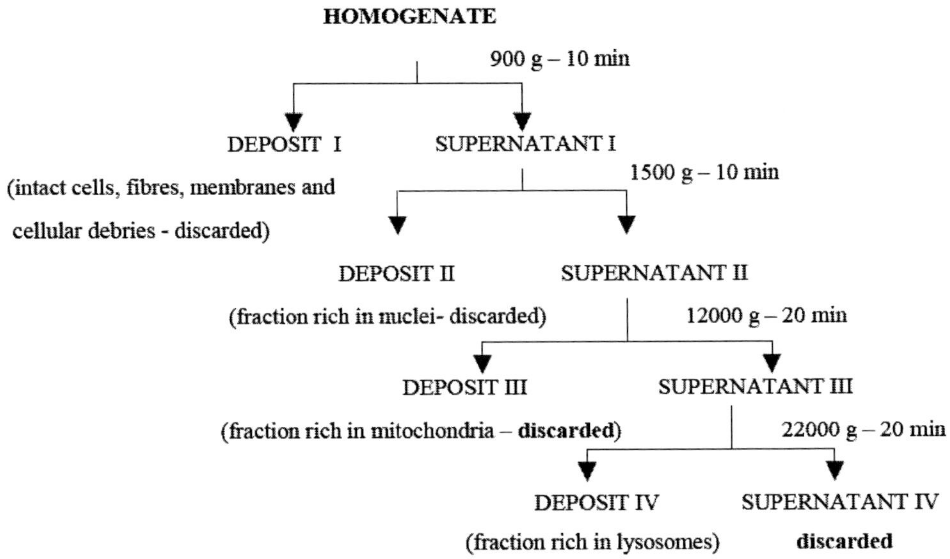

I.3. Determination of the Acid Phosphatase Activity

The sediment originated from centrifugation at 22.000g, which contained most of the lysosomes, was re-suspended into the same quantity of the buffered solution as the initial homogenized material, and divided into 2 tubes. A 4.10^{-3} M digitonine solution was added to one of them and both were incubated in ice for 20 minutes. After this, they were centrifuged once more at 22.000g, for 20 minutes, to settle the intact lysosomes. The stability of lysosomal components was assessed by determining the acid phosphatase activity measured according to Andersch and Szcypinski (1947). This classical technique uses p-nitrophenyl-phosphate as enzyme substrate, which is transformed in p-nitrophenol and phosphoric acid. The addition of sodium hydroxide stopped the reaction and the formed p-nitrophenol (yellow anion) was quantitatively determined by photometry, at 405nm. The quantity of measured p-nitrophenol / time is directly proportional to the acid phosphatase activity, which was expressed in thousand units per milligramme of total proteins, determined according to Bradford (1976). The experimental isolation conditions of the lysosomes was controlled by the digitonine test, which, according to Ishisaka *et al.*(1998), breaks down all the lysosomes, because it attaches to the cholesterol of its membranes, labilizing them.

The results obtained on the biomarker experiment were compared with the controls, using ANOVA and the Tukey-Kramer multiple comparison test. Values were considered to be significant if $p<0.05$.

II. The Sea-Urchin Embryonic Development Test

Adult sea-urchin were collected at an area considered as free from industrial and domestic waste, at the coast of Salvador, Bahia. The animals were cleaned and kept over night in one aquarium containing filtered (20μm) natural sea-water (36 ppt), taken from the same area. Immediately prior to each test, gametes were collected from the aboral surface of males and females (3-6 individuals), after they start spawning under stimulation of a 0.5M KCL solution, injected into the coelomic cavity (Kobayashi, 1990). Pooled eggs were suspended in filtered (GF/C1.2μm) sterilized (129°C; psi1.5/cm^2) seawater. Fertilization was accomplished by transferring 1ml of sperm to 100mL of a dense egg suspension (Nascimento et al., 2002). Countings on first division embryos were made 2h later, to quantify the suspension volume necessary to be added to the test tubes, to provide 20 embryos/ml, according to the test protocol developed by Araújo and Nascimento (1999).

The obtained 100%-WSF from each biodiesel sample was dosed in a geometric dilution series of six loadings (0%, 4.6%, 10.0%, 22.0%, 46.0% and 100%), in triplicate vessels containing the same sea water used for fertilization, to which the embryos were pooled. Simultaneously, a reference toxicant-test (DSS- Dodecil Sodium Sulphate) was conducted as described by Pereira et al. (2001) to determine the embryos sensitivity and health, with the purpose to assure the test precision. After an exposure of 36h, two 10ml-samples were removed from each test vials, preserved in 5% buffered formalin and later examined under a compound microscope. The first one hundred larvae or late embryos were counted as normal or abnormal, and the responses to the different treatment were recorded as the percentage of embryos failing to develop or developing in an abnormal manner (Kobayashi, 1981; Nascimento, 2005). Dissolved oxygen, salinity, temperature and pH were maintained at 5.9-6.5ppm, 36ppt, 25+- 2^0C and 7.0 to 8.5 respectively, according to the test protocol developed by Araujo and Nascimento, (1999).

Sea-urchin embryo responses to biodiesel-WSFs were expressed as a percentage net risk of abnormality, using Abbot's formula (Finney, 1971). Effects were calculated based on concentration-response curves and analyzed by the Trimmed Spearman Karber (Hamilton et al., 1977) computer statistical method. This was done to provide EC_{50} values, equivalent to the biodiesel-WSFs concentrations that may cause abnormalities in 50% of the exposed embryos. Means of the EC_{50}-results and their standard deviation have been calculated using the Graphpad Instat, version 3.0 (Graphpad Software Inc., 1997).

III. The Microalgae Test

The microalgae test was carried out according the methodology specified in the ISO 10253:1995(E), which is standardized to determine toxic effects caused by chemical contaminants on marine microalgae growth. The test involved the exposure of a mono-specific *S.costatum* inoculum, in exponential growth phase, to dilutions of the different biodiesel-WSFs.

The inocula for the *Skeletonema costatum* culture were obtained from the algae bank at the Institute of Biology, Federal University of Bahia, where they have been maintained under appropriate controlled conditions (Leite, 1986). Modified Conway medium was used for this microalgae cultivation (Griffith et al, 1973). The algae from exponential growing phase cultures were quantified by counting the cells in a Neubauer chamber, under an optical microscope. Calculations have been made to provide an inoculation of 1000 cells/ml in each flask test (closed Erlenmeyer flasks, 150ml capacity), containing 100ml of the culture medium (filtered and autoclaved natural sea-water from a pollution-free reference area, salinity 28, enriched with Conway solutions).

Serial dilutions (0%, 4.6%, 10.0%, 22.0%, 46.0% and 100%), of biodiesel-WSFs have been applied to this algae culture growth test. The flasks were incubated for 96h at a temperature of 24^0 C in a rotary shaker, under continuous illumination (4000-4800 lux) provided by fluorescent lamps (40W each). The culture growth inhibition was measured by the reduction on cellular number counted in Neubauer chamber, compared to an identical control (0% WSF). Simultaneously, a reference toxicant (DSS) test was conducted as described by Pereira et al. (2001). Each test was repeated three times and was fully randomized with regard to vials location during incubation and the order of fluorescence measurements and cell counts. Effects were expressed as IC_{50} values (equivalent to the biodiesel-WSFs, which causes 50% culture growth inhibition), estimated by Trimmed Spearman Karber (Hamilton et al., 1977), based on concentration-response curves traced from data obtained by optical counter (Neubauer chamber) and checked by coulter counting (Coulter Counter model ZI 991 3044-B). The coefficient of variation (CV=sd / X.100) among tests was calculated to determine the precision and repeatability.

To determine possible significant differences in toxicity among the various biodiesel-WSFs, the comparisons of IC_{50} results were done by ANOVA (Graphpad Software Inc., 1997), followed by a Tukey test (parametric test of multiple comparison).

IV. Chemical Analysis

The water-soluble-fractions, obtained from the biodiesel samples diluted in freshwater and in seawater (36ppt), were stored in closed amber-glass flasks and kept inside a refrigerator at 4^0 C. All the samples were analyzed by gas chromatography (GC) at day 1, 60 and 120. Water soluble fractions were prepared also from all neat biodiesel samples, maintained under the same conditions for 180 days, and chemically analyzed in day 1 to provide a comparison of the three kinds of biodiesel degradation at conditions of low temperature and in a sealed environment.

All measurements were carried out using a Varian CP-3900 (Automatic Headspace Sampler, TELEDYNE TEKMAR HT3™ capillary gas Chromatograph). The analyte was adsorbed onto the SPME fiber and then thermally desorbed in the inlet of a Varian CP-3900 GC, equipped with a split/split less injection port and flame ionization detection (FID) system. The injector and detector temperatures (150°C and 250°C, respectively) were held constant during the analysis. The fused silica capillary column used for separation was a 0.30m, 0.32 mm, i.d. 3μm film thickness Varian CP-3900 (100%-methylsilicone). The GC oven was programmed as follows: the initial temperature of 50 °C was held for 5.0 min, increased to 250°C at 10 °C/min. Nitrogen was used as carrier gas at a constant flow rate of 2 mL/min and the injector was operated in split mode (20:1 split ratio).

The 10 mL vials, containing 2 mL of WSF from the three types of biodiesel B_{100} (originated from WCO, palm oil and castor oil) were capped with Teflon lined septum caps. The WSF were then exposed to the headspace of the vial, and the volatile compounds were adsorbed. The retention time of methanol (Rt = 1.30 min) was determined by direct injection of neat methanol. The actual amount of methanol present in the WSF was red as 1.10^{-4} % mass/mass. This determination was performed using the described headspace SPME, and the GC conditions. The kind of capillary column used for methanol determination, was fused silica (100%-methylsilicone), which provides reliable analysis and is suitable for a wide range of applications. Methanol peaks were clearly recognized by the use of a standard samples as control. It is well known for the determination of biodiesel degradation that methanol measurement by GC analysis is more accurate than results given by other methods (Li et al., 2007).

Results and Discussion

Biodiesel is considered as a clean fuel, when compared to fossil-fuels (Fukuda et al., 2001; Zhang et al., 2003). However, most of the studies related to its environmental effects are based on engine emissions. The data related to its toxicity or to the toxic effects caused by the effluents generated during its production processes are scarce. Data of chemical characterization of Biodiesel exist (ANP, 2007), but the toxicity evaluation cannot be analytical, since chemical analyses determine concentrations of isolated contaminants, but not their possible synergic or additive interactions that can increase its effects on biota (Nascimento et al., 2006). Chemical analysis, besides being expensive and often technically complicated to be run, do not take into account the effect of possible existing interactions between the environment and the different types of toxic agents (Sarkar et al., 2006). Biological responses to contaminants exposition are the only secure way to estimate environmental damages.

The use of biomarkers is growing in importance and increasingly being used as preventive toxicological tools, because they constitute earlier warning of chronic effects, which include long-term noticeable alterations on biota, under low contaminant exposure. They are able to detect the primary biochemical defense responses such as the induction of stress proteins (Nascimento et al., 1998; Leite et al., 2005), as far as the alteration of inducible enzymes involved in metabolism. The search for simpler and cheaper biomarkers has led to studies of lysosomes, which are able to detect the earliest cellular responses resulting from exposure to toxic pollutant agents, mainly those having lipophilic characteristics (Köhler et al. 2002; Martins et al., 2005).

Lysosomes are cellular multifunctional organelles, rich on hydrolytic enzymes, responsible for the breakdown of all the constituents of the cells and macromolecules derived from the extracellular space, via endocytosis. Besides these primary functions, they are involved in cell defense mechanisms, in the protection against toxic agents and infection by viruses and bacteria (Cuervo and Dice, 2000). Physical-chemical modifications, which lead to the loss of integrity of the membranes and diverse components of the lysosomes are almost always associated with cellular dysfunction, inflammatory and degenerative diseases, as well as apoptosis and cell death (van Nierop et al., 2006). Within this scope, lysosomes proved to be sensitive biomarkers and directly linked to the biochemical mechanisms of response and defense against aggressors (Ramos et al, 2008). The findings of Hwang et al., (2004) and Okay et al., (2006) have shown that there are significant alterations in the stability of lysosomes isolated from the tilapia liver, when exposed to hydrocarbons in the environment. Furthermore, they represent a simple, low cost technical approach (Ramos et al., 2008).

In the present research, the acid phosphatase activity was used as a biomarker of the lysosomal membranes integrity in fishes exposed to different biodiesel WSFs. The higher the enzymatic activity of the lysosomal fraction from hepatocytes of the exposed fishes, the less stable were the isolated lysosomes.

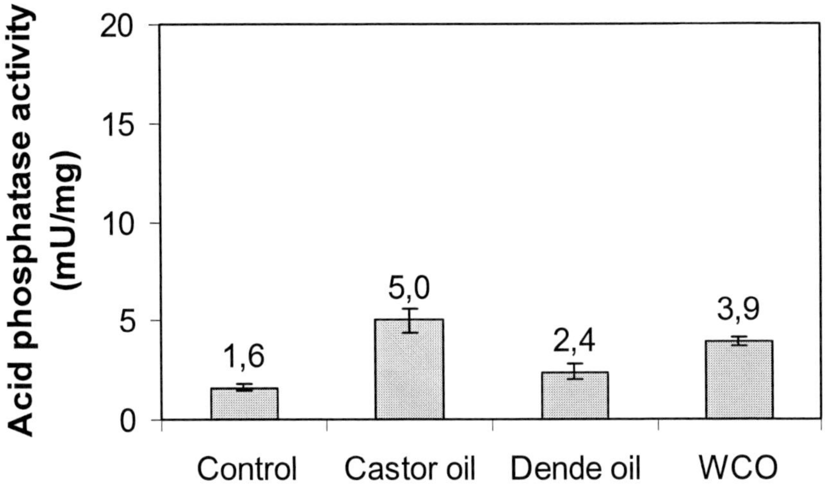

Multiple range comparison test

Values joined for the same line do not differ significantly (p> 0,05)

Biodiesel (WSF)	Castor oil	WCO	Dende oil	Control
Acid Phosphatase	5,0	3,9	2,4	1,6
(mU/mg)means	!-----------!	!----------!	!----------------------------	!

Figure 1. Values of acid–phosphatase activity, measured in the liver lysosomal fraction of fishes (Tilapia), exposed to the WSFs from castor oil, palm (dende)-oil, and WCO-biodiesel fuels.

The results (Figure 1) indicated that, among the analyzed samples, WSF of castor oil biodiesel was the most toxic (5,0±0,61mU/mg), followed by WCO biodiesel (3,9±0,20 mU/mg) and palm (dende) oil biodiesel (2,4±0,44 mU/mg). The toxicity of the analyzed WSFs was evidenced by the high values of acid phosphatase activity in liver tissues of exposed Tilapia, when compared to the control fishes (1,6±0,20mU/mg).

The migration of the enzymes to the external medium, after breakage of the lysosomes has been used previously (Petrovic et al., 2004) as one of the safest parameters for the evaluation of the physical and chemical stability of their membranes. The loss of the lysosomal enzymes was related to noxious effects of various xenobiotics (Köhler et al., 2002; Martins et al., 2005). Assessment of lysosomal stability, measured by techniques related to autophagia or endocytosis of stains, such as neutral red and cresil violet, was proved to reflect ecotoxicological responses, associated with environmental changes (Koukouzika and Dimitriadis, 2005; Martins et al., 2005). The formation of heterophagosomes containing the stains depends on cellular signals located in the plasmatic membranes. These signals are extremely sensitive to fluctuations in the extra-cellular environment (Strömhaug et al, 1998). These characteristics indicate the lysosomal membrane stability as a good biomarker (Ramos, 2005).

Techniques that make use of the various components of the lysosome are increasingly being used to monitoring contaminant effects to the most diverse species of the aquatic biota. Among them, the most reliable are those that evaluate the activity of marker enzymes such as acid phosphatase, β-N-acetil-hexosaminidase (EC 3.2.1.52), cytocromo C oxidase (EC 1.9.3.1), neutral esterase (EC 3.1.1.1), cathepsin B (EC 3.4.22.1) e β-glucuronidase (EC 3.2.1.31) (Strömhaug et al., 1998, Petrovic et al., 2004, Marigómez et al., 2005 and van Nierop et al., 2006). Therefore, given the simplicity of the method, the sensitivity of the response and its relatively low cost, the lysosomal integrity, measured through acid phosphatase activity can be an important tool for the evaluation of aquatic pollution by toxic xenobiotics.

Compared to the results obtained by Ramos (2005), who determined the acid phosphatase activity as a biomarker in liver of Tilapia exposed to WSF of different gasoline formulations, the biodiesel-WSFs toxicity, analyzed as part of the present research, have showed to be less toxic. Despite this, biodiesel shall not be considered as a bio-compatible fuel. Even though biodiesel consists mainly of fatty acids, the accumulation of these greases is damaging to the environment (Novak and Kraus, 1973). Furthermore, the possible reversion of the transesterification process, in the presence of water, can result in methanol production, which may appear during biodiesel degradation and, specially, in the biodiesel-WSFs. Methanol, which is toxic to biota (Roel et. al., 2000), was pointed out as the main suspect of the toxicological results obtaining in this research.

While chemical analysis cannot measure the interactions among contaminants, toxicity tests play an important role in detecting potentially hazardous contaminants and products well ahead of their use, as well as in estimating their potential effects on the natural environment (Nascimento et al., 2000a). However, predictions of the likelihood of ecological harm caused by industrial products are more feasible if the tests involve a key process such as reproduction, and/or key species of the ecosystem at risk (Callow, 1994). The invertebrate early-life-toxicity tests and the microalgae tests, by focusing on the first links in food webs of coastal marine ecosystems, provide results that are more realistically predictive of harmful effects at ecosystem level (Beiras and His, 1995), since they are mostly related to chronic biological responses.

Results of the present research, involving ecotoxicological tests on *Echinometra lucunter* embryos, showed that the castor oil biodiesel-WSF, processed by methylic route and basic (NaOH) homogene catalysis, presented greater toxicity, (CE_{50}-36h = 2,45%), followed by WCO biodiesel-WSF (CE_{50}-36h = 5.97 %) and palm (dende) oil-biodiesel-WSF (CE_{50}-36h=16.26%). Considering these results it was verified that biodiesel from castor oil was also the most toxic for the embryos of *Echinometra lucunter*.

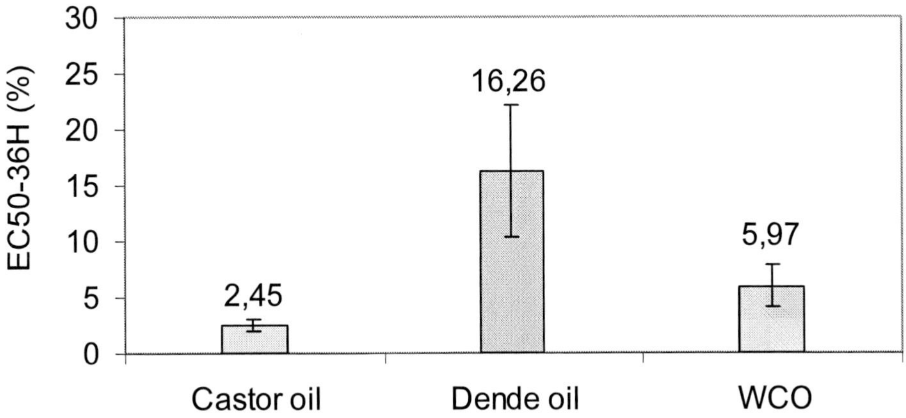

Multiple range comparison test

Values joined for the same line do not differ significantly (p> 0,05)

Biodiesel (WSF)	Dende oil	WCO	Castor oil
EC_{50}-36H (%)	16,26	5,97	2,45
	!-----------!	!-----------------	-----------------!

Figure 2. Effective concentrations of castor oil, palm (dende) oil and WCO biodiesel-WSFs, causing 50% abnormal development on exposed embryos of *E. lucunter*.

In relation to the tests based on culture inhibition growth of *S. costatum*, the IC_{50} values for the WSF of the different biodiesels samples were lower than for the sea-urchin embryos, being, respectively 1.29%, 8,27% and 15,81% for castor oil, WCO and, palm (dende) oil-biodiesel. Even presenting a similar response trend to the biodiesel-WSFs, the microalgae growth test results pointed out the palm oil biodiesel-WSF as the less toxic, while for the sea-urchin embryos, the less toxic among the three analyzed biodiesel-WSFs, was the WCO biodiesel-WSF. In a recent work, based on exposure of the microalgae *Tetraselmis chuii* to different WSF of castor oil, WCO, and palm oil biodiesel, it was showed that they have respectively inhibited (CI_{50}-96h) the growth of the cultures in concentrations of 48.32%, 56,74 and 93,52 (Pereira, personal communication). Especially concerning the data on *S. costatum*, the results showed the biodiesel-WSFs as very toxic to microalgae.

An ecotoxicological study comparing the diesel fuel to biodiesel (rapeseed oil transesteryfied with methanol) effects on microbial community of aerated soils, showed that the biodiesel fuel is toxic at concentrations above 12% w/w, while the diesel fuel exhibits toxic properties at concentrations higher than 3% w/w. (Lapinskienè et al., 2006).

The present research provided CG-chromatograms results of all the WSFs used to estimate the toxicity of the different biodiesel fuels. The WSFs were obtained and chemically analyzed at different periods (1, 60, 120 days after storage as WSFs and as WSF obtained at at day 1, after 180 days of storage as neat biodiesel samples.). The different types of biodiesel, from where the analyzed WSFs were isolated, have been obtained by transesterification, a reaction of a fat or oil with methanol to form esters and glycerol, through a number of consecutive, reversible reactions. A catalyst was used to improve the reaction rate and yield, so that, after the transesterification process, the products were a mixture of esters (biodiesel), glycerol, methanol, catalyst and tri-, di- and monoglycerides (Ma and Hanna, 1999).

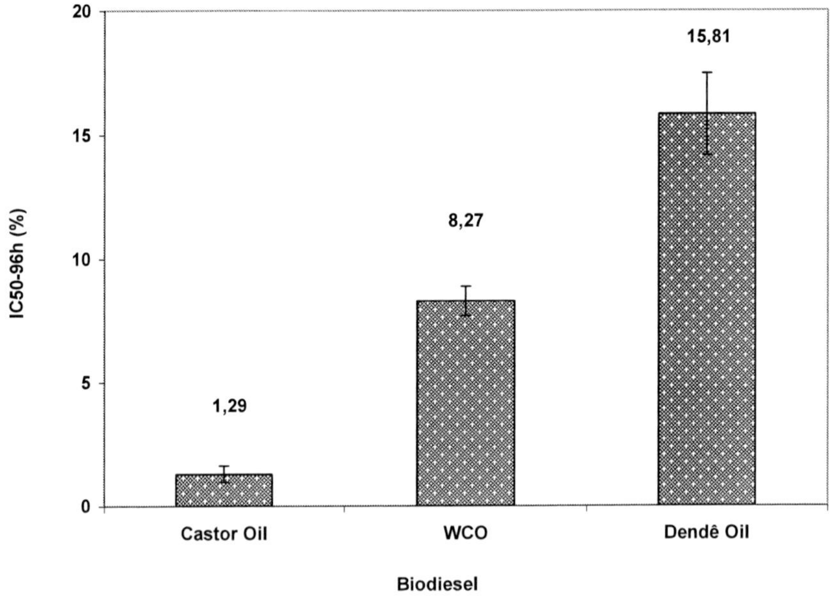

Multiple range comparison test

Values joined for the same line do not differ significantly (p> 0,05)

Organisms	*S.costatum/ Dendê Oil*	*S.costatum* WCO	*S.costatum* C.Oil
X=	15,81	8,27	1,29
	!--------------------!	!--------------------!	!--------------------!

Figure 3. Concentrations of castor oil, palm (dende) oil and WCO biodiesel-WSFs, causing 50% growth inhibition in cultures of *S. costatum.*

All the impurities and the presence of water are supposed to be avoided in order biodiesel to meet the necessary specifications (ANP, 2008) at refining. Because the reaction is reversible, excess alcohol was used to shift the equilibrium to the products side. However, it is reported that some of the un-reacted methanol may remain in the ester layer even when washed, if not removed by distillation or evaporation (Wimmer, 1992). Methanol is a toxic product to biota (Roel, 2000) and can be the responsible for the biological altered responses obtained in the present research.

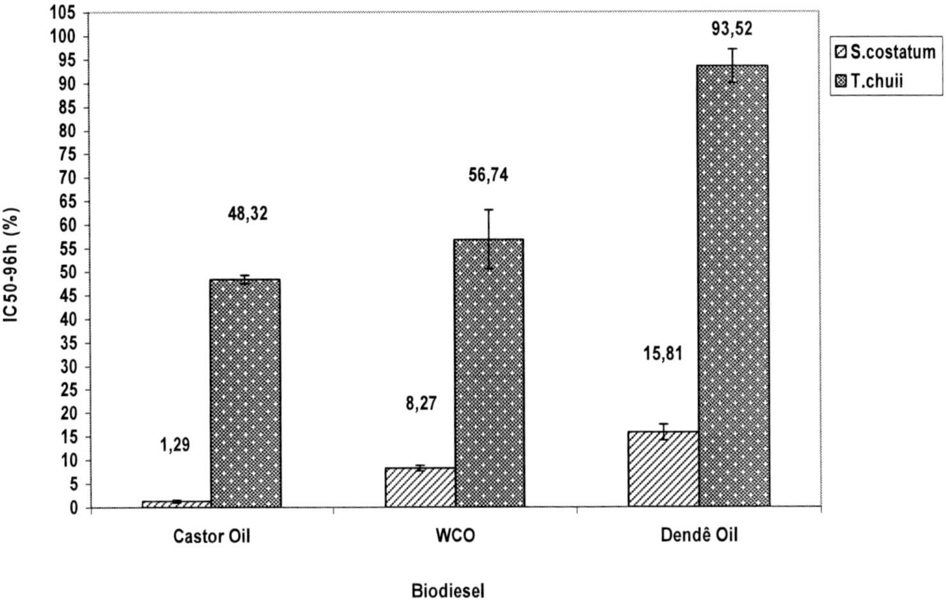

Multiple range comparison test

Values joined for the same line do not differ significantly (p> 0,05)

Values IC_{50}-96h	*T.chuii*/ Dendê Oil	*T.chuii*/ WCO	*T.chuii*/ C.Oil	*S.costatum*/ Dendê Oil	*S.costatum*/ WCO	*S.costatum*/ C.Oil
X=	93,52	56,74	48,32	15,81	8,27	1,29

Figure 4. Comparative responses (inhibition concentration) of *S. costatum* and *T.chuii* in cultures exposed to castor oil, palm (dende) oil and WCO biodiesel-WSFs.

GC-chromatograms carried out on the same biodiesel-WSFs samples, used for the ecotoxicological estimates, pointed out methanol as the more conspicuous contaminant. It can be observed for the seawater samples of the biodiesel-WSFs (Figures 5 to 7), that the methanol increases from day 1 up to day 120 in all the analyzed WSFs, what also happened for the freshwater-based WSFs, even though at a lower rate (Figure 9). Comparatively, for the WSFs obtained from each biodiesel, the degradation (and consequently the methanol concentration) was higher in the seawater than in the freshwater-soluble-fractions, possibly because the salts, present in seawater, may have catalyzed this process, leading to its acceleration. For all the analyzed samples, other oxidation by-products (in lower quantities as compared to methanol and not identified in this study) appeared in the WSF-chromatograms, increasing in concentrations with time (Figures 5 to 7).

The comparative results on the 1-day-WSFs obtained from neat biodiesel samples maintained for 180 days in sealed dark bottles and under low temperatures (4^0 C), showed for seawater, as for freshwater (Figure 8) that, by this period, the mean methanol concentration is lower in comparison to the value obtained for day-120; however, it is still significantly higher ($p<0.05$) for the castor oil-WSFs, what lead to the conclusion that this ester is hydrolyzed more easily than in the biodiesel-WSFs from palm (dende) oil or waste cooking oil, representing consequently a higher danger to the environment.

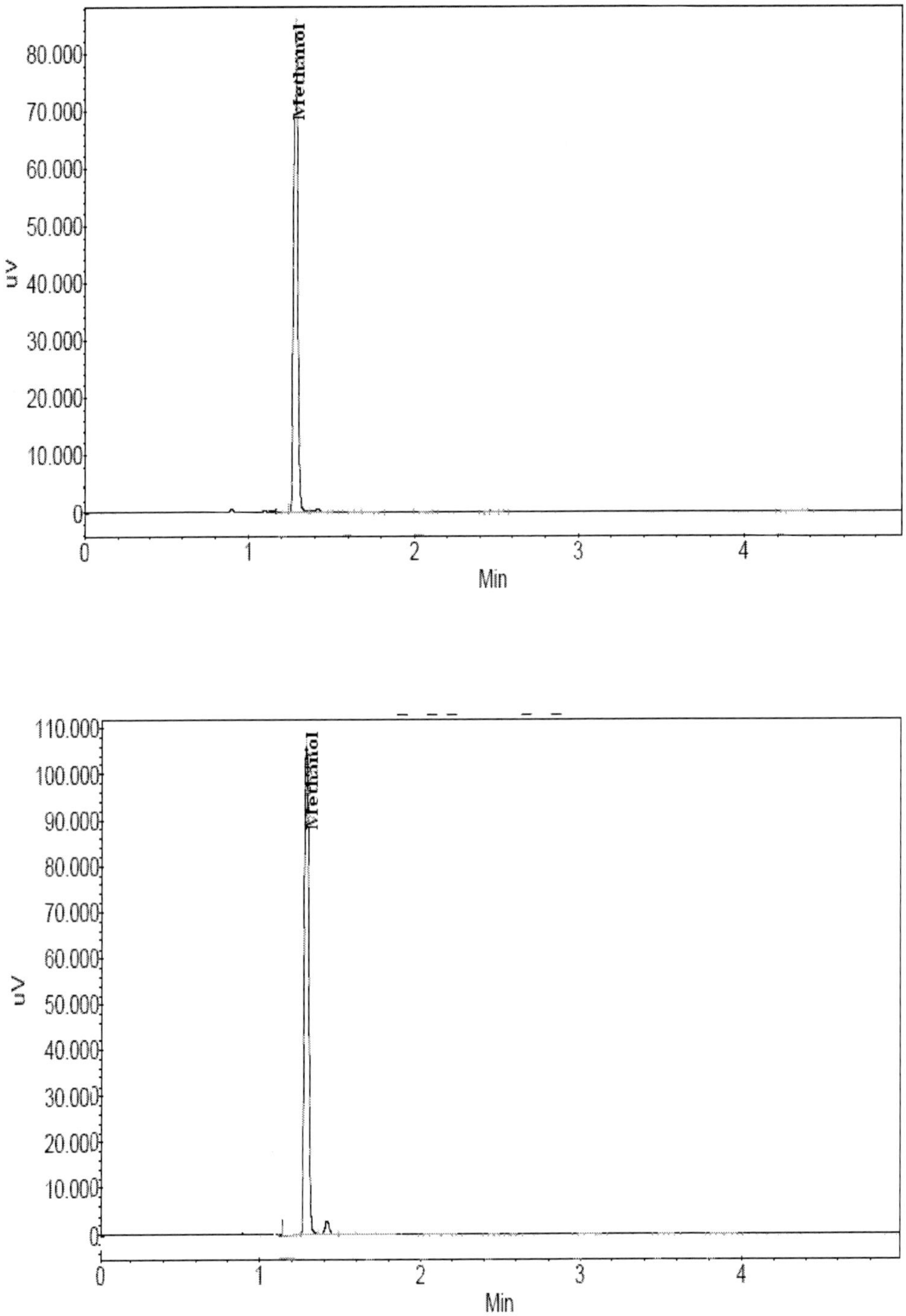

Figure 5. GC-chromatograms of castor oil biodiesel-WSFs in seawater at day 1 (above) and at day 120 (below).

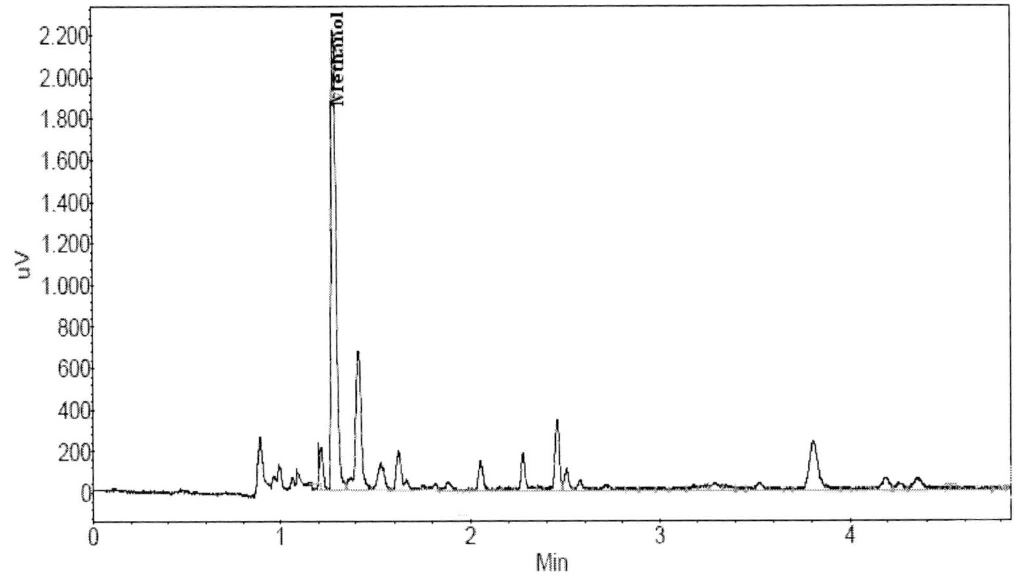

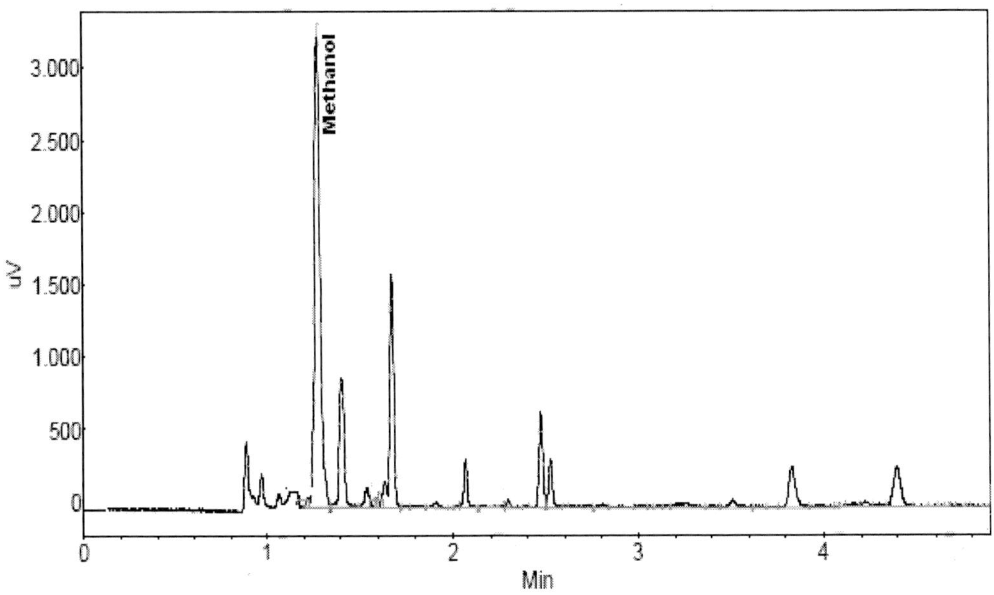

Figure 6. GC-chromatograms of palm (dende) oil biodiesel-WSFs in seawater at day 1 (above) and at day 120 (below).

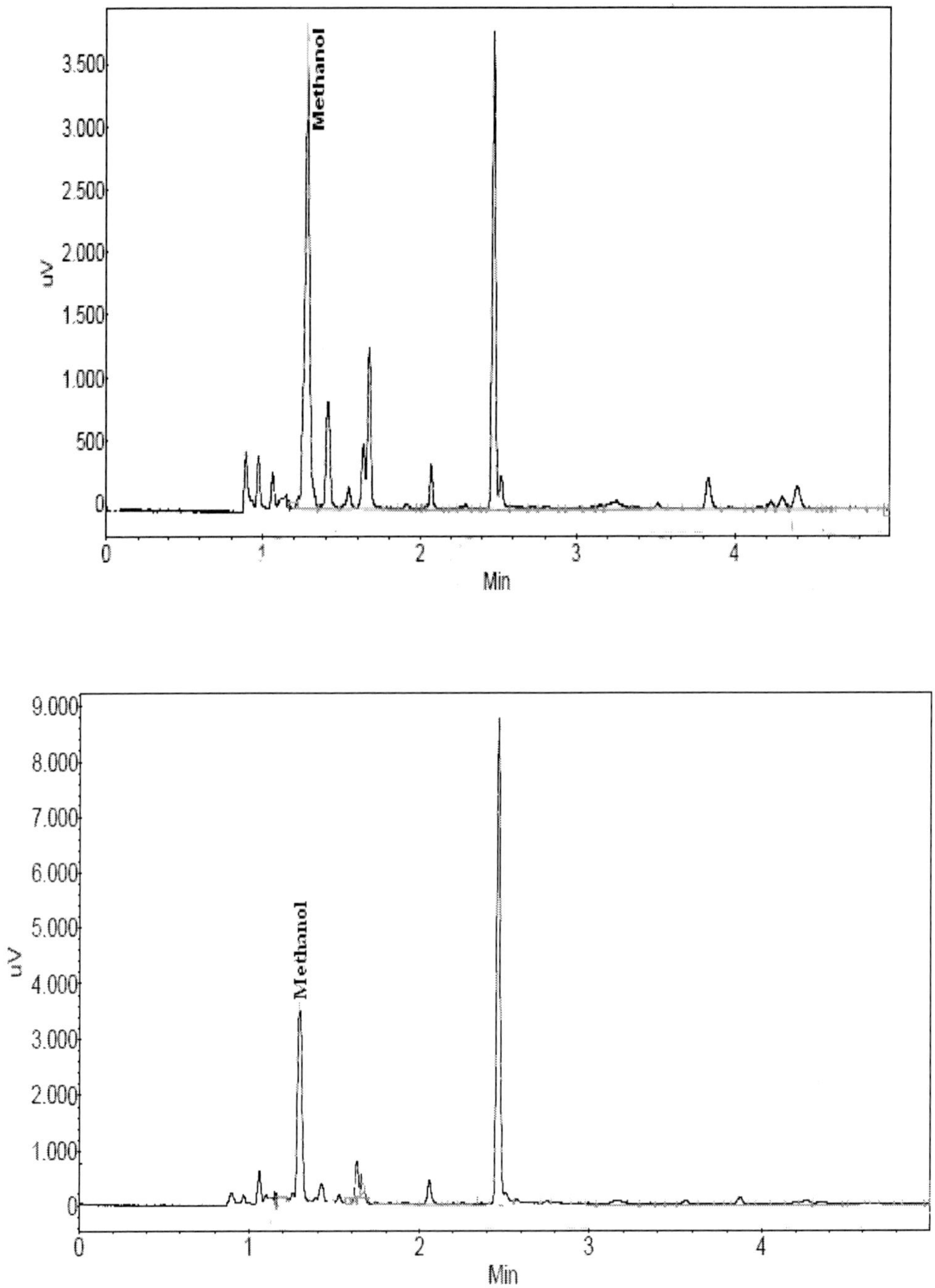

Figure 7. GC-chromatograms of waste-cooking-oil biodiesel-WSFs on seawater at day 1 (above) and at day 120 (below).

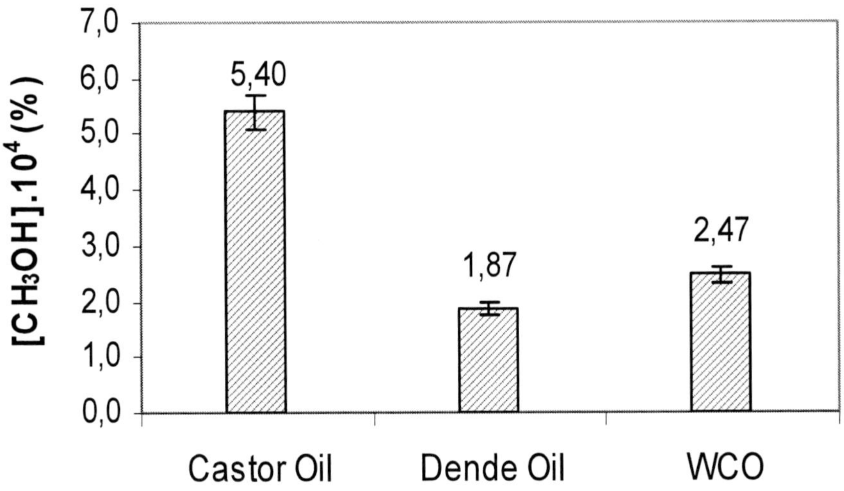

COMPARATION OF MULTIPLE MEANS

Values joined for the same line do not differ (p> significantly; 0,05)

Biodiesel	Castor oil	WCO	Dende oil
$CH_3OH].10^4$	5,40	2,5	1,9
	!------------!	!--------------------------------	--!

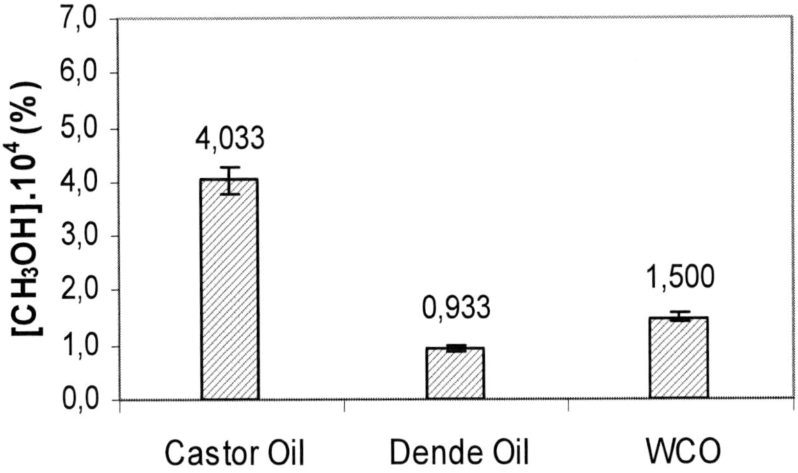

COMPARATION OF MULTIPLE MEANS

Values joined for the same line do not differ (p> significantly; 0,05).

Biodiesel	Castor oil	WCO	Dende oil
$CH_3OH].10^4$	4,03	1,5	0,9
	!------------!	!--------------------------------	--!

Figure 8. Concentrations of methanol (%) obtained by GC-chromatograms from castor oil, palm (dende) oil and WCO biodiesel-WSFs on seawater (above) and on freshwater (below), on day 1, after 180 days of storage (dark sealed bottles, 4°c) as neat biodiesel samples (B_{100}).

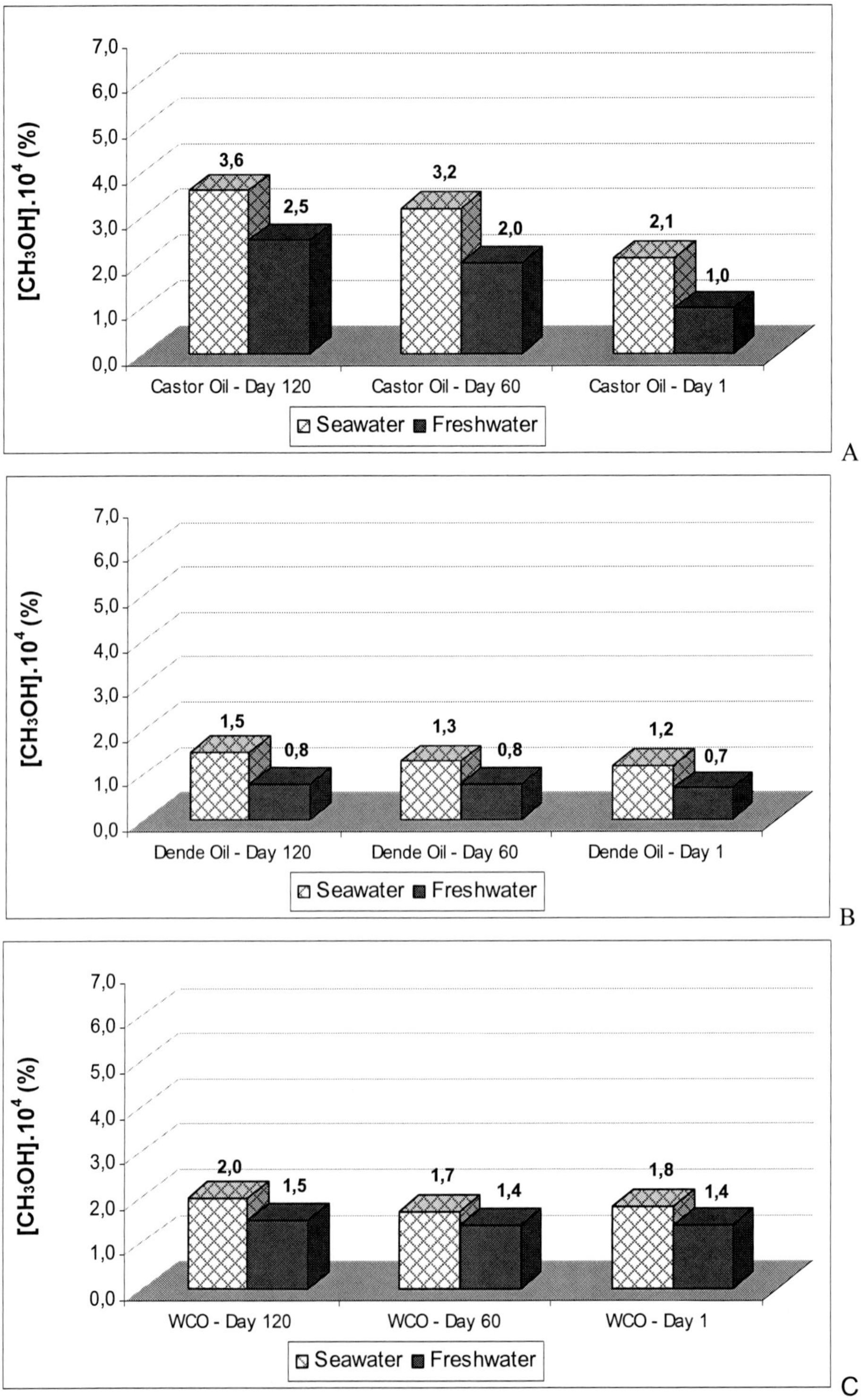

Figure 9. Comparison of Methanol concentrations obtained by GC-chromatograms from different biodiesel-WSFs on seawater and freshwater, analyzed at different periods (1, 60 and 120 days) of storage. A- castor oil biodiesel-WSF; B- palm (dende) oil biodiesel-WSF; C- WCO biodiesel-WSF.

It can be noticed that even maintained at laboratory-controlled conditions, the biodiesel samples were also degraded as far as their WSFs. Biodiesel is considered as having a good degradability. It is expected to degrade biologically over 98% within three weeks, while at the same conditions and in the same period, diesel will degrade only 50% (Williamson and Badr, 1998). However, the rate of degradation varies dependant mainly on temperature. Experimental results (Leung et al., 2006) demonstrated that the biodiesel degraded less than 10% within 52 weeks when stored at 4 to 20^0 C, while nearly 40% degradation was found for samples stored at 40^0 C, irrespective of the presence of the presence of water moisture or maintenance conditions (sealed or exposed to ambient air). However, due to hydrolysis, water content in biodiesel will enhance biodiesel degradation under temperatures lower than 40^0 C. In the present research the degradation process leaded to a crescent quantity of methanol with time, varying in intensity among the different biodiesels analyzed, and showing the most intense rate for the castor oil, and the least, for palm (dende) oil (Figure 8).

Considering the reversibility of the transesterification process by hidrolytic reactions, and the toxicological results of the present research, the biodiesels WSFs, (and possibly the biodiesel washing water discarded as effluent from the productive process) may became more toxic with time, due to the increased presence of methanol, which constitutes an environmental risk, not being yet properly addressed.

In most of the Brazilian plants especially in the small ones, the biodiesel washing water generated at the end of the biodiesel process route, is not treated before discarded. This wastewater can reach the pluvial collecting system or, by runoff, reach fresh water or coastal ecosystems. If biodiesel is to be used as a source of energy, these findings may represent a potential challenge for aquatic ecosystems, especially on coastal environments, considering that in countries like Brazil, 80% of the population lives in or around coastal areas.

Conclusion

Considering the importance of biodiesel as one of the options to increase the consumption and production of renewable fuels for sustainable development, it is important to control the negative direct or indirect effects of its production, use and trade, which will determine its ultimate economic, environmental and social impacts.

Given that few investigations have been performed on biodiesel-WSF toxicity, the findings of this study provide an important foundation for further detailed investigation into the effects of those products on aquatic ecosystems, especially close to urban developments.

The ecotoxicological bioassays, carried out on test-organisms occupying diverse biological niches in aquatic ecosystems exposed to biodiesel-water soluble fractions, provided valuable information about the interactive biological effects of the soluble chemical compounds present in biodiesel fuels. This study proved that there is an environmental risk associated with the runoff of biodiesel and its water-soluble-fractions, which should be addressed, considering the perspectives, in next future, of this biofuel to, worldwide, substitute fossil fuels to some extent.

In view of the increasing amount of methanol formed with time (up to 120 days) during the biodegradation process of the analyzed biodiesel samples, it would be advisable to consider the necessity of effluent treatment as part of the biodiesel plants.

REFERENCES

Al-Widyan, M., Tashtoush, G., Abu-Qudais, M. (2002).Utilization of ethyl ester of waste vegetable oils as fuel in diesel engines. *Fuel Process. Technol.* 76: 91–103.

ANP-Agência Nacional de Petróleo, Gás Natural e Biocombustíveis. (2007). Dados Estatísticos. http://www.anp.gov..br.

ANP-Agência Nacional de Petróleo, Gás Natural e Biocombustíveis. (2008). Novas especificações para o biodiesel. http://www.anp.gov..br.

Andersch, M.A., Szcypinski, A.J. (1947). Estimation of alkaline and acid phosphatase activities in normal human serum by using p-nitrophenol phosphate. *Am. J. Clin. Pathol.* 17: 571-574.

Agarwal, A. K., Das, L. M. (2001). Biodiesel development and characterization for use as a fuel in compression ignition engines. *J. Eng. Gas Turbines Power*, 123: 440-447.

Agarwal, D., Sinha, S. Agarwal, A .K. (2006). Experimental investigation of control of Nox emissions in biodiesel-fueled compression ignition engine. *Renew. Energy*, 31: 2356-2369.

Araújo, M.M.S. Nascimento, I.A. (1999). Testes ecotoxicológicos marinhos: análise de sensibilidade. Coimbra: *Ecotoxicol. Environ. Restor.*, 1(1): 41-47.

Balcao, V. M., Paiva, A.L., Malcata, F. X., (1996). Bioreactors with immobilized lípases. State of the art. *Enz. Microb. Technol.* 18 (6): 392-416.

Basri, M.; Heng, A.C., Razak, C.A.N, Wan Yunus, W.M.Z., Ahmad, M., Rahman, R.N.A., Ampon, K., Salleh, A.B., (1997). Alcoholysisof palm oil mid fraction by lípase from *Rhizopus rhizopodiformis*. *J. Amer. Oil Chem. Soc.,* 74 (2):113-116.

Beiras, R., His, E., (1995). Toxicity of fresh and freeze-dried hydrocarbon polluted sediments to *Crassostrea gigas* embryos. *Mar. Pollut. Bull.*, 30 (1):47-49.

Bradford, M.M. (1976). A rapid and sensitive method for the quantities of protein utilizing the principle of protein-dye-biding. *Anal. Biochem.*, 72: 248-254.

Bodin, N., Burgeot, T., Stanisière, J.Y., Bocquené, G., Menard, D., Minier, C., Boutct, I., Amat, A., Checrcl, Y. And Budzinski, H. (2004). Seasonal variation of a battery of biomarkers and physiological indices for the mussel *Mytilus galloprovincialis* transplanted into the northwest Mediterranean sea. *Comp. Biochem. Physiol. C. Toxicol. Pharmacol.* 138: (4): 411-427.

Bünger, J., Krahl, J., Franke, H.-U., F., Munach, A., Hallier, E., (1998). Mutagenic and cytotoxic effects of exhaust particulate matter of biodiesel compared to fossil diesel. *Mutat. Res.*, 415: 13-23.

Burkhardt, H., Dyson, R., Morton-Mar, J. (2006). Handbook on Critical Global Issues and Viable Solutions. *Council on Global Issues*. Toronto. Canada 22pp.

Callow, P., (1994). Ecotoxicology: What are we trying to protect? *Environ. Toxicol. Chem.* 13:7-22.

Carrareto, C., Macor, A., Mirandola, A., Stoppato, A., Tonon, S. (2004). Biodiesel as alternative fuel: experimental analysis and energetic evaluations. *Energy* 29: 2195-2211.

Cardone, M., Prati, M. V., Rocco, V., Seggiani, M., Senatore, A., Vitolo, S. (2002). *Brassica carinata* as an alternative oil crop for the production of biodiesel in Italy: engine performance and regulated and unregulated exhaust emissions. *Environ. Sci. Technol.*, 36: 4656-4662.

Carlier, P., Hannachi, H., Mouvier, G., (1986). The chemistry of carbonyls in the atmosphere- a review. *Atmos. Environ.*., 20: 2079-2099.

Carter, W.P.L., (1995). Computer modeling of environmental chamber measurements of maximum incremental reactivities of volatile organic compounds. *Atmos. Environ.*, 29: 2513-2527.

Connemann, J. Fisher, J. (1998). *Biodiesel in Europa 1998: Biodiesel processing technologies.* International Biofuels Congress, Brazil. 15pp

Corrêa, S. M ., Arbilla, G. (2006). Aromatic hydrocarbons emissions in diesel and biodiesel exhaust. *Atmos. Environ.,* 40: 6821-6826.

Cuervo, A.M., Dice, J.F., (2000). When lysosomes get old. *Exp. Gerontol.*, 35: 119-131.

Cruz, A. C. S., Nascimento, I. A., Couto, B. C., Pereira, S. A., Leite, M. B., Bertoletti, E., Zagatto, P. A. (2007). Estimation of the critical effect level for pollution prevention base don oyster embryonic development toxicity test: the search for reliability. *Environ. Intern. Special Issue,* 33: 589-595.

Doornbosch, R., Steenblik, R. Biofuels:Is the cure worse than the disease? OECD-Orgaization for Economic Co-operation and Development Report. SG/SD/RT(2007)3 56pp

EMBRAPA, 2005. <http:// sistemasdeproducao.cnptia.embrapa.br>.

Evangelista, A. J. A., Nascimento, I. A., Pereira, S. A., Leite, M. B., Martins, L. K., Fillmann, G. (2005). *Assessing the potential toxicity of marine sediments found in petroleum industry areas: a new approach base don shrimp pos-larvae responses.* Ciencias Marinas, Mexico, 31(1 A): 43-55.

Fent, K., (2001). Fish cell lines as versatile tools in ecotoxicology: assessment of cytotoxicity, cytochrome P4501A induction potential and estrogenic activity of chemical and environmental samples. *Toxicol. In Vitro.*, 15:477-488.

Finney, D.J. (1971). Proibit Analysis. 3 ed. London. Combridge University Press, 333 pp..

Fukuda, H., Kondo, A., Noda, H. (2001). Biodiesel fuel production by transesterification of oils. *J. Biosci. Bioeng.*, 92: 405-416.

Gaffney, J.S., Marley, N.A, Martin, R.S., Dixon, R.W., Reyes, L.G., Popp, C.J., (1997). Potential air quality effects of using ethanol-gasoline fuel blends: a field study in Albuquerque, New Mexico. *Environmental Science and Technology*, 31: 3053-3061.

Global Footprint Network, (2003). Humanity`s Footprint 1961-2003: demands vs. biocapacity. http://www.footprintnetwork.org/

Graboski, M. S., Mccomick, R. L., (1998). Combustion of fat and vegetable oil derived fuels in diesel engines. *Prog. Energy Combust Sci.* 24:125-164.

Graphpad Software Inc. (1997). Instat guide to choosing and interpreting statistical tests. Graphpad Software, Inc. San Diego. California, U.S.A. Available in http//www. Graphpad.com.

Griffith, G.W., Kenslow, M.A.M., Ross, L.A., (1973). A mass culture method for Tetraselmis sp. A promising food for larval crustaceans. *Nat. Mar. Fish. Service. Gulf Coastal Fisheries Center*. Contribution, p.354.

Haas, M. J. (1997). Fuels as solvent for the conduct of enzymatic reactions. Patent (5,697,986). USA.

Haas, M. J., McAloon, A. J., Yee, W.C., Foglia, T.A. (2006). A process model to estimate biodiesel production costs. *Bioresour. Technol.,* 97: 671-678.

Hamilton, M.A., Russo, R.C., Thurston, R.V. (1977). Trimmed Spearman-Karber Method for Estimating Median Lethal Concentrations in Toxicity Bioassays. *Environ. Sci. Technol.*, 11 (7): 714-719.

Hwang, H-M., Wade, T.L., Sericano, J.L. (2004). Destabilized lysosomes and elimination of polycyclic aromatic hydrocarbons and polychlorinated biphenyls in eastern oysters (*Crassotrea virginica*). *Environ. Toxicol. Chem.* 23: 1991-1995.

IEO-International Energy Outlook, (2007). EIA-Energy Information Administration, 2007 Report. Washington, D. C. http://www.eia.doe.gov/oiaf/ieo/pdf/0484(2007).pdf

International Energy Authority. (2004). Biofuels for transport: an international perspective (IEA, Paris, France,2004, chap. 6). www.iea.org/textbase/nppdf/free 2004/biofuels 2004.pdf

IPCC–Intergovernmental Panel of Climate Change. (1996). WMO/UNEP/IPCC Assessment Report http://www.ipcc.ch/.

IPCC–Intergovernmental Panel of Climate Change. (2007). WMO/UNEP/IPCC. Fourth Assessment Report (AR4 Synthesis Report). 52pp http://www.ipcc.ch/.

Ishisaka, R., Utsumi, T., Yabuki, M., Kanno, T., Furuno, T., Inoue, M. And Utsumi, K. (1998). Activation of caspase-3-like protease by digitonina-treated lysosomes. *FEBES Letts*. 435: 233-236.

Iso, M., Chen, B.X., Eguchi, M. (2001). Production of biodiesel fuel from triglycerides and alcohol using immobilized lipase. *J. Mol. Catal. B: Enzymatic* 16 (1), 53-58.

Kalam, M. A., Masjuki, H. H., (2002). Biodiesel from palm oil- an analysis of its properties and potential. *Biomass and Bioenergy*, 23:471-479.

Kose, O., Tuter, M., Aksoy, H. A., (2002). Immobilized *Candida antarctica* lipase catalyzed alcoholysis of cotton seed oil in a solvent-free medium. *Bioresour. Technol.* 83 (2):125-129.

Koukouzika, N., Dimitriadis, V.K. (2005). Multiple biomarker in *Mytilus galloprovincinalis* from Greece Coast: "Lysosomal membrane stability, neutral red retention, micronucleus frequency and stress on stress". *Ecotoxicol.* 14, pp. 449-463.

Knothe, G., Dunn,R. O. Bagby, M. O. (1997). Biodiesel: the use of vegetable oils and their derivatives as alternative diesel fuels. In: Saha, B. C., Woodward, J. (Eds). *Fuels and Chemicals from Biomass*. American Chemical Society, Washington, D. C. pp 172-208.

Knothe, G., Gerpen, J, V., Krah, J, Ramos, L.P., (2006) *Manual do Biodiesel*. São Paulo, Editora Edgard Blücher. p. 340.

Köhler, A., Wahl, E., Söffker, K. (2002). Functional and morphological changes of lysosomes as prognostic biomarkers of toxic liver injury in a marine flatfish (*Platichthys flesus*). *Environ. Toxicol. Chem.* 21 (11):2434-2444.

Kobayashi, N. (1981). Comparative toxicity of various chemicals, oil extracts and oil dispersants to Canadian and Japanese sea urchin eggs. *Publs. Seto. Mar. Biol. Lab.,* 26, 1(3): 123-133.

Kobayashi, N. (1990). Marine pollution bioassay by sea urchin eggs, an attempt to enhance sensitivity. *Publs. Seto. Mar. Biol. Lab* 34(.4-6):225-237.

Lapinskienè, A., Martinkus, P., Rebzdaitė, V., (2006). Eco-toxicological studies of diesel and biodiesel fuels in aerated soil. *Environ. Pollut.*, 142 (3): 432-437.

Leite, M.B.N.L., Ramos, M.L.N., Nascimento, I.A. (2005). Avaliação de toxicidade de formulações de gasolina: proteínas de estresse como bioíndicadores. In: V Congresso

Brasileiro de Pesquisas Ambientais de Saúde, São Paulo. *Anais do V Congresso Brasileiro de Pesquisas Ambientais de Saúde*. 357-361.

Leite, M.B.N.L., (1986). Isolamento, seleção e levantamento das exigências ambientais para o cultivo em massa de clorofíceas. Salvador: Tese (Mestrado em Botânica). Instituto de Biologia da Universidade Federal da Bahia, Brasil.

Leung, D.Y.C., Koo, B.C.P., Guo, Y. (2006). Degradation of biodiesel under different storage conditions. *Bioresour.Technol.*, 97: 250–256.

Li, H.; Zhan, H.; Fu, S.; Liu, M.; Chai, X.-S. (2007).Rapid determination of methanol in black liquors by full evaporation headspace gas chromatography. *J.Chromatog.* A, 1175: 133–136.

Liebreich, M. (2008). Food prices increases: is it fair to blame biofuels? *New Energy Finance.* Research Note 3pp.

Lott, M., (2002). QSS Group Inc., 4500 Forbes Boulevard, Suite 200, Lanham, MD 20706.

Ma, F., Hanna, M. A. (1999). Biodiesel production: a review. *Bioresour. Technol.*, 70:1-15.

Martins, L.K., Nascimento, I.A., Fillmann, G., King, R., Depledge, M.., Readman, J., Evangelista, A.J.A. (2005). Lysosomal responses as diagnostic tool for the detection of oil chronic pollution at Todos os Santos bay. *Environ. Res.*, USA, 98 (2): 387-396.

Marigómez, I., Lekube, X., Cajaraville, M.P., Domouhtsidou, G., Dimitriadis, V. (2005). Comparison of cytochemical procedures to estimate lysomal biomarkers in mussel digestive cells. *Aquat. Toxicol.* 75: 86-95.

Millennium Ecosystem Assessment, (2005). Ecosystem and Human well-being: biodiversity synthesis. World Resources Institute, Washington, D. C. 100pp. www.mill ennium assessment.org/en/documents/document.354.aspx.pdf .

Mittelbach, M., (1990). Lipase catalyzed alcoholysis of sunflower oil. *J. Amer. Oil Chem. Soc.*, 67 (3): 168-170.

Modi, M.K., Reddy, J.R.C., Rao, B.V.S.K., Prassad, R.B.N., (2007). Lipase-mediated conversion of vegetable oil into biodiesel, using ethyl acetate as acyl acceptor. *Bioresour. Technol.*, 98: 1260-1267.

Nascimento, I. A. dos Santos, J. M., da Cruz, A. C., Pereira, S. A. Leite, M. B. N. L., Nascimento, M. A. (2008). Eco-compatibilidade na matriz energética brasileira: O biodiesel pode ser considerado como combustível ambientalmente correto? Anais do VI Congresso Brasileiro de Planejamento Energético . Salvador. Bahia. pp. 375-385.

Nascimento, I.A., Pereira, S.A., Leite, M.B.N.L. (2006). Biomarcadores como instrumentos preventivos de poluição (Capítulo 17). In : Zagatto, P. A. and Bertoletti , E. (eds). *Ecotoxicologia Aquática: Princípios e Aplicações*. 1 ed. São Paulo: RIMA. 2006 1: 413-432.

Nascimento, N. C. (2005). Buscando a precisão em testes ecotoxicológicos com ouriço do mar *Echinometra lucunter* (Echinodermata, Echinoidea): caracterização de desenvolvimento larval e análise de sensibilidade à substâncias de referência. Dissertação – Universidade Federal da Bahia, Orientador: Iracema Andrade Nascimento.

Nascimento, I.A., Sousa, E.C.P.M., Nipper, M. (2002). Métodos em Ecotoxicologia Marinha: Aplicações no Brasil. Ed. Artes Gráficas e Indústria Ltda. São Paulo. 262pp.

Nascimento, I.A., Smith, D. H., Pereira, S.A., Araujo, M.M.S., Mariani, A.M., (2000a). Integration of varying responses of different organisms to water and sediment quality at sites impacted and not impacted by the petroleum industry. In: *State of Brazilian Aquatic Ecosystem: Aquat. Ecosys. Health Manag. Special Issue*. Canada, 3 (4):. 449-458.

Nascimento, I.A., Smith, D. H., Sampaio, M. G., Santos, G. V., Pereira, S.A. (2000b). Ecotoxicological diagnosis of Aratu Bay, Bahia, Brazil: a new approach to validate e reactive short term toxicity end-point by comparison with intertidal benthic activity. *Aquat. Ecosyst. Health Manag.*, Ontario Canada,. 3 (4): 485-497.

Nascimento, I.A, Oliveira, J. G.D., Pereira, S.A. , Leite, M.B.N.L.,Souza, L. E. P. M., Sampaio, S. A., Mariani, A. M. (1999). Validação de um índice de qualidade de águas costeiras para seleção de áreas para cultivo de *Crassostrea rhizophorae*: testes de crescimento . In: *Acuicultura 99*, 1999, Puerto La Cruz. Memórias Acuicultura em Armonia com El Ambiente. Caracas, Venezuela: Societal Venezuelan de Acuicultura/ World Aquaculture Society,1: 199-208.

Nascimento, I.A., Leite, M.B.N.L., Sansone, G., Pereira, S.A., Smith, D.H. (1998a). Stress protein accumulation as an indicator of impact by the petroleum industry in Todos os Santos bay, Brazil. *Aquat. Ecosys. Health and Manag.*, the Netherlands, 1 (1):101-108.

Nascimento, I.A., Smith, D. H., Pereira, S.A., Leite, M.B.N.L. (1998b). The use of different organisms and biological levels to determine water and sediment quality at industrial sites. *Cuadernos de Investigación Biológica*, Bilbao, España, 20 (1):339-349.

Nascimento, I.A. (1996). Stress protein accumulation in *Corbicula fluminea* (Mollusca: Bivalvia) as an indicator of environmental water quality, related to heat shock protein response. Biociências, Rio Grande do Sul, 4 (1): 138-146.

Novak, J. T., Kraus, D. L. (1973). Degradation of long fatty acids by activated sludge. *Water Res.* , 7: 843-851.

OECD (Organization for Economic Co-operation and Development). Agricultural Market Impacts of Future Growth in the Production of Biofuels. OECD Document No. AGR/CA/APM(2005)24/FINAL. Paris, 2006.

ORNL (Oak Ridge National Laboratory). Biofuel Feedstock Assessment. Available in: <http://www.ornl.gov> . October, 2008.

Okay, O.S., Tolun, L., Tüfekçi, V., Telli-Karakoç, F., Donkin, P. 2006. Effects of pyrene on mussels in different experimental conditions. *Environ. Intern.* 32: 538-544.

Paixão, J. F., Nascimento, I. A., Pereira, S. A., Leite, M. B. L., Rodrigues, I. L. P., (2007). Estimating the gasoline components and formulations toxicity to microalgae (*Tetraselmis chuii*) and oyster (*Crassostrea rhizophorae*) embryos: an approach to minimize environmental pollution risk. *Environ. Res.*, 103: 365-374.

Pang, X., Shi, X., Mu, Y., He. H., Shay, S., Chen, H., Li, R. (2006). Characteristics of carbonyl compounds emission from a diesel-engine using biodiesel-ethanol-diesel as fuel. *Atmos. Environ.*, 40: 7057-7065.

Pereira, S. A., Silva, C. C., Nascimento, I. A., Rodrigues, A. R., Valença, A. R. Impacto de poluentes petroquímicos na fotossíntese da microalga *Tetraselmis chuii*. Livro de Resumos da 53ª Reunião Anual da SBPC, Salvador-Ba, Jul/ (2001). p.345.

Petrovic, S., Semencic, L., Ozretic, B., Ozretic, M. (2004). Seasonal variations of physiological and cellular biomarkers and their use the biomonitoring of north Adriatic coastal (Croatia). *Mar. Pollut. Bull.*, 49: 713-720.

Pinto, A. C., Guarieiro, L. L. N., Rezende, M. J. C., Ribeiro, N. M., Torres, E. A ., Lopes, W. A ., Pereira, P. A ., de P., de Andrade, J. B. (2005). Biodiesel: an overview. *J. Braz. Chem. Soc.* 16 (6B):1313-1330.

Prokop, T., (2002). *Imperial Western Products*, 14970 Chandler St., Coachella, CA 91720.

Royon, D., Daz, M., Ellenrieder, G., Locatelli, S. (2007). Enzymatic production of biodiesel from cotton seed oil using t-butanol as a solvent. *Bioresource Technology*, 98: 648-653.

Ralph, P.J., Burchett, M.D. (1998). Impact of Petrochemicals on the Photosynthesis of *Halophila ovalis* Using Chlorophyll Fluorescence. *Mar. Pollut. Bull.*, 36 (6): 429-436.

Ramos, M.L.N. (2005). Desenvolvimento e aplicações de técnicas para a avaliação da toxicidade de 14 formulações de gasolina, utilizando *Oreochromis niloticus* (Linnaeus, 1758); Cichidae, Teleostei, e ovas do Gastrópode *Pomacea lineata* (SPIX, 1827) como organismos testes. Dissertação de Mestrado-Universidade Federal da Bahia.

Ramos, M. L. N., Rodrigues, L. E. de A., Nascimento, I. A. (2008). The lysosomal stability as a biomarker for the determination of pollution in aquatic environments. *Braz. Arch.Biol. Technol.*, 51: 1071-1077.

Righelato, R.; Spracklen, D. V. (2007). Carbon mitigation by biofuels or by saving and restoring forests? Science, 317: 902. www.sciencemag.org/cgi/content/full/317/5840/902/DC1.

Roel, A. R., Vendramim J. D., Frighetto R.T.S., Frighetto, N., (2000). Atividade tóxica de extratos orgânicos de *Trichilia pallida* Swartz (Meliaceae) sobre *Spodoptera frugiperda* (J. E. Smith), *An. Soc. Entomol. Bras.*, 29 (4): 67-98

Sarkar, A., Ray, D., Shrivastava, A.N. And Sarker, S. (2006). Molecular biomarkers: Their significance and application in marine pollution monitoring. *Ecotoxicol.* 27: 109-107.

Schuchardt, U., Sercheli, R., Vargas, R.M. (1998). Transesterification of vegetable oils: a review. *J. Braz. Chem. Soc.*, 9:199-210.

Searchinger, T., Heimlich, R., Houghton, R. A., Dong, F., Elobeid, A. Fabiosa, J., Tokgoz, S., Hayes, D., Yu, T.-H. (2008). Use of U.S. Croplands for biofuels increases Greenhouse Gases through emissions from land-use change. *Science*, 319: 1238-1240.

Sharp, C. A., Howell, S. A., Jobe, J., (2000). The effects of biodiesel fuels on transient emissions from modern diesel engines. Part II Unregulated emissions and chemical characterization. *SAE Techn. Paper* 2000-1-1968.

Sheehan, J., Camobreco, V., Duffield, J., Graboski, M., Shapouri, H., (1998). An overview of biodiesel and petroleum diesel life cycles. *Report of National Renewable Energy Laboratory (NREL) and US-Department of Energy (DOE).* Task No. BF886002.

Shimada, Y., Watanabe, H., Sugihara, A., Tominaga, Y. (2002). Enzimatic alcoholysis for biodiesel fuel production and application of the reaction to oil processing. *J. Mol. Catal. B: Enzym.* !7: 133-142.

Strömhaug, E., Berg, T.O., Fengsrud, M., Seglen, O. (1998). Purification and characterization of autophagosomes from rat hepatocytes. *Biochem. J.* 335, p. 217-224.

Turrio-Baldassarri, L., Battistelli, C. L., Conti, L., Crebelli, R., De Berardis, B., Iamiceli, A.L., Gambino, M., Iannaccone, S. (2004). Emission comparison of urban bus engine fueled with diesel oil and biodiesel blend. *Sci. Total Environ.*, 327: 147-162.

Wang, W.G., Lyons, D.W., Clark, N.N., Gautam, M., Norton, P.M., (2000). Emissions from nine heavy trucks fueled by diesel and biodiesel blend without engine modification. *Environ. Sci. Technol.*, 34: 933-939.

Wiltsee, G., (1998). Waste grease resource in 30 US metropolitan areas. In: *The Proceedings of Bioenergy 98 Conference*, Wisconsin, pp 956-963.

Williamson, A. M. Badr, O. (1998). Assessing the viability of using rape methyl ester (RME) as an alternative to mineral diesel fuel for powering road vehicles in the U.K. *Appl. Energy,* 59: 187-214.

Wimmer, L. (1992). Preparation of esters of fatty acids with short-chain alcohols. Austrian AT 349-571.

Winkaler, E.U., Silva, A.G., Galindo, H.C. E Martinez, C.B.R. (2001). Biomarcadores histológicos e fisiológicos para o monitoramento da saúde de peixes de ribeirões de Londrina, Estado do Paraná. *Acta Scient.* 23 (2): 507-514.

WCED-World Commission on Environment and Development, (2007). Report of the United Nations / WCED . General Assembly Resolution 42/187 (Dec. 1987). Retrieved Nov. 2007.

World Bank. (2004). Annual Review of Development effectiveness. ISBN 0-8213-6303-4. http://www.publications.worldbank.org/.

World Bank (2008). Rising food prices: policy options and World Bank response. http://www.publications.worldbank.org/.

USEPA-U.S. Environmental Protection Agency. (1984). *Drilling Fluid's Toxicity Test.* Federal Register, v.50, n. 165, Monday, August 26, p.34631-34636.

Van Nierop, K., Muller, F.J.M., Stap, J., Van Noorden, C.J.F, Van Eijk, M., Groot, C. (2006). Lysosomal destabilization contributes to apoptosis of germinal center B-lymphocytes. *J. Histochem. Cytochem.* 54 (12):1425-1435.

Yang, L., Lai, C-T., Shieh, W. K. 2000. Biodegradation of dispersed diesel fuel under high salinity condition. *Water Res.*, 34: 3303-3314.

Xu, Y., Du, W., Liu, D., (2005). Study of the kinetics of enzymatic interesterification of triglycerides for biodiesel production with methyl acetate as the acyl acceptor. *J. Mol. Catal. B: Enzimatic*, 32(5-6): 241-245.

Zhang, Y., Dubé, M.A., McLean, D.D., Kates, M., (2003). Biodiesel production from waste cooking oil: 1- Process design and technological assessment. *Bioresour. Technolo.*, 89: 1-16.

Zhang, X., Peterson, C., Reece, D., Haws, R., Mller, G.,. (1998). Biodegradability of biodiesel in the aquatic environment. *Trans. ASAE*, 41:1423–1430.

In: Industrial Pollution Including Oil
Editor: Harry Newbury and William De Lorne
ISBN 978-1-60456-917-9

Chapter 3

IMPACT OF VARIOUS CONCENTRATIONS OF CRUDE-OIL ON THE PHYSICOCHEMICAL PROPERTIES OFA SOIL IN NIGERIA

O. Obire*, E. C. Anyanwu and R. N. Okigbo
Department of Applied And Environmental Biology, Rivers State University of Science and Technoology, P.M.B. 5080, Port Harcourt, Nigeria
Department Of Botany, Nnamdi Azikiwe University, Awka, Nigeria

ABSTRACT

The effect of the addition of different concentrations (0% - control, 0.5%, 1%, 3% and 5%) of crude oil (Brass blend, black colour 'API 41.7 – 43.0') on the physicochemical characteristics of soils was investigated at 24h after pollution and at 18weeks after pollution. This was to compare the soil characteristics 24h after pollution and several weeks after pollution. The pH values of the control soils and polluted soils at 24 hours after pollution ranged from 6.4 to 7.9. Exchangeable acidity ranged from 0.18 to 0.25. Calcium, magnesium, potassium and sodium values ranged from 1.99 to 6.88, 0.27 to 0.35, 0.09 to 0.24 and 0.03 to 0.06 meq $100g^{-1}$ soil respectively. Available phosphorus, ammonium and nitrate values ranged from 24 to 28 ppm, 15.7 to 19 ppm and 14 to 28ppm respectively. Organic carbon, total nitrogen and carbon/nitrogen (C/N) ratio values ranged from 1.58 to 4.23%, 0.14 to 0.217% and 8.83 to 28.6% respectively. The pH values of the control and oil-polluted soil at 18 weeks after pollution ranged from 6.3 to 7.8. Exchangeable acidity ranged from 0.15 to 0.23. Calcium, magnesium, potassium and sodium values ranged from 1.94 to 6.38 meq $100g^{-1}$ soil, 0.28 to 0.37 meq $100g^{-1}$ soil, 0.11 to 21 meq $100g^{-1}$ soil and 0.02 to 0.14 meq $100g^{-1}$ soil respectively. Available phosphorus, ammonium and nitrate values ranged from 22 to 29 ppm, 15 to 28 ppm, and 24 to 30 ppm respectively. Organic carbon, total nitrogen and carbon/nitrogen ratio values ranged from 1.60 to 3.27%, 0.08 to 0.169% and 10.9 to 23.0 respectively. The percent (%) organic carbon was lowest in control as compared with the polluted soils at 24h. This is as a result of organic carbon being added to the polluted soils through the addition of the crude oil whose major constituents are carbon and hydrogen. While there

* E mail: omokaro515@yahoo.com; okigborn17@yahoo.com

was a general decrease in % organic carbon in all the polluted soils at the after 18weeks, there was an increase in the control soil. The % total nitrogen was higher in the control than 0.5% and in 3% polluted soils. Generally, there was a reduction in % total nitrogen in all the soils after 18weeks. The C: N ratio was lowest in the control. However, there was an increase in the control, 0.5%, and in 1% polluted soils and a reduction in 3% and 5% polluted soils at the end of the investigation.

INTRODUCTION

The increase in demand for crude oil as a source of energy and as a primary raw material for industries has resulted in an increase in its production, transportation and refining which in turn has resulted in gross pollution of the environment (Gutnick and Rosenberg, 1977). Bartha (1986) reported that generally both in the marine and the terrestrial environments, the low-level routine discharges (effluents, urban runoff, cleaning operations, oil treatment of roads for dust control etc.) also account for about 90% of the total anthropogenic petroleum hydrocarbon pollution.

Snowden and Ekweozor (1987) reported that petroleum hydrocarbons get into the environment through accidental discharges due to oil industry activities. Oil industry activities as reported by Obire and Wemedo (1996) include among others drilling, exploration and production processes. Oil pollution of the terrestrial environment also comes from oil-well blowouts and seepages, pipeline overflow and stationary spills from storage tank in refinery areas and from sales and uses of crude oil and refined petroleum products (Obire and Wemedo, 1996). Incidentally contamination may lead to addition of oil to the soil as a deliberate policy of waste disposal management (Flowers *et al.*, 1984; Moffat and Linden, 1995; Obire and Wemedo, 1996).

Plice (1948), Baker (1970), Rowell (1977) also reported that oil spillage cause extensive damages ranging from the destruction of terrestrial flora and fauna, biomagnification of the toxic components of the petroleum, conversion of arable land to barren land and the destruction of the aesthetic quality of environment. As reported by Fagbani *et al.* (1988); Portmann *et al.* (1989) and NDWC (1995), in some cases, oil spills contaminates and also kill several hectares of mangrove swamp and in most cases seedlings are killed.

Petroleum toxicity has been identified as a contributing factor to the stressed conditions of a soil-polluted by crude oil. Crude oil even in small amounts can delay germination of a seed while in larger amounts, it will prevent germination of the seeds altogether. Phytotoxic and other stress imposing properties of crude oil were suggested as causes of death to plants.

Plice (1948) reported that pure toxicity of crude oil plays a part in decreasing seed viability and that volatile oil fractions have high wetting capacity and penetrating power in such a way that if they come in contact with plants seed, they enter the seed coat easily and then kill the embryo. The extent of contamination of the soil after spillage determines the rate at which oil moves as a body through the soil had taken place. The movement of the oil through and out of the soil largely depends upon the type of oil, organic content, texture and moisture, the depth to the water table and the content of the soil (Rowell 1977).

The soil harbours both macro- and micro-organisms as well as plant roots and it is also one of the most dynamic sites in nature. Many of the biochemical reactions concerned in the mineralization of soil organic matter and in the nutrition of plants particularly agricultural crops are mediated by microorganisms whose activity is governed by the soil environment. In the presence of adequate supply of oxygen, soil fertility (nutrient recycling and mineralization) being a growth linked process involving both macro- and micro- organisms requires the presence of mineral elements such as C, Ca, Mg, K, S, Fe, N, P, in addition to various trace elements. Any factor whose presence in the soil environment adversely affects these mineral elements also adversely affects plant growth. In soils affected by crude oil, oxygen demand by the aerobic microbes using the oil carbon for growth is very high and because of the high demand, the oxygen can be depleted rapidly (Rowell, 1977). Bossert and Bartha (1984) reported that as oil gradually sinks down into the soil, air and pore spaces are filled leading to exclusion of air. However, Udo and Fayemi (1975) observed poor plant growth in oil-contaminated soils and suggested that suffocation of the plants caused by exclusion of air by the oil or exhaustion of oxygen by increased microbial activity was partly responsible. Nutrient is one factor that can hinder biodegradation of pollutants if not handled properly. Therefore, inadequate provision of the nitrogen, phosphorus, potassium and sulphur (which is probably the most important and the most easily modified of all the factors) could limit the rate of hydrocarbon degradation in the terrestrial environment (McGill and Nyborg, 1975).

According to OTA (1990), in any case when oil is spilled in large quantities, the microorganisms in that environment will be limited in their ability to degrade the petroleum due to lack of nutrients, but when nutrients are added to the environment, the organisms will regain enough ability to overcome the limitations and biodegradation will then take place unhindered.

The aim of this investigation is to assess the impact of the addition of various concentrations of crude oil on the physicochemical properties of a soil in the Niger Delta as regards the soil fertility status. Results obtained will help in recommending processes that will augment the native soil fertility status, improve rate of oil recovery and crop yield as to sustain agricultural development in the Niger Delta.

Materials and Methods

Study Area and Source of Crude Oil

The study is a fallow patch of land in Nkpolu-Oroworukwo, Mile 3, Diobu area of Port Harcourt, Rivers State that was fenced to prevent animal and human encroachment. The land was marked out and prepared into unpolluted and crude oil-polluted plots from which soil samples for the study were collected and analyzed. Crude oil (Brass blend) with a specific gravity (API) of 41.7 – 43.0 and black in colour was obtained from Nigeria Agip Company (NAOC), Obrikom flow station, Rivers State, in large sterile plastic containers.

Experimental Design and Treatment Level and Application of Crude Oil

The experimental design was the Randomized Complete Block Design (RCBD). Each plot or block unit was 30cm x 30cm. Since microbially influenced agricultural soil fertility is in the range of 0 – 15 cm depth (Toogood, 1977) then the volume of soil per plot was 30 x 30 x 15 = 13,500cm^3. The study involved 5 treatments, each replicated three times. Assignment of treatment to plot was carried out using the Table of Random Numbers of Gomez and Gomez (1984). The volume of crude oil applied to individual plots in various concentrations (0.5%, 1%, 3% and 5%) was calculated in millilitres using a 1:1 soil volume oil volume relationship (Amadi *et al.*, 1993). The appropriate amount of crude oil was measured and applied evenly on the surface of each plot and allowed to sink gradually for 30 minutes except for the unpolluted plots (0% crude oil) which served as controls.

Since microbiological oxidation of any organic substrate requires that certain specific conditions be met (McGill and Nyborg, 1975) the following subsequent treatments were carried out in order to speed up the rate of crude oil degradation in the soil; All experiment and control plots were tilled once a week in order to make oxygen available to the microorganisms involved in crude oil degradation process.

All experiment and control plots were watered to 60% water holding capacity (Margesin and Schinner, 1997) before oil application and subsequently as required.

Collection of Soil Samples for Analysis

Soil samples (400g) of surface soil (0 – 15cm depth) was collected from each plot consecutively after tilling with a sterile scoop and transferred into sterile polythene bags. Soil sample collections were made from 3 – 4 random points per plot and then bulked to form composite samples. All soil samples for future analyses were stored at 4 ± 2°C according to ISO and OECD standards (Torstensson *et al.*, 1998). Soil samples from each plot was collected before pollution, 24 hours after pollution and 18 weeks after pollution.

Physico-Chemical Analysis of Soil Samples

The physico-chemical properties of the soil samples were determined before and after oil application. Soil samples to be analysed for physico-chemical properties were air-dried, sieved through 2mm mesh, and stored appropriately (Laverty, 1977). The parameters determined include soil particle size, water holding capacity, pH, temperature, organic carbon, total nitrogen, available phosphorus, exchange capacity, exchangeable cations – sodium, potassium, magnesium and calcium, ammonium ion (NH_4^+), and nitrate (NO_3^-).

Particle size analysis was carried out using the hydrometer method (Bouyoucos, 1951). Soil pH was measured in 1:2.5 soil: water suspension, following 30 minutes shaking, using a pH meter 7020, Electronic Instrument Ltd., Kent) and (Laverty, 1977; Flowers *et al.*, 1984).

The temperature of the soil samples was determined using a mercury thermometer. This was done by inserting a thermometer into the tilled soil for a period of five minutes and the readings taken immediately the thermometer was removed from the soil.

Organic carbon was determined in accordance with titration procedures as described in the method of Walkey and Black (1934).

Total nitrogen, ammonium and nitrate content were determined by the Kjeldahl digestion and steam distillation of KCl extracts of soil according to the method of Bremmer (1965).

Available phosphorus was determined colorometrically in NH_4F + HCl extract of soil, using a 1.5 soil solution ratio. This is the method of Bray and Kurtz (1945).

Exchangeable acidity was determined by the method of Dewis and Freitas (1970) as cited by Allen *et al.* (1974).

Exchangeable cations were determined after extraction of soils with 1N ammonium acetate for 1 hour. Sodium and potassium in the filtrate were determined using the flame photometer in accordance with the manufacturer's instructions. Magnesium and calcium were determined using titration procedures.

Determination of Water-Holding Capacity

The water-holding capacity of the soil was determined and the 60% water-holding value was calculated from the corresponding value.To determine the water-holding capacity of the soil for this study, $100cm^3$ of the soil was first determined by pouring the homogenized, air-dried, and 2mm – sieved soil sample into a $100cm^3$ measuring cylinder. The later was taped until all visible air spaces were filled and the soil adjusted to the $100cm^3$ of the soil.

The $100cm^3$ soil was put into a funnel with cotton wool. The funnel was tapped carefully and persistently until all visible air spaces were filled. The funnel was allowed to stand in the open end of 100ml measuring cylinder. One hundred millilitres of water was quickly poured into the funnel containing the soil sample. The water was allowed to drain through the soil until no water dripped into the measuring cylinder. To ensure that no water was left in the funnel, the experimental set up was allowed to stand overnight.

The volume of water, which passed through the soil, was then measured by reading the level of water in the measuring cylinder. The amount of water retained was calculated by subtracting the volume of water drained into the measuring cylinder from the volume of water added to the soil.

The 60% water-holding capacity of the $100cm^3$ soil used for the determination was later translated to 60% water-holding capacity of the total soil volume of each experimental plot.

RESULTS

The present investigation revealed that at various oil concentrations the measured properties showed a decrease in their values except in the control (0% oil pollution) where an increase was observed in the following parameters: Soil pH, magnesium, sodium, ammonium, nitrate and organic carbon while the others parameters measured showed a decrease in their values. The particle soil analysis revealed that the soil texture is sandy loam having 36% clay, 10% silt and 54% sand. The results of the soil chemical properties of the control and oil-polluted soils for various concentrations of oil at 24 hours after pollution and at 18 weeks after pollution are as shown in Table 1. The data tabulated are the means obtained from triplicates of each experiment.

The pH values of the control soils and polluted soils at 24 hours after pollution ranged from 6.4 to 7.9. That is, the soils were slightly acidic and slightly alkaline. Exchangeable acidity ranged from 0.18 to 0.25. Calcium, magnesium, potassium and sodium values ranged from 1.99 to 6.88, 0.27 to 0.35, 0.09 to 0.24 and 0.03 to 0.06 meq $100g^{-1}$ soil respectively.

Table 1. Mean values of chemical characteristics of control soil and soils polluted with various concentrations of crude oil

	Concentration of crude oil pollution of soil samples									
	0% (control)		0.5%		1%		3%		5%	
Soil parameters	24h	18wks	24h	18wks	24h	18wks	24h	18wks	24h	18wks
pH	6.4	7.1	7.9	7.8	6.7	6.3	7.8	7.4	7.6	7.0
Exchangeable acidity	0.21	0.20	0.25	0.23	0.22	0.23	0.24	0.21	0.18	0.15
Ca^{2+}	1.99	1.94	6.88	6.38	4.81	4.55	4.65	4.21	4.63	3.79
Mg^{2+}	0.32	0.35	0.31	0.28	0.33	0.30	0.27	0.30	0.35	0.37
K^+	0.15	0.13	0.16	0.11	0.24	0.21	0.09	0.11	0.13	0.19
Na^+	0.05	0.08	0.06	0.03	0.04	0.02	0.06	0.14	0.03	0.12
Available phosphorus	27	29	26	26	28	25	24	22	25	22
NH_4^+	16.0	15.0	15.9	26.0	15.7	28.0	19	26.0	17	21.0
NO_3^-	28	30	27	29	20	31	21	24	14	24
% Organic carbon	1.58	1.80	2.42	1.60	4.23	3.08	4.00	2.93	4.13	3.27
% Total nitrogen	0.179	0.165	0.150	0.080	0.217	0.169	0.140	0.127	0.207	0.164
C:N Ratio	8.83	10.9	16.1	20.0	19.5	19.9	28.6	23.0	20.0	20.0

Available phosphorus, ammonium and nitrate values ranged from 24 to 28 ppm, 15.7 to 19 ppm and 14 to 28ppm respectively. Organic carbon, total nitrogen and carbon/nitrogen (C/N) ratio values ranged from 1.58 to 4.23%, 0.14 to 0.217% and 8.83 to 28.6% respectively. The pH values of the control and oil-polluted soil at 18 weeks after pollution ranged from 6.3 to 7.8. Exchangeable acidity ranged from 0.15 to 0.23. Calcium, magnesium, potassium and sodium values ranged from 1.94 to 6.38 meq $100g^{-1}$ soil, 0.28 to 0.37 meq $100g^{-1}$ soil, 0.11 to 21 meq $100g^{-1}$ soil and 0.02 to 0.14 meq $100g^{-1}$ soil respectively. Available phosphorus, ammonium and nitrate values ranged from 22 to 29 ppm, 15 to 28 ppm, and 24 to 30 ppm respectively. Organic carbon, total nitrogen and carbon/nitrogen ratio values ranged from 1.60 to 3.27%, 0.08 to 0.169% and 10.9 to 23.0 respectively.

This study has revealed the differences in soil chemical characteristics and hence the fertility status of soils as influenced by various concentrations of crude oil pollution.

Exchangeable acidity, magnesium, potassium, sodium and total nitrogen values were generally low in all the polluted soils. The pH was lowest in the control soils as compared with the polluted soils. However, there was an increase in the pH of the control soil while there was a general decrease in the pH of all the polluted soils at the end (18 weeks) of the investigation. The control and 5% oil polluted soils recorded the lowest values of exchangeable acidity. This explains in part the toxicity of the soil as a result of release of aluminium and hydrogen ions into the soil by certain materials or clay particles, and this causes unfavourable soil condition for the growth of microorganisms, plants and animals. This result is in agreement with the findings of Schwendinger (1968) which says that at above

3% concentration, oil is dangerous both to the plant growth and microorganisms. Calcium, available phosphorus, ammonium nitrate and organic carbon showed a consistent increase in the values with slight variations. Calcium was lowest in the control at 24h of investigation and there was a reduction in calcium with increasing concentration of crude oil pollution of the soils. There was also a general decrease in the concentration of calcium in all the samples at the end of the investigation. Calcium uptake by plants and microorganisms has a dual function in soil. It serves as an essential nutrient as well as the dominant base that keeps the soil neutral in reaction. Depletion of calcium in the soil leads to the replacement of the calcium by hydrogen ions which makes the soil to be acidic. This explains why the control is more acidic than the polluted soils 24h during the investigation. In general, no remarkable difference was observed in exchangeable acidity of the soils polluted with various concentrations of crude oil at the end of the experiment as compared to the values obtained at the on-set of experiment. This was due to stability in exchangeable cations in oil-polluted soils most especially calcium. The results obtained for the exchangeable cations during this investigation are in agreement with the results obtained by McGill and Nyborg (1975). At 24h after pollution, except for the 5% polluted soil, magnesium was highest in the control soil. There was also an increase of magnesium and sodium in the control soil, 3% and in 5% soils but a reduction in the 0.5% and 1% polluted soils at the end of the investigation. Potassium decreased in control, 0.5% and in 1% but increased at 3% and 5%. Calcium uptake by plants and microorganisms has a dual function in soil. It serves as an essential nutrient as well as the dominant base that keeps the soil neutral in reaction. Depletion of calcium in the soil after 18weeks can lead to the replacement of the calcium by hydrogen ions which makes the soil to be acidic and this leads to deterioration of soil structure, enhanced solubility of Fe, Al and Mn to toxic levels, less solubility of phosphorus, inhibition of activities of soil organisms leading to accumulation of organic matter in the soil (Alexander, 1977). Available phosphorus was highest in the control and while there was an increased in the control, there was a general decrease in phosphorus in the polluted soils at the end of 18weeks. Plants and microorganisms require the presence of available forms of phosphorus for their growth. Addition of petroleum as a source of carbon leads to deficiencies of phosphorus as increase in microbial abundance to enhance the utilization of petroleum will draw from available store putting great demand on the phosphate supply; and deficiency of this phosphorus may lead to depressed crop yields (Cooke, 1975; Alexander, 1977).

Ammonia was lowest in the controls while nitrate was highest. There was a general increase in nitrate concentration in all the soils. The percent (%) organic carbon was lowest in control as compared with the polluted soils at 24h. This is as a result of organic carbon being added to the polluted soils through the addition of the crude oil whose major constituents are carbon and hydrogen. While there was a general decrease in % organic carbon in all the polluted soils at the after 18weeks, there was an increase in the control soil. The % total nitrogen was higher in the control than 0.5% and in 3% polluted soils. Generally, there was a reduction in % total nitrogen in all the soils after 18weeks. The C: N ratio was lowest in the control. However, there was an increase in the control, 0.5%, and in 1% polluted soils and a reduction in 3% and 5% polluted soils at the end of the experimentation. The range of pH, potassium, ammonium, nitrate, organic carbon, and total nitrogen recorded in this study is however higher than the range reported by Obire and Nwaubeta (2002). While the range of calcium, magnesium, sodium, available phosphorus and C: N ratio is lower than the range reported by Obire and Nwaubeta (2002).

The results obtained are in agreement with the findings of Cook and Westlake (1973) which says that in oil-polluted soils, nitrogen and phosphorus quickly become limiting to biodegradation of oil by microorganisms and plant growth while C/N ratio is widened. This would leads to a serious competition for the limited nutrient elements thereby resulting to a delay in the rate of soil recovery and decrease in crop yield.

REFERENCES

Alexander, M. (1977). *Introduction to Soil Microbiology*, 2nd edn, John Wiley and Sons, Inc., New York, p. 467.

Allen, S. E., Grimshaw, H. M., Parkinson, J. A. and Quarmby, C. (1974) Chemical Analysis of Ecological Materials. 1st edn, Blackwell Scientific Publications, London.

Amadi, A., Dickson, A. A. and Maate, G. O. (1993). Remediation of oil- polluted soils: 1. Effect of organic and inorganic nutrient supplements on the performance of maize (*Zea mays L*). *Water, Air and Soil Pollution. 66*: 59-76.

Baker, J. M. (1970). Studies on salt marsh communities. The effects of a single oil spillage. pp. 16 - 43. *In*: E.B. Colwell (ed.). *The Ecological Effects of Oil Pollution on Littoral Communities. Institute of Petroleum,* London.

Bartha, R. (1986). Biotechnology of pollutant biodegradation. *Microbial Ecology 12*: 155-172.

Bossert, I. and Bartha, A.S. (1984) "Hydrocarbons in soil ecosystem" cited by A. Amadi, S.D. Abbey and A. Nma (eds) Effects of oil spill on soil properties. *Water, Air and Soil Pollution. 86*: 1-11.

Bray, R. H. and Kurtz, L. T. (1945) Determination of total organic and available forms of phosphorus in soils. *Soil Science* 39: 39-45.

Cook, F. D. and Westlake, D. W. S. (1973) Biodegradability of Northern Crude Oils. *Environmental - Social Program, Northern Pipelines Task Force, Report 74-1.*

Cooke, C.B.E. (1982) Fertilizing for Maximum Yield. 3rd Edn, Granada Publishing Ltd., Britain, 465 pp.

Fagbami, A. A., Udo, E. J. and Odu, C. T. I. (1988). Vegetation damage in an oil field in the Niger Delta of Nigeria. *Journal of Tropical Ecology. 4*: 61-75.

Flowers, T. H., Pulford, I. D. and Duncan, H. J. (1984). Studies on the breakdown of oil in soil. *Environmental Pollution. (Series B) 8*: 71-82.

Gomez, K. A. and Gomez, A. A. (1984). Statistical Procedures for Agricultural Research. 2nd edn. Singapore, John Wiley and Sons Inc. 680pp.

Gutnick, D. L. and Rosenberg, E. (1977). Oil tankers and pollution: A microbiological approach. *Annu. Rev. Microbiol. 31*: 379 396.

Laverty, D. H. (1977). Analytical procedures on soils subjected to oil spills. *In*: J. A. Toogood (ed.) *The reclamation of agricultural soils of oil spills.*

Margesin, R. and Schinner, F. (1997). Efficiency of indigenous and inoculated cold-adapted soil microorganisms for biodegradation of diesel oil in Alpine soil. *Appl. Environ. Microbiol. 63 (7)*: 2660-2664.

McGill, W.B. and Nyborg, M. (1975). Reclamation of wet forest soils subjected to oil spills. *Alberta Inst. of Pedology, Canada, Publ. No. G - 75 - 1.*

Niger Delta Wetlands Centre (1995). Review of initial assessment of environmental issues in the Niger Delta and "Niger Delta Biodiversity". *Report by the Niger Delta Wetlands Centre, Port Harcourt,* Nigeria.

Obire, O. and Nwaubeta, O. (2002). Effects of refined petroleum hydrocarbons on soil physicochemical and bacteriological characteristics. *J. Appl. Sci. Environ. Mgt. 6(1)*: 39 – 44.

Obire, O. and Wemedo, S. A. (1996). The effect of oilfield wastewater on the microbial population of a soil in Nigeria. *Niger Delta Biologia. 1(1)*: 77 - 85.

Office of Technology Assessment (OTA) (1990) Bioremediation for Marine Oil Spill. Workshop. Washington D. C., Government Printing Press 4: 1-30.

Plice, M. J. (1948). Some effects of crude petroleum on soil fertility. *Soil Science Society Proceedings. 13*: 413 - 416.

Portmann, J., Biney, C., Chidilhe, A. and Zabi, S. (1989). State of the marine environment: West and Central African Region. *UNEP Regional Seas Reports and Studies. No. 108.*

Rowell, M. J. (1977). The effect of crude oil spills on soils: A review of literature. *In*: J.A. Toogood (ed) *The Reclamation of Agricultural Soils after Oil Spills. Part 1: Research, Alberta Inst. of Pedology, Publ. No. M - 77 - 11, pp. 1 - 33.*

Schwendinger, R.B. (1968). Reclamation of soil contaminated with oil. *J. Inst. Petrol. 54: 182 – 197.*

Snowden, R. J. and Ekweozor, I. K. E. (1987). The impact of a mineral oil spillage in the estuarine Niger Delta. *Marine Pollution Bulletin.* 18 (11): 505-599.

Toogood, J. A. (1977). Burning as a reclamation technique. *In*: J. A. Toogood (ed) The Reclamation of Agricultural Soils after Oil Spills. *Part 1: Research, Alberta Inst. of Pedology, Canada, Publ. No. M- 77-11, pp. 89 - 91.*

Torstensson, L., Pell, M. and Stenberg, B. (1998). Need of a strategy for evaluation of arable soil quality. *Ambio. 27(1)*: 4 - 8.

Udo, E. J. and Fayemi, A. A. A. (1975). The effect of oil pollution of soil on germination, growth and nutrient uptake of corn. *J. Environ. Qual. 4*: 537 - 540.

Walkley, A. and Black, I. A. (1934) Methods of soil analysis. *Soil science* 37: 29.

In: Industrial Pollution Including Oil
Editor: Harry Newbury and William De Lorne
ISBN 978-1-60456-917-9

Chapter 4

RISK ASSESSMENT OF AIRBORNE PARTICLES AND VOLATILE ORGANIC COMPOUNDS FROM INDUSTRIAL AREAS

Andrea Mueller*[*1], Gunnar Wichmann[3], Laura Massolo[2], Andrés Porta[2], Uwe Schlink[1], Alicia Ronco[2]* and *Olf Herbarth[1,4]

[1]Helmholtz Centre for Environmental Research – UFZ,
Department of Human Exposure Research and Epidemiology Permoserstr. 15, 04318 Leipzig, Germany
[2]Centro de Investigaciones del Medio Ambiente,
Facultad de Ciencias Exactas, Universidad Nacional de La Plata,
47 y 115; (1900)-La Plata,
Argentina. CICPBA-CONICET
[3]ENT-Research-Laboratory, ENT-Department, University of Leipzig,
Liebigstr. 26, 04103 Leipzig, Germany
[4]Department of. Environmental Medicine and Hygiene, Faculty of Medicine,
University of Leipzig, Liebigstrasse 27, 04103 Leipzig, Germany

ABSTRACT

Particulate matter (PM) and bound chemicals like polycyclic aromatic compounds (PAHs) as well as volatile organic compounds (VOC) in air are potential mediators for adverse health effects. The presented study includes a determination of exposure to VOC, determination of PM and thereon bound PAHs as well as investigations of mutagenic, cytotoxic and immunomodulating effects of PM in three regions in La Plata (Argentina), influenced either by industrial or traffic emissions, and in a control area. La Plata was selected as an example for the influence of industrial emission since Argentina's main oil refinery is located nearby.

* Corresponding author: Andrea Müller. Department of Human Exposure Research and Epidemiology, Helmholtz-Centre for Environmental Research –UFZ, Permoserstr. 15 D 04318 Leipzig, Germany, Phone: ++49 341 235 1541. Fax ++49 341 235 1786. Email: a.mueller@ufz.de

Following sampling using high volume collectors with cascade impactors to separate six size fractions of PM (from 10 to less than 0.49 μm) PM-absorbed compounds were hexane-extracted. Additionally, 28 different VOC (aromatics, alcanes, terpenes, and halocarbons) were collected by passive sampling and analyzed for typical fingerprints.

For an assessment of toxicity and mutagenicity, extracts of PM were tested for mutagenicity (AMES-test: *Salmonella typhimurium* TA 98/S9+) and cytotoxicity (*Tetrahymena pyriformis*). In addition, cell respiration and ATP production of *T. pyriformis* were determined as parameters for oxidative stress and immunomodulatory effects on human leukocytes were analyzed.

For risk assessment, the lifetime cancer risk (LCR) was calculated for PAHs using the concentration of benzo[a]pyrene (BaP) and BaP equivalents as well as for VOC by concentration of benzene.

Highest total PAH concentrations were found in the industrial area. The size distribution of particles demonstrates the greater relevance of smaller particles since more than 50 % of PAHs were absorbed to particles <0.49m. The highest mutagenic potency and cytotoxicity as well as immunomodulatory effects were also associated with particles <0.49 μm. These effects correlated with PAH concentrations especially of high condensed PAHs with 4 and 5 rings. Investigations of the cell respiration showed an uncoupling of the respiratory chain by extracts of fine particles.

Regarding the VOC concentration a dominance of alkanes and aromatic compounds was observed, even though a different chemical distribution was seen. In the industrial area significant higher concentrations of hexane, heptane and benzene occurred.

According to the higher concentration of benzene and BaP also the highest level for LCR was estimated for the industrial area compared to the others.

Due to the complexity of effects related to the different exposure scenarios, both, exposure analyses and an effect assessment by various bioassays allowed us to estimate the risk of industrial air pollution.

1. Introduction

Exposure to ambient air particulate matter (PM) and bound chemicals as well as exposure to volatile organic compounds (VOC) is associated with various diseases. Several epidemiological studies have shown an association of particular matter with asthma, acute respiratory symptoms, chronic cough, bronchitis, and respiratory and cardiovascular mortality (Brunnekreef and Holgate 2002, Schwartz, 2004,).

PM is classified by its aerodynamic diameter as coarse (2.5–10 μm), fine (0.1 - 2.5 μm), and ultrafine (< 0.1 μm). Fine and ultrafine PM are produced by combustion (motor vehicles, power plants), whereas coarse particles are generated by mechanical processes which produce dust from non-combustive sources. Coarse particles depose predominantly in the nose and throat and are usually cleared by exhalation or mucociliary clearance. PM with an aerodynamic diameter <10 μm (PM_{10}), and moreover especially those <2.5 μm ($PM_{2.5}$), are a significant threat to human health (Ferin et al., 1992; US EPA 1996 a, b; Hauser et al., 2001). $PM_{2.5}$ and especially smaller particles, the sub-micrometer (fine) including the ultrafine particles <100 nm ($PM_{0.1}$), are respirable and therefore can deposit and accumulate in the deeper airways and thus can have immediate and prolonged impact on the epithelial and immune cells of the lung and on the whole organism. Indeed, increased $PM_{2.5}$ exposure was shown to be associated with numerous health effects. Dockery and colleagues (1992) demonstrated a correlation between fine particle concentrations in ambient air and human mortality. New evidence regarding the risk for human health caused by fine and ultrafine particles in ambient air aroused an increasing concern (Junker et al., 2000; Schwartz and Neas, 2000; WHO, 2000; Herbarth et al., 2001). Also they are suspected to be associated with an increased incidence of lung cancer (Pope et al., 2002). In numerous studies a higher mutagenicity, genotoxicity and toxicity of fine particles could be demonstrated (Hsiao et al., 2000; Kawanaka et al., 2004; Healey et al., 2005). A reason for the observed higher mutagenicity and increased risk for exposed human beings might be the drastically greater surface per volume of smaller particles and the therefore increased capacity to adsorb chemicals.

Polycyclic aromatic hydrocarbons (PAHs) and volatile organic compounds (VOCs) are widely distributed in the atmosphere. They arise from natural and anthropogenic processes and represent an important group of air pollutants. Anthropogenic emission sources of PAHs and VOCs in the atmosphere include traffic and domestic heating. Petroleum refineries and petrochemical plants are other contributing sources of emissions, mainly originating from production processes, storage tanks and waste areas (Brenner et al., 1989; Watson et al., 2001; Cetin et al., 2003).

Indoor VOC exposure is thought to be of greater concern in the community because indoor concentrations of many pollutants are often higher than those typically encountered outdoors. At the same time, indoor VOC concentration may be the dominant contributor to personal exposure because most people spend over 80% of their time indoors, either in the home or in the work place (Guo et al., 2004; Wang et al., 2007).

Measurement of VOCs in the indoor environment has received substantial research attention for several years because indoor VOC levels may pose potential health effects to occupants of dwellings. While some VOCs may be present at concentrations that are not considered acutely harmful to human health with short-term exposure, long-term exposure may result in mutagenic and carcinogenic effects (Son et al., 2003). The occurrence and concentrations of VOCs in residences can be affected by outdoor atmospheric conditions, indoor sources, indoor volume, human activities, chemical reactions, ventilation rates, and seasonal factors (Son et al., 2003; Schlink et al., 2004).

For actions reducing adverse exposure in the urban environment it is important to know the emission of the dominant sources. For that reason, techniques of source apportionment are currently of growing interest in urban air pollution sciences. Early attempts at source identification used indicator substances, such as acetylene (Whitby and Altwicker, 1978), aromatic organic compounds (Edgerton et al, 1989) as indicators of motor vehicle emissions.

To recognize the effect of a source, some authors model the dispersion of the pollutants (Peace et al., 2004) for the situation of interest. In the case of multiple sources the scenario is too complex for such source models. Receptor models provide an alternative framework for quantifying source contributions (Hopke, 1985).

Receptor models contrast with source models (Seigneur, 1998) that combine source emission rates with meteorological transport and chemical changes to estimate concentrations at a receptor. Some receptor models are statistical, establishing empirical relationships between variables by multivariate mathematical techniques. In the chemical mass balance (CMB) modeling approach developed by the U.S. Environmental Protection Agency (USEPA, CMB8 User Manual, 2001) multiple linear least squares regressions are used to identify the probable sources of VOCs (Srivastava, 2004).

Chemically bound compounds to particulate matter such as polycyclic aromatics hydrocarbons (PAHs) and their derivatives, particularly highly condensed compounds with five or six aromatic rings, are known to be mutagenic and carcinogenic (Tuominen et al., 1988; Zwoździak et al., 2001). Some of these compounds and their mixtures can produce DNA damage, such as inducing mutations in bacterial and mammalian cells, chromosomal effects and oncogenic transformations (Lewtas, 1988 and 1993).

Previous studies of airborne particulate matter and chemically bound compounds have indicated the presence of high concentrations of PAHs in urban air from areas subject to domestic and industrial emissions (Harrison and Jones, 1995; Monarca et al., 1997).

Mutagenic activity and cytotoxicity were found in organic extracts from dust particles — total suspended particulate matter <100 µm (TSP); particulate matter < 10 µm (PM_{10}) and particulate matter < 2.5 µm ($PM_{2.5}$) (Spinosa De Martinis et al., 1999; Ye et al., 1999; Černa et al., 2000; Müller et al., 2001; Zwoździak et al., 2001).

Moreover, higher mutagenic and toxic effects have been detected in connection with smaller particle size fractions in bioassays (Monarca et al., 1997; Hsiao et al., 2000). Increasing concern has been aroused by new evidence of the risk to human health posed by fine and ultrafine particles in the air (Junker et al., 2000; Schwartz and Neas, 2000; Hauser et al., 2001; Herbarth et al., 2001).

The presented study includes the determination of exposure to VOC, determination of PM and thereon bound PAHs as well as investigations of mutagenic, cytotoxic and immunomodulating effects of PM from different size fractions in three regions in La Plata (Argentina), influenced either by industrial or traffic emissions, and a control.

2. Characterisation of the Study Area

The study region in La Plata, Argentina, was selected as an example for the influence of industrial emissions and their influence on human health since Argentina's mail oil refinery is located near the city.

La Plata and its neighboring areas are located northeast of Buenos Aires province (Argentina) and have a population of around 700,000 inhabitants. Three sampling zones were considered: industrial (I), urban (U), control (C) areas (Figure 1).

The industrial area holds the country's main oil refinery (total crude oil distillation capacity: 38,000 m^3/day^{-1}) located next to six petrochemical plants producing diverse compounds such as aromatics (benzene, toluene, xylenes), aliphatic solvents (n-pentane, n-hexane, n-heptane), polypropylene, polybutene, maleic anhydride, cyclohexane, methanol, methyl tertiary butyl ether, and petroleum coke. This industrial complex is approximately 10 km north-northeast of the main urban sector of the city.

The urban area is characterized by heavy traffic; the number of vehicles registered is approximately 200,000. The rural control area with low traffic is placed 15 km away from the urban and industrial sectors.

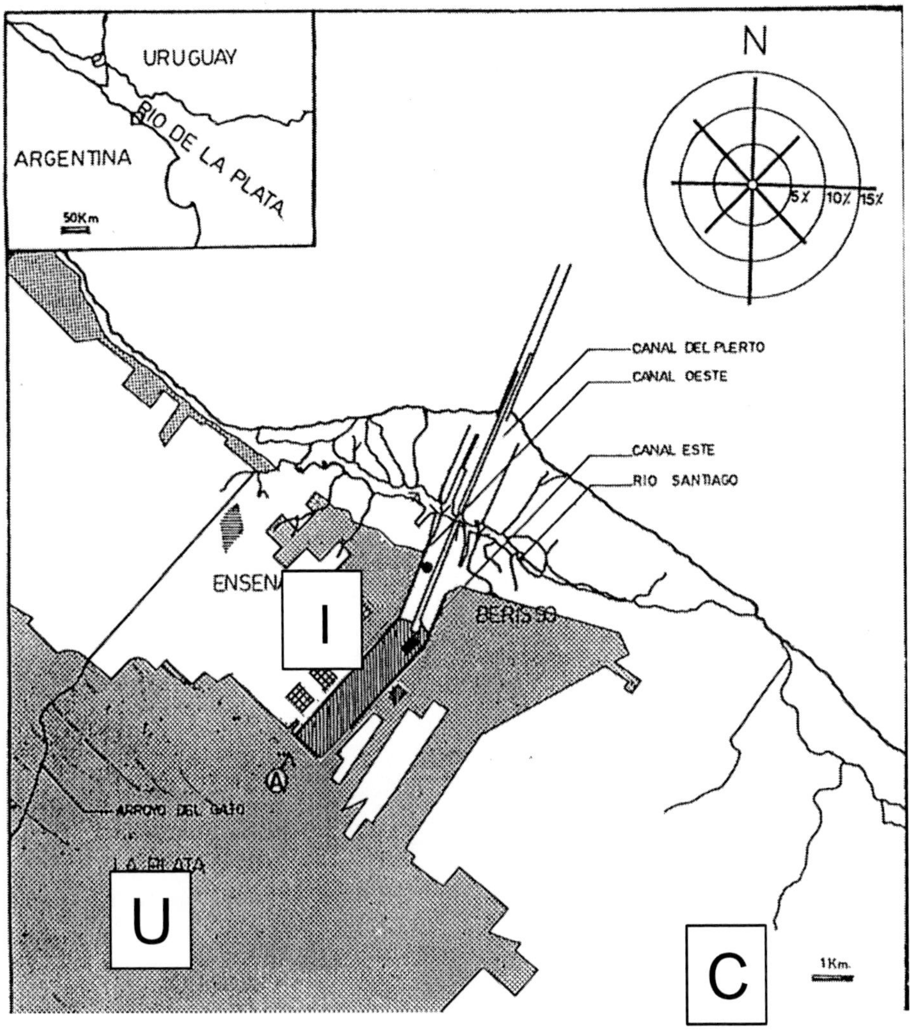

Figure 1. Sampling points in La Plata, Argentina. Samples were collected using a high volume sampler in three different areas: influenced by petrochemical industry (I), urban area influenced by traffic (U) and control area (C). Passive sampling monitors to analyse concentration of VOC were placed indoors and outdoors in representative places in all regions.

3. Analyses of Air Contaminants

3.1. Analyses of Air Particulate Matter and Bounded Polycyclic Aromatic Compounds

3.1.1. Methods

For the study samples of airborne particulate matter (PM_{10} and smaller fractions) were selected from three locations shown in figure 1. The sampling was carried out with high volume collectors using cascade impactor with a 10 μm inlet (Sierra and Wedding and Associates Inc.) to separate six fractions by their aerodynamic size:

fraction 1: 10-7.2 μm,
fraction 2: 7.2-3 μm,
fraction 3: 3-1.5 μm,
fraction 4: 1.5-0.95 μm,
fraction 5: 0.95-0.49 μm
fraction 6: < 0.49 μm.

The concentration of particles was determined by gravimetric methods (Marple and Lui, 1974; 40 CFR, 1987). Dust samples were extracted using hexane in the accelerated extraction system ASE 200 (DIONEX, Canada) under a pressure of 14 mPa and a temperature of 150 °C. The extracted sample was reduced by rotary evaporator to ~ 0.5 ml and nitrogen flow near dryness and re-dissolved with 1 ml hexane. The sample work-up procedure is as follows: 2.5 g silica gel 60 (Merck, Germany) inserted into the clean-up column was first conditioned by eluting with 20 ml of hexane/dichloromethane (1:1). The extract (1 ml hexane) was applied to the top of the clean tube and eluted with 4 ml hexane/ dichloromethane (6:1). After discarding three milliliters, the elution of the PAH mix was carried out with 16.5 ml hexane/dichloromethane (6:1). After adding 1-octanol as a keeper, the PAH fraction was evaporated again following the described method and re-dissolved with 1 ml acetonitrile.

For testing in bioassays (chapter 4) portions of the organic extracts without 1-octanol were reduced to dryness and re-dissolved in dimethylsulfoxide (DMSO).

The chemical analysis was performed by HPLC (Jasco, Germany) which is equipped with two pumps (980-PU), degasser (DG 980-50), autosampler (AS-950), column oven (CO-200) and fluorescence detector (FP 920). The column oven temperature was set to 30 °C, the injection volume was 10 μl and the chromatographic separation was performed on a C18 column from Jasco (l=2,500 mm, d=3 mm, 5 μm) with an acetonitrile / water gradient of 60:40, hold 3 min, ramp to 100:0 within 30 min and hold 10 min, with a flow rate of 0.5 ml min^{-1}. Before the next injection the column was equilibrated with the initial conditions for constant retention times.

The fluorescence detection of the 15 investigated PAHs was performed by following the wavelength program to detect selectively the compounds of interest. The detector started at (excitation/emission) 275/350 nm to detect naphthalene, acenaphthene, fluorene and phenanthrene. After 14 min, it changed to 260/420 nm to detect anthracene, fluoranthene and pyrene. 7 min later, it switched to 265/380 nm to detect benzo(a)anthracene and chrysene. After 27 min, it changed to 290/430 nm to detect benzo(b)fluoranthene, benzo(k)fluoranthene, benzo(a)pyrene, dibenz(ah)anthracene and benzo(ghi)perylene, and at 38.7 min it switched to 300/500 nm to detect indeno(1,2,3-cd)pyrene.

To analyze the PAH composition (15 components, Table 1) the HPLC system was calibrated using an external standard method with a mixture of 16 PAHs in acetonitrile containing 10 ng μl^{-1} of each compound (Dr. Ehrenstorfer GmbH, PAH-Mix 9). Acenaphthylene, although contained in the standard, is only weakly fluorescent and thus not detected. The detection limit for the HPLC system ranged from 1 to 7 pg μl^{-1} depending on the compound (Table 1). For recovery tests the NIST standard reference material 1649a, containing 12 components of the 16 US EPA PAHs, was applied. Filter reagent blanks and spiked samples were analyzed for quality control. The average recovery of the total procedure (extraction, clean-up and chemical analysis) for the PAHs ranged from 90 % to 119 %.

The solvents used were liquid chromatography grade (LiChrosolv) from E Merck, Germany. The mixture of 16 standard PAH components was supplied by Dr. Ehrenstorfer GmbH, Germany (PAH-Mix 9, L20950900AL). (Rehwagen et al. 2005).

Table 1. Polycyclic aromatic compounds analysed

Non carcinogenic compounds	Detection limit [pg μl^{-1}]	Carcinogenic compounds (according to IARC 1983, 2002)	Detection limit [pg μl^{-1}]
Acenaphthene	0.88	Naphthalene	1.32
Fluorene	1.53	Benzo(a)anthracene	1.61
Fluoranthene	2.80	Benzo(a)pyrene	1.28
Pyrene	5.07	Dibenz(a.h)anthracene	0.68
Anthracene	2.01	Benzo(b)fluoranthene	2.57
Phenanthrene	1.79	Benzo(k)fluoranthene	1.64
Chrysene	2.56	Indeno(1.2.3-cd)pyrene	5.21
Benzo(g.h.i)perylene			6.80

3.1.2. Concentration of Particulate Matter and Polycyclic Aromatic Compounds

The concentration of PAHs in association with airborne particles was investigated during a period of four weeks in summer and four weeks in winter between 2000 and 2002. Table2 shows the arithmetic mean and the median as well as the 25th and 75th percentile of the concentration of PM, and the concentration of each analyzed compound.

The highest concentration of PM was found in the urban area. However, the sum of the PAHs in the industrial area (mean 7.1 ng m^{-3}) was obviously higher than in the other two regions (4.8 ng m^{-3} and 1.9 ng m^{-3}, Table 2).

The concentration in the industrial area was approximately twice as high as in the urban area of La Plata City. However, no significant differences were observed for any of the analyzed compounds, although concentrations several times higher compared to the control area in La Plata occurred in both locations (Table 2). This finding could be caused by an impact of the refinery and the petrochemical complex on the urban area.

Greater differences between the industrial and the control area of La Plata were found for benzo(a)anthracene (BaA) and chrysene.

The mean concentrations of benzo(a)pyrene (BaP) as an indicator of carcinogenic risk (WHO, 1987) in the city of La Plata (0.48 ng m^{-3}) were similar to results found in Las Palmas, Gran Canaria (0.34 ng m^{-3}) (Castellano et al., 2003) and twice as high as in Sao Paulo, Brazil (0.13–0.28 ng m^{-3}), Los Angeles, USA (0.21 ng m^{-3}) and Melbourne, Australia (0.17 ng m^{-3}) (Venkataraman et al., 1994; Panther et al., 1999; Vasconcellos et al., 2003).

The mean of BaP concentration in the industrial area of 0.91 ng m^{-3} was comparable to an industrial region in Hong Kong (Guo et al., 2003).

Table 2. Concentration (mean and confidence intervals CI 25-75%) of air particulate matter (PM_{10}) (in µg m^{-3}) and of each PAH bound on particles (in ng m^{-3}) for the investigated areas in La Plata

	Industry Mean CI 25-75%	Urban Mean CI 25-75%	Control Mean CI 25-75%
Particulate matter	19.30 18.47-48.24	32.27 23.64-40.55	11.03 10.63-11.32
Naphthalene	0.10 0.02-0.58	0.057 0.01-0.43	0.104 0.02-0.24
Acenaphthene	0.01 0.01-0.09	0.005 0.00-0.01	0.000 0.00-0.01
Fluorene	0.01 0.01-0.03	0.021 0.02-0.02	0.011 0.00-0.01
Phenanthrene	0.10 0.06-0.23	0.126 0.11-0.15	0.053 0.02-0.09
Anthracene	0.03 0.01-0.05	0.016 0.01-0.02	0.006 0.00-0.01
Fluoranthene	0.27 0.01-0.34	0.201 0.17-0.46	0.108 0.00-0.22
Pyrene	0.30 0.21-0.35	0.266 0.25-0.34	0.139 0.02-0.19
Benzo(a)anthracene	0.67 0.10-1.19	0.263 0.24-0.30	0.104 0.00-0.19
Chrysene	0.68 0.24-1.49	0.401 0.37-0.45	0.152 0.02-0.30
Benzo(b)fluoranthene	1.06 0.56-2.96	0.64 0.55-0.748	0.283 0.08-0.47
Benzo(k)fluoranthene	0.48 0.21-1.03	0.241 0.21-0.29	0.126 0.03-0.18
Benzo(a)pyrene	0.91 0.26-2.00	0.487 0.40-0.50	0.155 0.03-0.35
Dibenz(a,h)anthracene	0.06 0.03-0.23	0.034 0.03-0.05	0.015 0.01-0.04
Benzo(g,h,i)perylene	1.20 0.60-2.72	0.981 0.90-1.25	0.262 0.10-0.78
Indeno(123)cd-pyrene	1.18 0.46-2.19	0.612 0.61-0.70	0.324 0.08-0.47
PAH sum	7.12 2.80-15.41	4.781 4.18-5.01	1.874 0.42-3.51

3.1.3. Distribution of PAHs According to Particle Size

In this study the concentration of PAHs associated with particles of six different size fractions was analyzed. Figures 2A-C show the distribution of PAHs depending on particle size. More than 50% of all PAHs were associated with the smallest particle size fraction (< 0.49 μm) in all locations studied. However, in the industrial area of La Plata as well as in the control areas a similar proportion of particles with an aerodynamic diameter between 0.49–0.95 μm (filter 5) was observed. In these places approximately 25% of PAHs were associated with particles of fraction 5, but with the highest concentration in the industrial area. For the sampling location in the city, 15% of PAHs were bound to these particles. This distribution demonstrates different sources of PAHs. Whereas in the cities traffic is mainly responsible for the emission of PAHs, in the other areas other sources of combustion exert an additional influence, such as the plants in the industrial area and the combustion of wood for heating processes in the control areas. It can be supposed that a higher proportion of PAHs on particles of less than 0.5 μm originates more from traffic, whereas the PAHs bound to the next particle fraction were emitted from other combustion sources. These findings correspond to the results of the ratios of several PAHs used to distinguish emission sources.

How reported also in various other studies, the highest concentration of particle mass and bound chemicals was associated with the smallest size fraction (less than 0.49 μm) (Pierce and Katz, 1975; Sirce et al., 1987, Schnelle-Kreis et al., 2001). The distributions observed also in this study indicated PAHs on particles less than 1 μm result from combustion and other high temperature sources and little attributed to re-suspension of surface dust and soil particles (Sheu et al., 1997; Kiss et al.,1998). In the urban area PAHs were mainly emitted into the atmosphere from sources such as motor vehicle exhaust, whereas in the industrial area combustion processes related to the petrochemical industry were responsible for PAH emission.

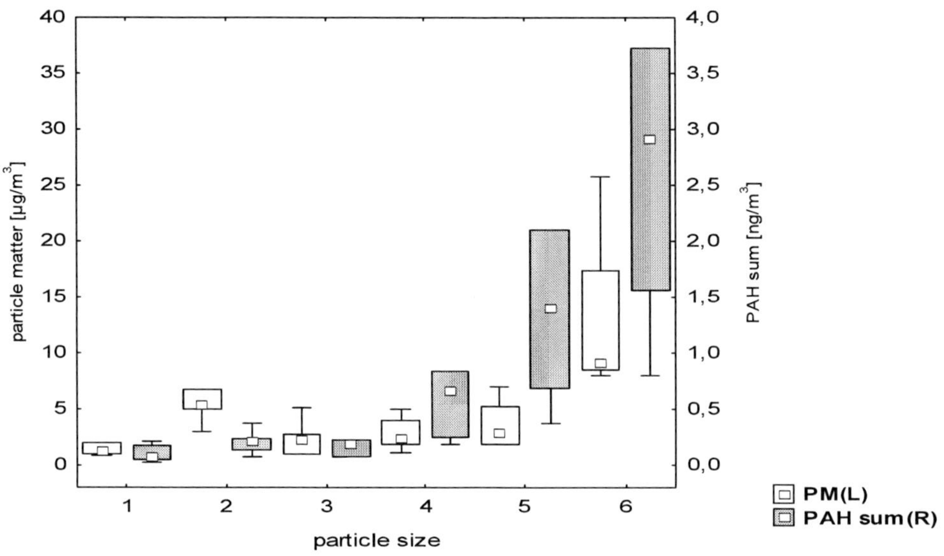

Figure 2A. Size distribution of PM [μg m-³] and sum of PAHs [ng m-³] in the industrial area.

Figure 2. (Continued).

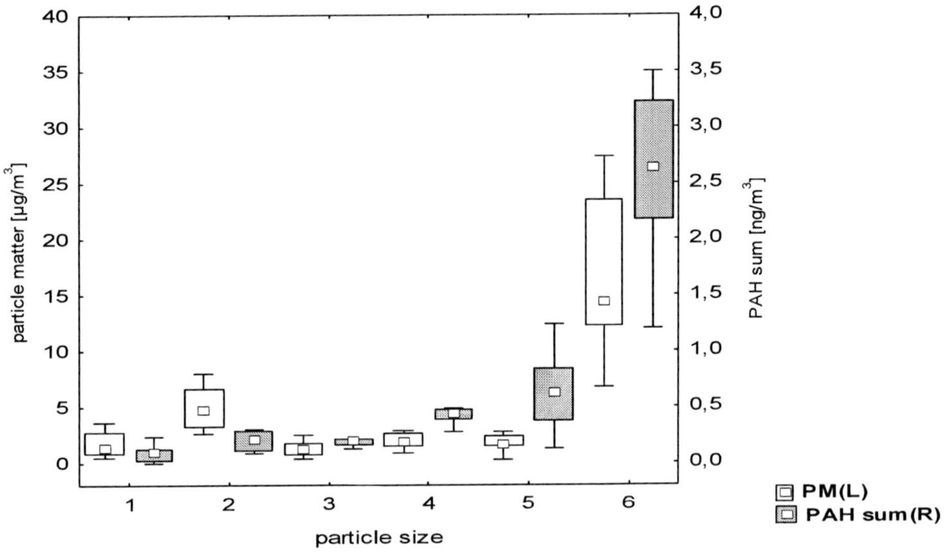

Figure 2B. Size distribution of PM [µg m-^{3}] and sum of PAHs [ng m-^{3}] in the urban area.

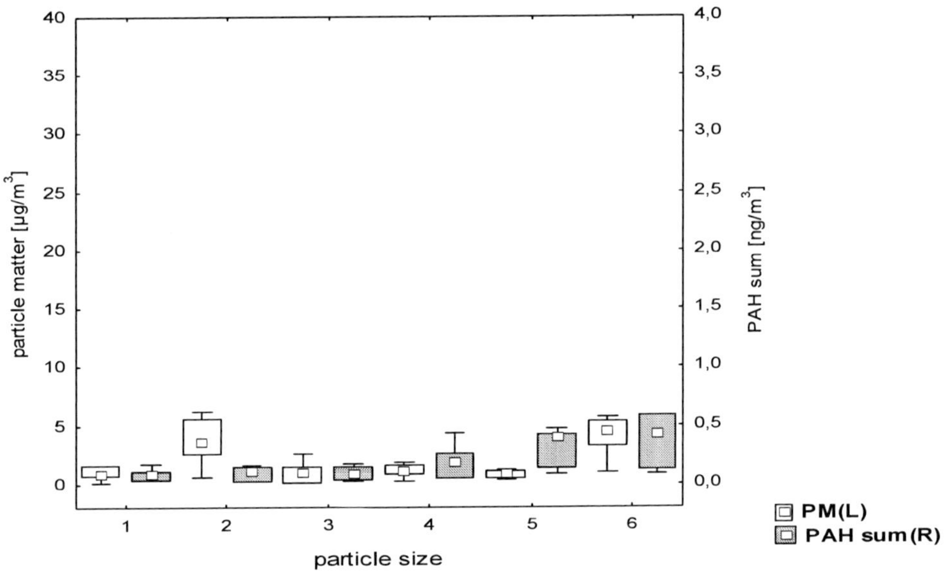

Figure 2C. Size distribution of PM [µg m-^{3}] and sum of PAHs [ng m-^{3}] in the control area.

Figures 2A-2C. Size distribution of PM [µg m-3] and sum of PAHs [ng m-3]. A- industrial area, B- urban area, C-control area. Particle size fractions 1: 10-7.2 µm, 2: 7.2-3 µm, 3: 3-1.5 µm, 4: 1.5-0.95 µm, 5: 0.95-0.49 µm 6: < 0.49 µm.

The similarity of PAH profiles from combustion processes did not allow to distinguish between industrial and urban sources. Otherwise, due to the vicinity of the city to the industrial places the emission of both places could interact one other. Therefore, no significant differences in concentration of PM, PAHs as well as in effects on the cells were observed (Rehwagen et al. 2005).

PAHs bound to air dust from urban, industrial and control areas were characterised regarding different sizes of particles. The highest percentage of PAHs was associated with the smallest particles in all locations studied.

These findings demonstrate a higher risk especially to fine respirable particles.

3.2. Volatile Organic Compounds (VOC)

3.2.1. Methods for Analyses of VOCs

Sampling was carried out in indoor and outdoor air in winter time between 2000 and 2002, with passive sampler (3M OVM 3500,1986) monitors in the selected places shown in figure 1. A sampling period of four weeks was selected, since an integrative sampling represented a mean load relative to personal exposure. Using this procedure, it can be assumed that the human exposure is more accurately assessed compared with a short time measurement. Indoor samples were collected in kindergartens and in the homes of selected children attending the kindergartens (n = 66), while outdoor samples were collected in representative sites of the three sectors (n = 22) parallel to the indoor measurement. The outdoor monitors were placed at rain-protected positions. Indoors, the passive monitors were placed in the middle of the room at a 1.5-2 m height with a minimum distance of 50 cm to the ceiling (Begerow et al., 1995; Rehwagen et al., 2003).

The study comprised the analysis of 29 VOCs. According to their structure and composition they belong to the following groups: *alkanes* (hexane, heptane, octane, nonane, decane, undecane, dodecane, tridecane); *cycloalkanes* (methylcyclopentane, cyclohexane, methylcyclohexane); *aromatic hydrocarbons* (benzene, toluene, ethylbenzene, m+p-xylene, styrene, o-xylene, 4-ethyltoluene, 3-ethyltoluene, 2-ethyltoluene, naphthalene); *chlorinated hydrocarbons* (chlorbenzene, trichlorethylene, tetrachlorethylene); *terpenes* (a-pinene, b-pinene, 2-carene, 3-carene, limonene). After exposure, the VOCs were desorbed from the adsorption layers (charcoal pads) by means of 1.5 ml of carbon disulfide (with low benzene, from Merck) containing 1% of methanol. The VOC analysis was performed on a PerkinElmer gas chromatograph with a masspectrometrical detector, equipped with an RTX-1 column (60 m 0.32 mm I.D., 1.0 μm film thickness; Restek). The oven temperature was held at 43 °C for 5 min, and then programmed to 200 °C at a rate of 2.5 °C min^{-1}. The injector temperature was held at 250 °C as well as the transfer line temperature. An electron energy of 70 eV was used for ionization. The source temperature was held at 200 °C. A sample volume of 1 μl was splitless injected. Integrated areas of selected fragment ions from each of the 29 VOCs were obtained with the software Turbomass, Version 4.4 (PerkinElmer) (Rehwagen et al., 2003).

Recovery coefficients were determined by direct injection of a known amount of the standard into a 3M sampler and subsequent extraction with carbon disulfide containing 1% methanol. The recovery was between 98% and 102%. For the studied VOCs, the background amounts contained in the monitors and in the carbon disulfide were determined for each charge of the samplers and chemicals and were subtracted from the sample results. The detection limits for the components were estimated as the three-fold standard deviation of five replicate measurements of monitor blanks. For components with blank values too low to be registered, it is usual practice to use the three-fold standard deviation of replicate measurements of a low-level standard solution (Begerow et al., 1995). The average concentration of each component over the sampling interval (in µg m^{-3}) was calculated according to the equation adopted from 3M (1986), using the adsorption coefficients in charcoal provided by Pannwitz (1991). The detection limits were found to be between 0.01 and 0.05 mg m^{-3} regarding a sampling interval of 4 weeks.

A non-parametric method (Mann–Whitney *U*-Test) was used for statistical analysis of results, according to the non-symmetrical distribution of the data (Sokal, 1981; Zar, 1998). Statistical analysis and plots of median values were prepared using Statistica 7.0. A *p*-value below 0.05 was regarded to be statistically significant.

3.2.2. Concentration and Distribution of VOC

Outdoor and Indoor Concentration Levels in the Investigated Areas

The concentration per compound groups are given in Table 3, showing the number of measurements, mean, median, and maximum concentration for outdoors and indoors in the different study areas. Significant differences ($p < 0.05$) could be observed for all compound groups of VOCs, both indoors and outdoors, between the industrial and the other areas, with the exception of terpenes.

Higher concentrations of outdoor VOCs were observed in the industrial area: between three and six times higher than in the other areas.

As various past studies have already shown, VOC concentrations are typically higher indoors than outdoors, due to accumulation if there is poor ventilation (Adgate et al., 2004, Schlink et al., 2004). Also in the study presented here, the median indoor concentration was higher than those outdoors. Similar values indoors were found for the urban and control area; the higher concentrations also indoor were found for the industrial area.

The main compounds outdoors as well indoors belong to the group of *alkanes* and *aromatics*. A detailed analysis of these components shows a different spatial distribution, specific to the investigated areas, which will be described in the following chapters.

The industrial area outdoors exhibited concentrations of *cycloalkanes* of 13.08 – 34.5 µg m^{-3}, which is between 5 and 15 times higher than in the other areas due to emissions of the petrochemical plant producing 85,000 ton/year of cyclohexane (IAPG, 1999). These caused significantly higher concentrations of cycloalcanes indoors (Cl 25-75% 6.3 – 47.3 µg m^{-3}) in the industrial area, too.

For *chlorinated hydrocarbons*, significant differences occurred only indoors in the industrial area. Chlorinated hydrocarbons can be contained in cleaning products and products for disinfection. The concentration of these compounds varied depending on differing human activities (Tang et al., 2005).

Terpenes, which can be emitted from furniture or other wood products and are contained in various cleaning and cosmetic products in households, were generally higher indoors (Gosselin et al., 1984; Hodgson et al., 2000). The concentrations ranged between 0.6 µg m^{-3} and 60 µg m^{-3}, which were lower than reported by other studies from European cities (Sareela et al., 2003, Schlink et al., 2004). No significant differences between the studied areas were seen.

Table 3. Concentration of VOC families and outdoor/indoor ratio for all studied areas in µg m^{-3}. *p*-value shows significant difference between indoors and outdoors; $*p < 0.05$, $**p < 0.01$. Median values of each compound was used to calculate the ratio

	Outdoor			Indoor		
	median	mean	max	median	mean	Max
Industrial	n = 4			n = 21		
alkanes	58.17	66.67	109.74	67.07	81.90	247.76
cycloalcanes	18.63	21.22	34.56	14.35	16.67	47.33
aromatics	47.96	53.35	71.53	52.32	85.49	380.19
Chlorinated	0.78	1.76	5.03	2.55	2.69	8.00
Terpenes	0.32	0.35	0.71	8.01	10.39	43.34
Sum VOC	125.57	143.35	216.80	152.24	197.14	610.73
Urban	n = 14			n = 22		
alkanes	11.59	14.96	47.17	28.62	54.22	274.40
cycloalcanes	3.17	3.02	4.49	4.17	7.96	56.68
aromatics	19.65	22.94	48.16	28.55	38.32	122.92
Chlorinated	0.86	1.31	6.23	0.43	1.46	10.98
Terpenes	0.31	0.48	1.28	4.48	6.89	34.01
Sum VOC	36.38	42.72	90.11	84.11	108.84	367.58
control	n = 4			n = 23		
alkanes	7.51	7.08	9.20	24.92	43.95	343.63
cycloalcanes	1.46	1.49	1.98	4.70	8.07	67.13
aromatics	10.79	11.16	14.77	21.28	44.53	258.40
Chlorinated	2.75	2.68	4.52	0.38	1.12	5.58
Terpenes	0.11	0.27	0.80	6.15	14.08	60.39
Sum VOC	23.86	22.67	27.63	62.15	108.75	659.91

Alkanes

Compounds of alkanes are the main group of VOCs outdoors and indoors. By analyzing the compounds of this group, different spatial distributions outdoors and indoors were found. Whereas the exposure of alkanes outdoors was dominated by C6 - C8 alkanes (hexane, heptane and octane), especially in the industrial area, higher concentrations of C9 - C13 compounds (nonane to tridecane) appeared indoors (Table 4).

The median outdoor concentrations of hexane, heptane and octane in sampling places closer to the industrial area (sum C6-C8 (Cl 25-75%) ranged between 27.7 and 87.4 µg m^{-3}) were between 6 and 17 times higher than in the other areas. These compounds are generated from the petrochemical industries located in the region. Because of the strong outdoor exposure, also the indoor concentration of the C6-C8 alkanes was significantly higher in the industrial area (sum C6-C8 (Cl 25-75%) ranged between 17.7 and 131.1µg m^{-3}).

N-Hexane is a neurotoxin, a narcotic, and an irritant of the eyes, skin, and mucous membranes (Hathaway et al. 1991). N-Hexane also causes productive and embryotoxic effects and is cytotoxic in mammalian and human test systems (NIOSH 1991). A concentration on n-hexane of 28.8 µg m^{-3} outdoor respectively 25.2 µg m^{-3} indoor in the industrial area were measured which demonstrate the risk in this area.

Table 4. Concentrations of alkanes in µg m^{-3} and indoor/ outdoor ratio for all studied areas. *p*-value shows significant difference between indoors and outdoors; *$p < 0.05$, **$p < 0.01$. Median values of each compound was used to calculate the ratio

	Outdoor			Indoor		
	median	mean	max	median	mean	Max
Industrial	n=4			n=21		
Hexane	28.82	32.95	56.50	25.21	30.62	94.42
Heptane	11.86	13.14	18.82	9.74	10.86	25.13
Octane	5.73	6.83	12.13	5.57	5.77	11.52
Nonane	4.62	4.00	7.43	6.05	9.40	33.43
Decane	4.03	4.35	6.47	9.15	5.87	34.84
Undecane	2.06	2.65	4.76	3.83	8.86	37.29
Dodecane	0.95	1.22	2.11	1.58	4.04	25.42
Tridecane	0.72	0.91	1.53	1.00	3.20	25.64
Urban	n=14			n=22		
Hexane	4.64	8.05	39.86	5.33	7.33	27.59
Heptane	1.71	1.65	2.23	1.96	3.37	15.97
Octane	0.64	0.69	2.32	1.11	2.09	17.66
Nonane	1.16	1.21	2.68	3.33	13.26	87.29
Decane	1.58	1.77	5.18	4.62	16.18	90.71
Undecane	0.84	0.88	1.91	3.26	8.83	76.94
Dodecane	0.38	0.38	0.60	1.19	2.35	20.10
Tridecane	0.34	0.32	0.58	0.57	0.81	2.74
Control	n=4			n=23		
Hexane	2.54	2.45	3.31	5.47	9.52	44.22
Heptane	0.89	0.89	1.19	1.89	2.60	8.16
Octane	0.34	0.35	0.45	1.03	1.73	6.57
Nonane	0.72	0.69	0.89	3.17	8.99	72.65
Decane	1.35	1.50	2.73	6.63	12.04	126.19
Undecane	0.74	0.75	1.14	3.59	8.56	70.71
Dodecane	0.22	0.22	0.27	1.22	3.35	18.25
Tridecane	0.21	0.22	0.30	0.85	2.08	11.85

The concentration of n hexane outdoor in the city and control area ranged between 2.5 and 4.6 µg m^{-3} and is mainly caused by traffic. The indoor concentration of hexane in the urban and control areas showed similar concentrations near 5 µg m^{-3}.

Indoor sources of hexane can be a wide range of household products like cleaners and polishes, paint, greases and lubricants (Sack et al., 1992). The mean concentration of C9 – C13 alkanes was generally higher indoors in all places, without significant differences between the investigated areas, which indicated mainly indoor sources.

These compounds are associated with human activities indoors like renovation, painting and cleaning (Lau et al., 1997; Watson et al., 2001).

Aromatic Compounds

Table 5 shows the concentrations of aromatic compounds in both outdoor and indoor air for all studied areas. Benzene, toluene, and xylenes were the predominant aromatic compounds in all cases. Concerning the outdoor concentrations of benzene, toluene and xylenes (BTX), the industrial area exhibits significantly higher concentrations (mean BTX 47.1 μg m^{-3}) than the other areas (Table 5).

Table 5. Concentrations of aromatic compounds in μg m^{-3} and indoor/outdoor ratio *p*-value shows significant difference between indoors and outdoors; *$p < 0.05$, **$p < 0.01$. Median values of each compound was used to calculate the ratio

	Outdoor			Indoor		
	median	mean	max	median	mean	Max
Industry	n=4			n=21		
Benzene	12.56	12.38	16.03	15.74	15.86	25.15
Toluene	18.55	17.31	20.26	20.41	32.70	219.44
Ethylbenzene	1.81	2.17	3.46	1.97	4.06	18.79
m+p-Xylene	11.47	14.64	24.71	9.52	16.12	69.63
Styrene	0.15	0.18	0.38	0.25	0.29	0.85
o-Xylene	2.46	2.77	4.10	2.82	4.74	17.13
4-Ethyltoluene	1.98	2.10	2.77	2.00	5.67	55.17
3-Ethyltoluene	0.81	0.88	1.27	0.80	2.44	26.09
2-Ethyltoluene	0.75	0.80	1.01	0.77	2.59	29.36
Naphthalene	0.13	0.13	0.16	0.46	1.02	9.56
Urban	n=14			n=22		
Benzene	2.90	3.00	5.35	2.96	3.59	12.74
Toluene	6.32	7.60	16.96	11.01	15.35	89.00
Ethylbenzene	1.02	1.25	2.29	1.27	1.36	2.85
m+p-Xylene	6.77	7.81	15.01	6.33	6.59	18.75
Styrene	0.13	0.17	0.42	0.19	0.24	0.74
o-Xylene	1.16	1.55	4.33	1.42	1.56	4.76
4-Ethyltoluene	1.36	3.12	3.06	1.18	2.80	28.24
3-Ethyltoluene	0.45	0.54	1.34	0.53	1.15	11.72
2-Ethyltoluene	0.45	0.50	1.00	0.68	1.31	11.62
Naphthalene	0.11	0.49	5.55	0.29	4.38	77.51
Control	n = 4			n=23		
Benzene	1.71	1.60	1.80	3.28	5.17	13.17
Toluene	3.85	4.21	6.26	9.24	18.45	87.83
Ethylbenzene	0.44	0.47	0.63	0.83	2.74	25.26
m+p-Xylene	2.55	2.89	4.25	3.69	10.22	97.16
Styrene	0.08	0.07	0.09	0.22	0.35	1.09
o-Xylene	0.44	0.50	0.73	0.86	2.77	25.63
4-Ethyltoluene	0.60	0.76	1.34	1.11	2.17	9.42
3-Ethyltoluene	0.24	0.32	0.63	0.43	0.88	3.68
2-Ethyltoluene	0.22	0.28	0.50	0.47	0.95	5.63
Naphthalene	0.06	0.06	0.08	0.28	0.83	7.14

This result is due to industrial emissions, considering that the regional petrochemical plants produce 250,000 ton/year of BTX (IAPG, 1999). The urban area shows outdoor levels of BTX (18.8 μg m^{-3}) two times higher than the control sectors, which is related with more intensive traffic in the city.

The analysis of the benzene: toluene ratio shows for the urban and control sectors values of 0.47 and 0.44 respectively. As reported in various studies, a value near 0.5 indicates vehicular exhaust as the dominant source for aromatic VOCs (Fernandes et al., 2002; Zou et al., 2003). The higher ratio obtained for the industrial area (0.67) points out the presence of additional sources of these compounds, such as the petrochemical industry.

Benzene has raised the focus of interest because of its known genotoxicity and carcinogenicity even at typical ambient concentrations (class I, IARC; 1995). The mean outdoor concentration in the industrial area (12.6 μg m^{-3}) was significantly higher than in the other areas, due to the industrial emission.

Outdoor sources of benzene in urban and control region are vehicular emissions (Watson et al., 2001; Na et al., 2004). The concentration in these areas ranged between 2.8 and 1.5 μm m^{-3}, and are higher than in some European cities like Helsinki (Edwards et al., 2001), Munich and Leipzig (Schlink et al., 2004), but are lower than other highly polluted cities like Athens (Chatzis et al. 2005), Caracas and Santigo de Chile (Gee and Sollars, 1998), and Seoul (Son et al. 2003). According to a European Union directive (December 2000) benzene concentrations in ambient air should not exceed 5 μg m^{-3} as a running annual average, with a long-term target of less than 1 μg m^{-3} (OJ, L 313, 2000).

The indoor concentrations of aromatic compounds were significantly higher in the industrial area resulting from the outdoor exposure, especially the concentration of the carcinogenic compound benzene, which was five times higher than in the other areas (Table 5). A median concentration of 15.7 μg m^{-3} indoors presents a significant health hazard for these inhabitants. According to the WHO (2000), an exposure to 1 μg m^{-3} produces a lifetime risk of 4 x 10^{-6} for leukemia. In the other areas, sources for indoor exposure of benzene can be influenced by traffic outside, but also by human activities like smoking. The urban and control areas have similar median indoor values of benzene.

3.3. Source Apportionment and Identification of VOC Fingerprints

Using the just presented VOC data (chapter 3.2), this chapter presents (i) the identification of typical spectra, (ii) the interpretation of spectra with sources, and (iii) an apportion of the observed pollution to these sources. A signature of each source is its specific composition of organic compounds, which is called the 'fingerprint' of the source (Xue et al, 2004). That means, the spectra are re-scaled, so that they represent only the relative composition and not the absolute concentration level. This procedure excludes the intensity of the pollution. What we present in this chapter is just the composition pattern of compounds that is a superposition of the fingerprints of the sources, irrespective of the proximity to the source.

An additional issue is the intrusion of the mixture of pollutants indoors. To quantify the amount of the intrusion we included indoor air samples into the study.

For the identification of the fingerprints we use a recently published new approach (Lee and Seung, 2001), which is much easier to understand and to calculate than earlier techniques. Additionally, modifying and extending the algorithm, we calculate not only the impact of the sources but also the contribution of other sources that are not identified yet.

3.3.1. Mathematical Approach

Classical statistical approaches to pattern recognition and data decomposition are principal component analysis (PCA) (Lucarelli et al, 2004), factor analysis (FA) (Pekey et al, 2004), and Empirical Orthogonal Functions (EOF), (Watson and Chow, 2002). All of them are based on an eigenvector decomposition of the correlation matrix. These eigenvectors are considered as latent variables (Fahrmeir et al, 1996) that are interpreted in terms of the individual sources.

Originally developed for the cognition sciences, these techniques, however, have considerable problems when used with air pollution data and have found only limited applications. Firstly, the identified eigenvectors are not unambiguous and can be transformed linearly, a fact that is utilised in the varimax rotation to improve the interpretability. Secondly, the calculated eigenvectors are orthogonal, which implies that the characteristic spectra of the sources are independent. For nearly all sources of air pollutants this is not valid, however. Therefore, the condition of orthogonal source-spectra will not be able to identify the true sources. The calculated eigenvectors and their eigenvalues are abstract because, expressed in terms of the original chemical concentrations, they typically contain negative elements.

In contrast, non-negativity constraints are typical requirements for models of mixtures of substances. The basic problem is to decompose a $(m \times n)$ matrix of (positive valued) observations X into a product of two matrices F ($(p \times n)$, the factors) and G ($(m \times p)$, the loadings) such that both have non-negative elements and p is a reasonable small value compared to m, n. Surely, there is no exact solution of this problem. A technique recently suggested by Lee and Seung (1997, 1999) is the non-negative matrix factorization (NMF) considering the above mentioned decomposition $X \approx GF$ .The calculation is simply an iterative multiplication using an update rule. The 'fingerprints', identified by the NMF algorithm, have a simple and meaningful interpretation. They are the basic elements, of which the observed spectra are weighted superpositions. As this is totally adequate to the constitution of urban air pollution we applied the NMF approach for the identification and apportionment of VOC sources in La Plata, Argentina.

3.3.2. Identification of Fingerprints

The application of the NMF algorithm requires an a priori specification of the number of sources. In several attempts to specify this parameter r we observed that three sources have a dominant impact, i.e. $r = 3$ was selected. The NMF produces two results: (a) the fingerprints of the sources, and (b) the contribution of each source to the composition of each sample.

Table 6. Source apportionment (percentage of the contribution of each source outdoors/indoors); the sites are (I) kindergarten in the industrial area, (U) down town kindergarten, (C) kindergarten in the control area, and (1-8) flats in the city; portions from 30% are printed in bold

#	site	period	Source industry		Source traffic		Source renovation		Sources unexplained	
			out	in	out	in	out	in	out	in
1	I	Mar '00	65	--	28	--	5	--	2	--
2	I	Aug '99	95	85	0	7	4	7	1	1
3	I	Mar '00	80	--	8	--	8	--	3	--
4	I	Aug '00	70	62	27	17	1	19	2	2
5	I	Mar '01	83	76	8	13	6	8	3	2
6	I	Aug '01	90	73	6	13	2	13	1	1
7	I	Aug '02	95	81	0	0	3	18	1	1
8	U	Aug '99	77	60	12	23	1	5	10	11
9	U	Mar '00	29	30	57	43	9	15	5	13
10	U	Mar '00	38	30	55	43	2	15	5	13
11	U	Aug '00	52	17	32	51	8	17	8	16
12	U	Aug '00	26	17	64	51	3	17	7	16
13	U	Mar '01	20	19	71	47	3	22	5	12
14	U	Mar '01	35	19	33	47	26	22	6	12
15	U	Aug '01	39	17	42	38	5	31	14	15
16	U	Aug '01	32	17	60	38	2	31	6	15
17	U	Aug '02	46	38	34	32	16	18	5	12
18	C	Aug '99	72	--	8	--	0	--	20	--
19	C	Mar '00	17	0	58	0	10	66	15	34
20	C	Aug '00	33	20	57	15	8	59	1	6
21	C	Mar '01	11	12	47	0	15	65	27	22
22	C	Aug '01	36	53	36	5	23	38	5	4
23	C	Aug '02	38	31	46	0	5	64	11	5
24	1	Aug '02	24	6	61	42	5	14	10	38
25	2	Aug '02	60	15	26	4	10	74	4	7
26	3	Aug '02	49	36	32	23	7	26	12	15
27	4	Aug '02	30	32	55	0	2	60	13	9
28	5	Aug '02	44	3	26	0	25	87	5	10
29	6	Aug '02	50	3	24	48	21	4	5	46
30	7	Aug '02	53	16	27	0	13	77	7	7
31	8	Aug '02	37	8	12	0	35	60	16	32

We identified the fingerprints of (1) industrial pollution, and (2) traffic pollution (see Figure 3a-c). Peculiarities of fingerprint 1 are the large fraction of hexane (ca. 28%) and the low difference between benzene and toluene, which make it attributable to the industrial emissions. We identified fingerprint 2 as traffic related, because of the large fractions of toluene and m- and p-xylene, and a typical benzene-toluene ratio of 1:5.6. Benzene, toluene, ethylbenzene and xylenes (BTEX) mainly originate from the same anthropogenic sources, with the greatest fraction deriving from vehicle emissions, and show typical source fingerprints (Rappenglück et al., 1998). The observed benzene-toluene ratio depends on the type of fuel used. In general, it amounts to 1:3 and can increase to up to 1:6 for aromatic-rich unleaded fuels (Guo et al., 2004).

In the outdoor air the dominant fingerprint largely corresponds to the location of the kindergartens (Table 6). In particular, the pattern of industrial pollution is 65-95% and the traffic pattern is always below 30% at the kindergarten in the industrial zone (I) (#1-7). The down town kindergarten (U) is dominated by the traffic fingerprint (#8-17).

The search for characteristic spectra of VOC arose in a third fingerprint (see Figure 3). As it can be seen (Table 6) this third fingerprint is mostly dominant for samples of indoor air. It contains high fractions of alcanes that are emitted with renovation activities. For that reason, to the third fingerprint we assign renovation and other indoor activities as a source.

Finally, as explained in the section 'Mathematical Approach', we combined the residual fractions into a spectrum of unexplained sources. For most samples, the fraction of unexplained sources is quite low, a fact that confirms the significance of the three identified fingerprints (especially for outdoor air). The source apportionment (Table 6) demonstrates that, for most sites and most time periods, the observed pollution is predominated by the local sources.

In the industrial area (#1-7) we recognised a very dominant fingerprint of the industrial pollution in outdoor air, as well as indoors. Patterns of traffic pollution as well as renovation are of minor importance. For the city (U) we see the dominance of the traffic pollution. This is outdoors complemented by industrial impacts. Interestingly, at site (U) the fingerprint of the traffic can often be observed also in the indoor samples, but the fingerprint of industry is rarer indoors. This may be due to the fact that traffic pollution persists all time, while industrial pollution occurs only with special meteorological conditions.

The kindergarten in the control area (C) is outdoors characterised by traffic pollution and, occasionally, industrial pollution, depended from the actual meteorological situation. Indoors it is dominated by a fingerprint of renovation and cleaning activities. This indicates that here outdoor pollutants only insignificantly penetrate the indoor environment. In this area pollutants of indoor activities are much more important for the indoor air quality than impacts of outdoor pollutants.

The pollutants of the petrochemical industries cover different parts of the urban area, depending on the meteorological and local conditions. In August 1999 the industrial emissions were distributed to the control area due to strong NNW winds. A similar situation seems to have been occurred in August 2002: industrial emissions are recognisable in the apartments (#24-31) and the urban and the control area kindergartens. In the apartments additionally a strong influence of VOC pattern by renovation activities were seen.

As to the central aim of this section, we may conclude that the industrial pollution is dominant in the industrial area, not only outdoors but also indoors. Furthermore, depending on the meteorological conditions, the industrial pollution can adversely affect the whole region of La Plata, including the control area that is ca. 15 km off down town and 25 km distant to the industrial area.

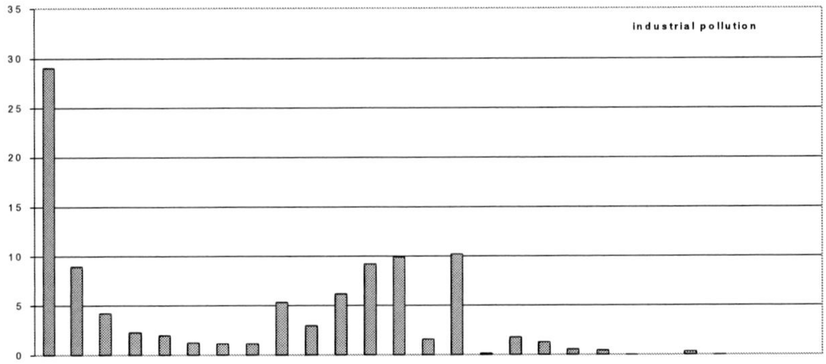

Figure 3A. Fingerprint for industrial pollution.

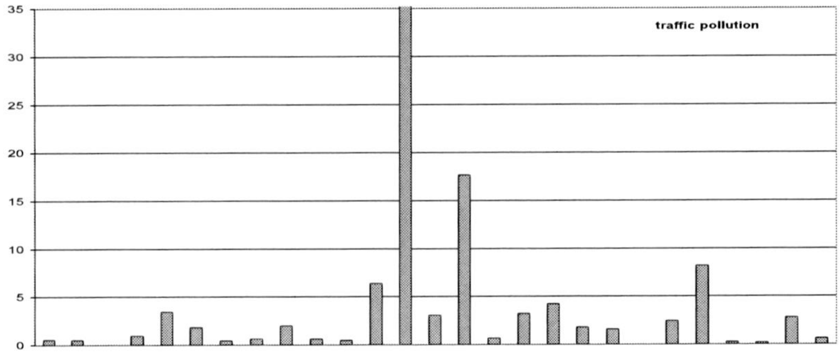

Figure 3B. Fingerprint for traffic pollution.

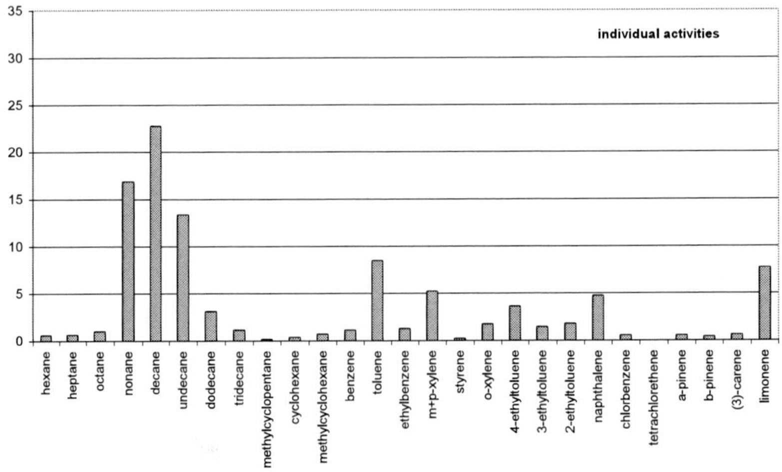

Figure 3C. Fingerprint for individual activities.

Figure 3A-C. The three sources and their fingerprints (composition in %).

Our results demonstrate the usefulness of the NMF algorithm in the identification of fingerprints and for source apportionment. Our extension of the algorithm enabled us to calculate, in addition to the source apportionment, the unexplained portion. The latter was found to be mostly lower than 30%, which means that the identified fingerprint really represent the most dominant sources. Fingerprints of industrial emissions are not only detectable outdoors but are also indoors noticeable. Under adverse conditions, i.e. winds from NNW, they may enter the rooms of the kindergartens. However, in contrast to industrial emissions, a more intense penetration was observed for traffic pollutions.

Outdoors we observed everywhere the fingerprints of two main sources: the industrial and the traffic emissions. The pattern of the latter dominates all residential areas. Indoors we identified a third fingerprint and assigned it to emissions with individual activities, such as renovation or cleaning. Sometimes this fingerprint was also detectable in the outdoor sample, indicating escaping indoor pollutants when windows opened and the outdoor sampler was mounted next to the window.

4. Risk Assessment of Airborne Particles and Bounded PAHs

The organic extracts of PM, sampled with a high volume collector using cascade impactors with a 10 μm inlet to separate 6 fractions by their aerodynamic size (see chapter 3.1.1) in all described study areas (figure 1), were tested for their mutagenicity using the Ames test with *Salmonella typhimurium* TA98 with metabolic activation (S9+). Additionally, cytotoxicity and changes in the energy-providing processes were investigated with cells of *Tetrahymena pyriformis.,* To verify immunmodulatory effects of chemical compounds bound to the sub-micrometer particles of different sampling locations three representative $PM_{0.5}$ samples from winter 2001 were selected for analysing the cytokine production in human blood leukocytes. Finally the life time cancer risk (LCR) was estimated for PAHs using the concentration of BaP and BaP_{eq} as well as for VOC by concentration of benzene.

4.1. Assessment of Mutagenicity

4.1.1. Methods for Analyis of Mutagenicity

The mutagenicity of air particulate extracts was assessed using of the Salmonella/Mammalian Microsomal Assay (Ames Test). Portions of the organic extracts without 1-octanol were reduced to dryness and re-dissolved in dimethylsulfoxide (DMSO) (chapter 3.1.1.). In order to obtain sufficient material for mutagenicity tests, sample extracts per site and season were pooled according to size fractions in three sub-samples as follows: coarse particles 10-3 μm, $PM_{2.5}$ 3-0.49 μm and $PM_{0.5}$<0.49 μm. Four dilutions and at least 2 replicates were tested for mutagenicity by the Salmonella/microsomal assay using the strain TA98 (kindly supplied by B.N. Ames, University of California, Berkeley, USA) with metabolic activation (+S9) and pre-incubation according to the standard plate incorporation protocol (Maron and Ames, 1983; USEPA, 1983). Benzo(a)pyrene (BaP) was used as positive control. Pure DMSO and extracts of filter blanks were also tested as negative controls. The results of mutagenicity were expressed as induced revertants mg^{-1} of particulate matter, calculated by linear regression analysis of the linear part of the dose-response curve and induced revertants m^{-3} on the basis of the content of particulate matter m^{-3} (Bernstein et al., 1982; McCubbin et al., 1991; APHA, 1998).

4.1.2. Mutagenicity of PM

Figures 4 A-C show the mutagenicity of particle extracts from the different investigated areas in La Plata. The high impact of particles less than 0.49 µm could be demonstrated very clearly (Figure 4C). Extracts of particle less than 0.49 µm from the urban and industrial area provoked the strongest mutagenic effect. Within the fraction of particles of 3-0.49 µm (figure 4B) the highest effects were caused by extracts from the industrial area. The mutagenicity of coarse particles (10-3 µm) can be estimated as very low in all areas (Figure 4A).

Several studies have reported results regarding the cytotoxic and mutagenic effects of airborne particulate matter. The most of them refer to investigations into extracts of TSP and PM_{10} (Spinosa De Martines et al., 1999; Černa et al., 2000; Reyes et al., 2000; Müller et al., 2001). However few data exists regarding the toxic and mutagenic effects of particles smaller than 2.5 µm, even though these particles pose a higher risk to human health (Hsiao et al., 2000). A higher contribution to mutagenic potency by organic compounds bound to fine particles < 1.5 µm has previously been reported in studies carried out in Greece (Viras et al., 1991, Tzounou et al., 1992) and Italy (Pagano et al., 1996; Monarca et al., 1997) tallying with the results reported here. Additionally, our results show a significant correlation of mutagenic effects with the levels of carcinogenic PAHs and BaP in respirable particle fractions smaller than 1.5 µm and 0.49 µm. This correlation was not observed for coarser particles (Table 7).

Table 7. Correlation factors r^2 for total PAHs, carcinogenic PAHs and Benzo(a)pyrene, and mutagenicity for each particle size fraction

Fraction No.	1	2	3	4	5	6
Size fraction [µm]	10-7.2	7.2-3	3-1.5	1.5-0.95	0.95-0.49	< 0.49
			Total PAH			
PAH	0,82	0,95	0,97	0,96	0,94	0,98
Carcinogenic PAH	0,75	0,92	0,95	0,98	0,97	0,95
BaP	n.s.	0,93	0,96	0,98	0,97	0,95
Mutagenicity			n.s.		0,88	0,88
			Carcinogenic PAH			
PAH	0,75	0,92	0,95	0,98	0,97	0,95
Carcinogenic PAH	0,82	0,97	0,98	0,97	0,96	0,98
BaP	n.s.	0,96	0,98	0,97	0,96	0,97
Mutagenicity			n.s.		0,92	0,85
			BaP			
PAH	0,75	0,93	0,96	0,98	0,97	0,95
Carcinogenic PAH	0,82	0,98	0,99	0,97	0,96	0,97
BaP	n.s.	0,97	0,98	0,97	0,96	0,97
Mutagenicity			n.s.		0,92	0,85

correlation factors r^2 were calculated by Statistica 6.0..
correlation were estimated as significantly for $p<0.05$.
n.s. – not significant.

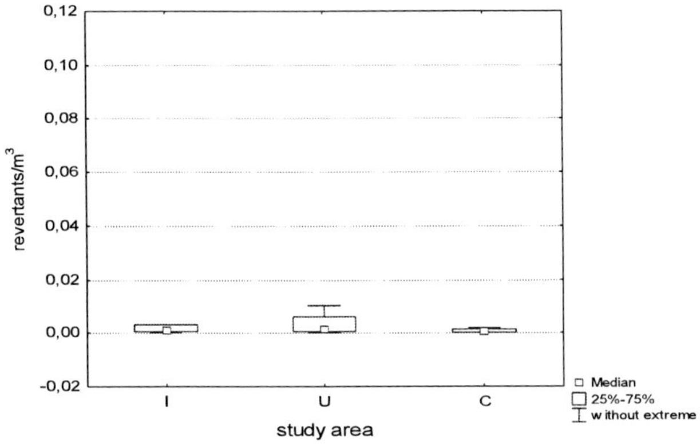

Figure 4A. Mutagenicity of coarse particles.

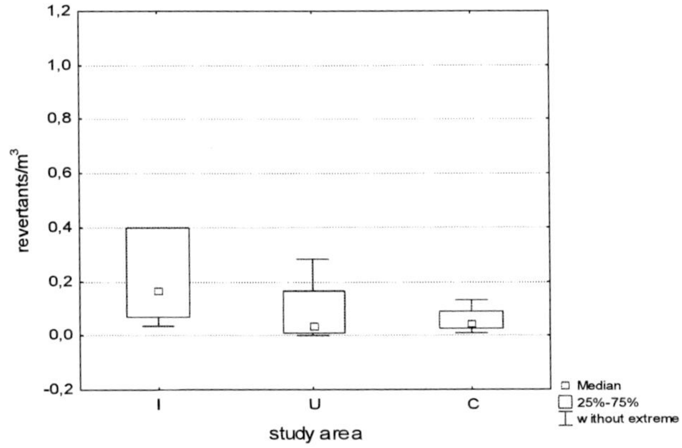

Figure 4B. Mutagenicity of PM $_{2.5.}$

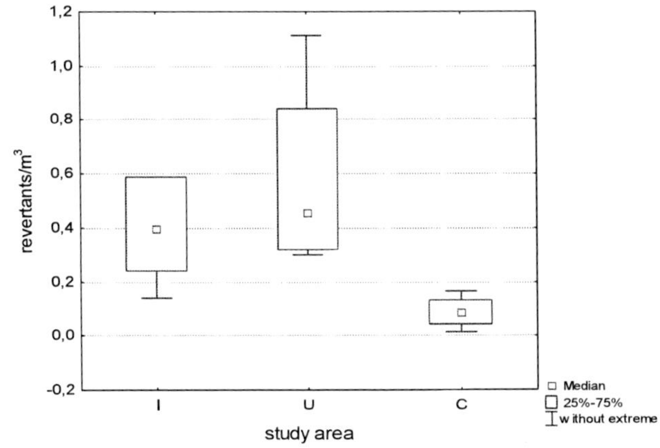

Figure 4C. Mutagenicity of PM $_{0.5.}$

Figures 4 A-C. Mutagenicity expresed as revertants/m3 from extracts of particle matter from different particle size in each study area industrial area (I), urban area (U), control area (C). A- coarse particles 10-3 μm, B- PM 2.5 3-0.49 μm C-PM 0.5 < 0.49 μm.

All the presented results of the bioassays demonstrate the high risk of fine particles. No significant different effects were observed between the industrial and urban area, even though different sources were responsible for the emission of particle matter (Massolo et al. 2002).

4.2. Assessment of Cytotoxicity

Cytotoxicity and changes in the energy-providing processes were investigated with cells of *Tetrahymena pyriformis,* a single-cell eukaryotic organism which is comparable to human cells in its sensitiveness. Therefore it is well established as suitable model for human respiratory epithelium-cell functionalities. (Sauvant et al. 1999, Müller et al.2001)

4.2.1. Methods of Cell Cultivation, Determination of Oxygen Consumption and ATP Concentration

Cell Cultivation

Tetrahymena pyriformis GL cells were cultured in 50-ml flasks using 5 ml of a proteose-peptone-trypton-salts medium (pH 7.4) as described by Gräf (1998) containing 0.5% proteose-peptone (Difco, Detroit, MI, USA), 0.5% tryptone (Difco) and 0.02% K_2HPO_4 (Merck, Darmstadt Germany) in darkness at 28°C without shaking. A 24-hour culture with a cell concentration of 0.2 x 10^5 *T. pyriformis* cells/ml was used as inoculum (Müller et al. 2006).

For toxicity testing, the cells of *T. pyriformis* were exposed to DMSO extracts of the particulate matter for 24 hours. All treated cultures had a concentration of 0.01% v/v of DMSO. The cell concentrations were measured after 24 hours, using a Multisizer™ 3 Coulter Counter (Beckman Coulter, Fullerton, USA).The results were expressed as the percentage of inhibition per cubic meter of air volume with respect to the DMSO control, calculated by linear regression analysis of the linear part of the dose–response curve (McCubbin et al., 1991, Bernstein et al., 1982).

Oxygen Consumption

The oxygen consumption was measured in a plexiglass chamber with an integrated Clark Sensor (Biolytik, Bochum, Germany) containing a 25 µm thick teflon-foil as membrane. Potassium chloride in water (3% w/v) was used as electrolyte. The calibration was performed with the culture medium in oxygen saturated as well as in reduced form. To determine the respiratory activity, the oxygen consumption of 2.00 ml of the cell suspension was measured at 28°C in a closed chamber. The homogeneity of the cell suspension was guarantied by use of a magnetic stirrer to avoid impoverished oxygen layers. Signals from the polarographic oxygen sensor were calculated by the program Oxylab version 1.92 (Biolytik). The oxygen consumption per cell was expressed as µmol O_2/min. The results were calculated as the percentage of the respiratory activity to the DMSO control.

ATP Concentration

ATP was extracted from the cultures by use of an ice-cold solution of perchloric acid (1.3 mM) containing EDTA (23 mM) to precipitate the proteins. After centrifugation (16,000 x g,

15 min, 4°C), a solution of potassium hydroxide (0.72 M) and potassium hydrogen carbonate (0.16 M) was added to neutralize the supernatants to a pH of 7.7. After a second centrifugation applying the same conditions, the supernatants were kept at -20°C.

The detection of ATP has been carried out using the Quantitative Monitoring ATP-Kit for free ATP (ThermoLabsystems, Helsinki, Finland) and was quantified by measurement in a luminometer (Wallac Victor 1420 Multilabelcounter, Perkin Elmer) as described by Loffhagen et al., 2003. In order to exclude errors by accidental inhibition of bioluminescence by components of the medium or toxin metabolites, the ATP concentration was measured using the constant-addition technique (Lundin and Thore 1975). The medium for ciliates was used as negative control medium. Each measurement was carried out in triplicates. The results were expressed as the percentage of the ATP concentration normalized to DMSO controls.

Statistical analyses were performed using STATISTICA 7.0. As the data did not follow a normal distribution function, the non-parametric Mann–Whitney rank sum test (*U*-test) was used to identify significant differences between groups of data. A *p*-value below 0.05 was regarded to be statistically significant.

4.2.2. Cytotoxic Effects of PM

Cells of *T. pyriformis* were strongly affected by the organic extracts of the particulate matter. As an example in figure 2 the effect of a 24-hour exposure of the extract of fraction 6 from the industrial area on the cell concentration, the oxygen consumption and the ATP concentration is demonstrated. In controls without DMSO the cell concentration achieved a level of 2.1×10^5 cells/ml. The oxygen consumption per cell was approximately 59 fmol/min, and the ATP concentration 22 fmol/cell (Figure 5). (Müller et al. 2006).

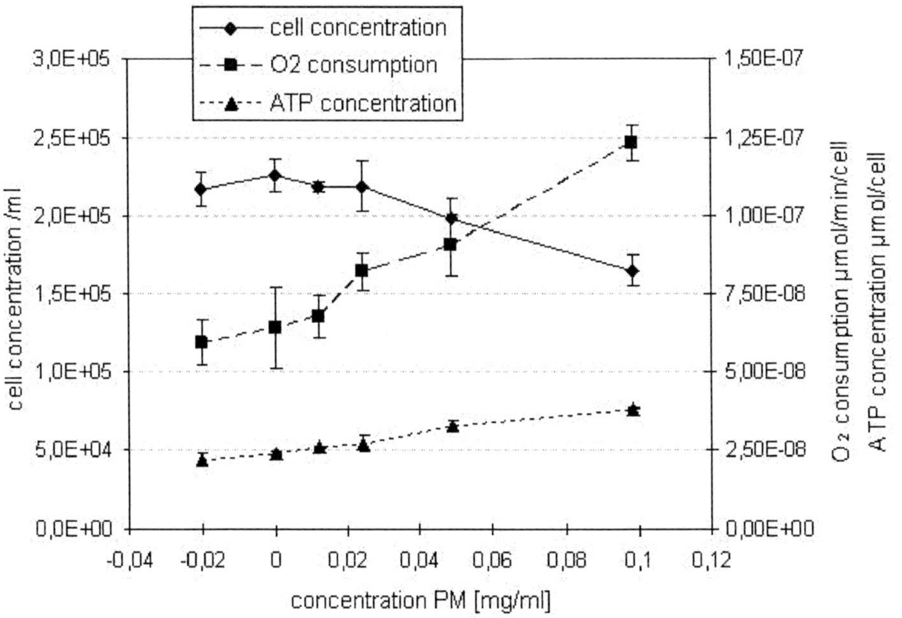

Figure 5. Cell concentration per ml, O2 consumption [µmol/ml/cell], and ATP concentration [µmol cell-1] of T.pyriformis after 24 hours of exposure with extract of fraction 6 from industry.

No significant differences between control with and without DMSO were observed. With increased concentration of particle concentration in the culture the cell concentration was inhibited significantly. In contrast, the oxygen consumption as well as the ATP concentration was significantly increased. For further comparison and correlation of the results according to the investigated areas and particle fractions, the results were calculated by linear regression analysis of the linear part of the dose–response curve (Bernstein et al., 1982, McCubbin et al., 1991) and expressed as the percentage of inhibition per cubic meter of air volume with respect to the DMSO control.

The highest effects on the cells were caused by the organic extracts from samples of the industrial and the urban area. Compared to the control area, a significant higher inhibition of cell proliferation and increase of oxygen consumption as well as the ATP concentration were observed in industrial and urban areas. In accordance with the particle mass concentration and the concentration of PAHs (see chapter 3.1.), no significant differences were observed between these two places. (Figure 6).

An increase of the observed effects was always associated with a decrease of particle size. Extracts from particles less than 1 µm (fractions 5 and 6) caused the strongest effects on the cell proliferation, as well as on their oxygen consumption and ATP concentration (Figure 7). All investigated extract samples caused an increase of the oxygen consumption and ATP concentration of *T. pyriformis* cells, even though a higher level of increased oxygen consumption was observed. This imbalance in the energy metabolism raised by the extracts of PM could indicate a partial decoupling in the respiratory chain. To visualize this phenomenon, a ratio between the inhibition of oxygen consumption and ATP concentration was calculated (table 8).

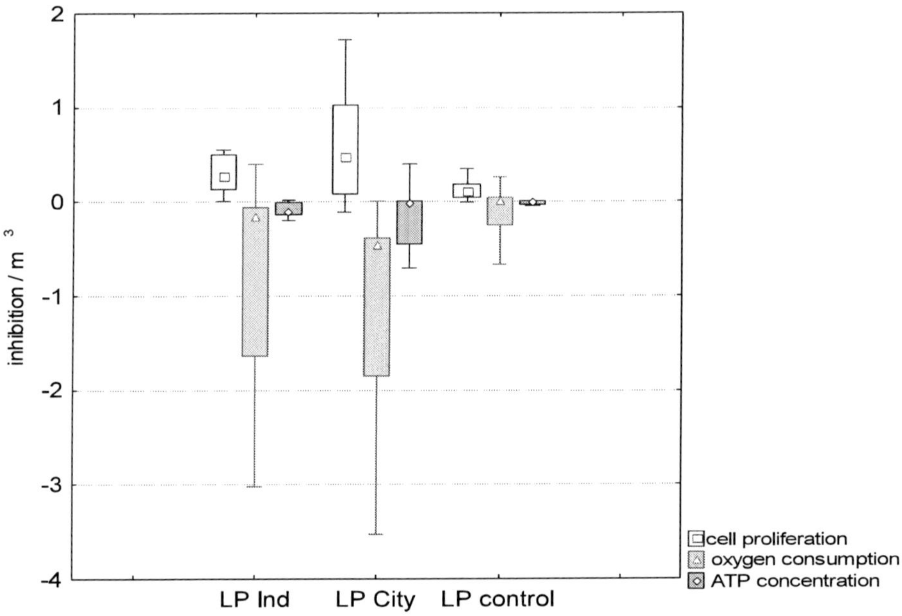

Figure 6. Inhibition of cell proliferation, oxygen consumption and ATP concentration in the investigated places of La Plata: LP Ind – industrial area, LP City – urban area, Control – control area.

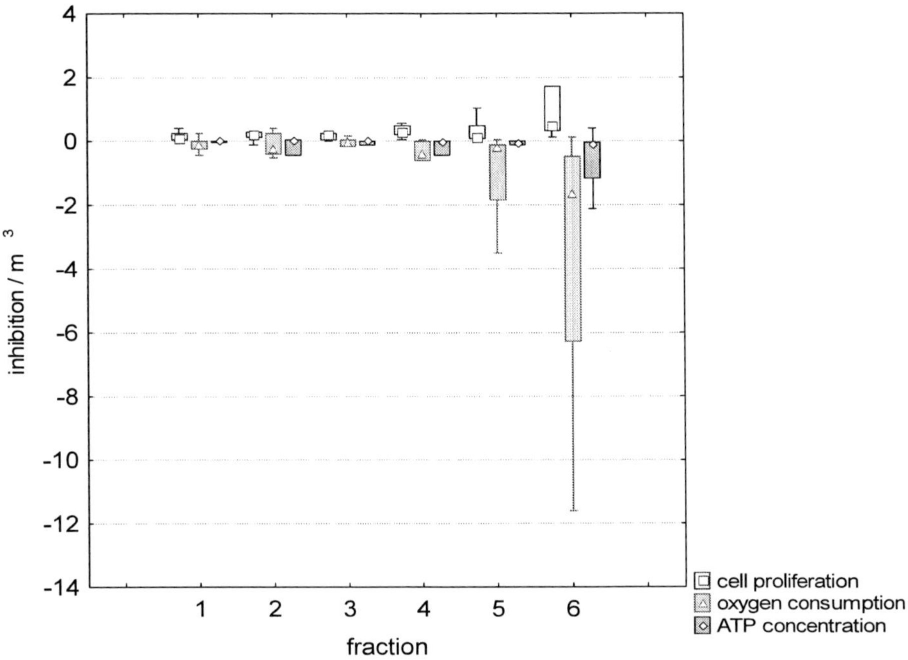

Figure 7. Inhibition of cell proliferation, oxygen consumption, and ATP concentration of T. pyriformis in dependence of the size of particles. Fraction 1: 10–7.2 μm; fraction 2: μm; fraction 3: 3–1.5 μm; fraction 4, 1.5–0.95 μm; fraction 5: 0.95–0.49 μm; and fraction 6, < 0.49 μm. Results are expressed as percentage inhibition per equivalent of cubic meter of ambient air.

These ratios were different for the investigated sampling locations and also for the different particle size fractions. An imbalance between oxygen consumption and ATP production expressed by a ratio more than one, was caused mainly by compounds adsorbed on particles less than 1 μm from the industrial and urban area (fractions 5 and 6). The observed effects on the cells seem to be associated with the concentration of PAHs. Nearly for all analyzed chemical compounds and the cell effects significant correlations showed up.

Table 8. Ratio between inhibition of oxygen consumption and ATP production for fractions of particulate matter collected at different sampling locations in La Plata. Inhibition was expressed as the percentage normalized to DMSO controls. As control results of culture without DMSO are shown

	Industry		Urban		Control	
	ratio	SD	ratio	SD	ratio	SD
control	0.98	0.005	0.99	0.004	0.98	0.003
Fraction 1	1.25	0.005	1.20	0.014	1.32	0.005
Fraction 2	1.36	0.010	0.88	0.026	1.18	0.030
Fraction 3	0.85	0.021	1.29	0.009	1.20	0.020
Fraction 4	1.32	0.023	1.04	0.019	0.91	0.006
Fraction 5	1.53	0.021	1.98	0.056	1.00	0.112
Fraction 6	1.35	0.083	1.88	0.022	1.09	0.006

How reported also in various other studies, the highest concentration of particle mass and bound chemicals was associated with the smallest size fraction (less than 0.49 μm) (Pierce and Katz, 1975). The distributions observed in this study also indicate that PAHs on particles less than 1 μm result from combustion and other high temperature sources and little attributed to re-suspension of surface dust and soil particles (Sheu et al., 1997; Kiss et al., 1998). In the urban area PAHs were mainly emitted into the atmosphere from sources such as motor vehicle exhaust, whereas in the industrial area combustion processes related to the petrochemical industry were responsible for PAH emission. The similarity of PAH profiles from combustion processes did not allow to distinguish between industrial and urban sources. Otherwise, due to the vicinity of the city to the industrial places the emission of both places could interact with one another. Therefore no significant differences in concentration of PM, PAHs as well as in effects on the cells were observed for these two places (Rehwagen et al., 2005).

In concurrence with the higher concentration of PAHs also the strongest effects on cells were caused by extracts from fine particles. In recent years numerous studies reported similar results about toxicity and mutagenicity obtained with different test systems (Massolo et al., 2002, Kawanaka et al., 2004, Hetland et al., 2004). It appears to be a strong correlation between the PAH concentration and biological effects. However, the coefficient of determination (r^2) demonstrates only a probability less than 50% for the analyzed PAHs to be the causative agents behind cytotoxicity and approximately 30% behind the changing in energy providing processes. PM contains a mixture of compounds including inorganic and organic substances. Due to the extraction methods used in this study, effects of inorganic compounds like salt and metals can be excluded. Otherwise, in PM occur additionally derivates from PAHs, e.g. nitrated polycyclic aromatic compounds (nitro-PAHs). PAHs and nitro-PAHs are typical indirect- and direct-acting mutagenic compounds in PM, respectively. Kawanaka et al. (2004) described a duplication of mutagenicity of PM with an aerodynamic diameter between 0.1 to 1.0 μm by nitro-PAHs, detected by the AMES test without a metabolic activation by S9.

In contrast to an inhibition of the cell proliferation, the respiratory activity and ATP concentration was significantly increased in general. An increase of oxygen consumption without net energy production in form of ATP indicates a decoupling of the respiratory chain. In case of a complete decoupling also the ATP concentration should be found to be reduced. In this study an increase of ATP concentration in the treated cells was found, but to a lesser extent. These results may indicate a partial decoupling. The ratio between oxygen consumption/ATP concentration suggests a partial decoupling processes, especially for extracts of particles < 1 μm from the industrial and urban areas. All observed effects were associated with the content of PAHs in PM.

There is evidence that PM and the organic fraction initiate oxidative stress on cells by presenting or by stimulating the cells to produce reactive oxygen species (ROS). Furthermore, the production of ROS has been maintained to play an important role in the primary cytotoxic effects of particle matter (Knaapen et al., 2004; Risom et al., 2005). Additionally, the organic fraction of air particles contains quinine radicals that may pass though redox-cycling producing hydroxyl radicals (Dellinger et al., 2001).

"Oxidative stress" has broadly been defined as an unbalance in favor of pro-oxidants and disfavor of antioxidants (Sies, 1993). It is imposed as a result of one of three factores: (I) an increase in oxidant generation, (II) a decrease in antioxidant protection and (III) a failure to repair oxidative damage. Oxidative stress is generated by a large variety of mechanisms, including mitochondrial respiration, inflammation, and metabolism of foreign compounds (Risom et al. 2005).

It is well documented that products of combustion like PAHs and nitro-PAHs can lead to production of O_2 radicals in cells. These O_2 radicals can directly reduce cytochrome c, resulting in enhancement of respiration (Butler et al., 1975, Mailer, 1990, Iuchi and Weiner, 1996, Zhao et al., 2003). Otherwise, an increase of oxygen radical concentrations leads to damages in the mitochondrial membrane which result in a significant decrease of intracellular ATP concentration (Ataullakhanov and Vitvitsky 2002, Brookes 2005). The observed shift of the ratio O_2 consumption/ATP production in the treated cells may therefore be a result of oxidative stress.

4.3. Assessment of Immunomodulation

The use of mitogen-stimulated human blood leukocytes was chosen for this purpose, since we were interested in the variability of potential immunomodulatory effects of chemical compounds bound to the sub-micrometer particles of different sampling locations on the cytokine production in ongoing human immune responses. Thus, we investigated in an *in vitro* test system the effects of $PM_{0.5}$-derived extracts on the vitality/proliferation and the cytokine-production of human peripheral blood mononuclear cells (PBMC) following suboptimal stimulation by phytohemagglutinin (PHA). Consequently, this study presents the effects of extracts from $PM_{0.5}$ from ambient air sampled in the rural control (C), the industrial (I), and the urban area (U) of La Plata (Figure 1) on the cytokine-production of PHA-stimulated human PBMC. The T cell cytokine interferon-γ (IFN-γ) was chosen as representative for the analyses. The obtained results for IFN-γ showed different effects of the chemical compounds extracted from $PM_{0.5}$ of the three regions of La Plata on the immune regulation which are presented here.

4.3.1. Methods for Immunological Investigations

Sample Handling

For immunological investigations, the DMSO extracts from the smaller particles $PM_{0.5}$ (fraction 6, see chapter 3.1.1.) were diluted in DMSO to yield eight dilutions with 50%, 25%, 12.5%, 6.25, 4.0%, 1%, and 0.1% of the respective extract in DMSO. These extract dilutions and the original extracts were further diluted in RPMI 1640 supplemented with 5% fetal calf serum (FCS; both from Biochrom, Berlin, Germany). As control, pure DMSO was diluted the same way to give a DMSO concentration of 1% v/v in the samples. These extract dilutions were thereafter applied in a volume of 50 µl into the cavities containing 100 µl of the diluted mitogen PHA (phytohemagglutinin) before the cells were seeded into the cavities (see below).

Cell Culture

PBMC were separated from buffy coats of six healthy blood donors by density gradient centrifugation with Ficoll-Paque™ (Amersham Pharmacia Biotech, Freiburg, Germany).

Thereafter, the PBMC were seeded into the 60 inner cavities of 96-well microtiter plates with flat bottom (Greiner, Nürtingen, Germany) with a final cell density of 1 x 10^6 PBMC/ml RPMI 1640 (Biochrom, Berlin, Germany) supplemented with 5% fetal calf serum (Biochrom, Berlin, Germany) in a total volume of 250 µl per well.

Dilutions of the extracts as well as DMSO and medium alone as controls were incubated in triplicate with PBMC which concomitant were stimulated by PHA (10 µg/ml final concentration). The final concentration of DMSO in the cultures receiving extract dilutions or pure DMSO was 0.2% v/v. After the incubation of 20 hours in humidified atmosphere at 5% CO_2 and 37°C, the supernatants were harvested and stored below – 20 °C until the analyses of produced cytokines in ELISA as described below.

MTT Assay

Immediately after removing 175 µl of culture supernatant from each well, 10 µl of a stock solution (5 mg/ml PBS) of MTT (3-[4,5-dimethyl-thiazol-2-yl]-2,5-diphenyl-tetrazoliumbromide, SERVA, Heidelberg, Germany) were added to each culture. After a 4-hour incubation period at 37°C, 100 µl stop solution (10 % w/v sodium dodecylsulfate in 50 % v/v N,N'-dimethylformamide, SERVA) were added to each well. After further incubation at 37°C over night, the optical density (OD) of the wells was read on an ELISA-Reader (Spectra Image, Tecan, Crailsheim, Germany) using a 570-nm filter.

The results for MTT metabolization of PBMC under influence of $PM_{0.5}$-extracts were calculated as the OD value in each of the determinations in triplicate normalized to the mean of control of the individual donor. These results were expressed as the percentage of vitality/proliferation compared to solvent control. Following this normalization of the individual values for MTT metabolization to the DMSO-containing control of the same donor, the resulting percent values were used for further analyses.

ELISA Assays

IFN-γ was measured in culture supernatants by indirect sandwich ELISA assays (OptEIA™ Kits, BD Biosciences, Heidelberg, Germany) using the TMB reagent set™ (BD Biosciences) following the manufacturer's instructions. After termination of the substrate reaction by addition of 1 M H_2SO_4, the optical density (OD) of the wells was read on an ELISA-Reader (Spectra Image, Tecan, Crailsheim, Germany) using a 450-nm filter with a 620-nm filter as reference. The lower limit of detection for the IFN-γ-ELISA was ≤ 4 pg/ml. The results for cytokine production of PBMC under influence of $PM_{0.5}$-extracts were calculated as the cytokine production in each of the determinations in the triplicate normalized to the mean of control of each individual donor. The results for the respective concentration of $PM_{0.5}$-extract were expressed as cytokine production compared to solvent control.

4.3.2. Immunomodulation by PM and Bounded PAHs

The DMSO extracts of $PM_{0.5}$ exerted differential effects on human PBMC regarding their proliferation as well as their cytokine production. At first, the three samples differentially influenced the MTT metabolizing capacity of PHA-stimulated PBMC. Figure 1 shows the effects of the three $PM_{0.5}$-derived extracts on PBMC proliferation on the basis of m^3 per ml cell culture (identical to $m^3/10^6$ PBMC). The $PM_{0.5}$-extract from the control area (C) caused only slight and insignificant effects (figure 8).

Contrary to these minor effects of chemicals adsorbed on $PM_{0.5}$ of the control area, much stronger effects were caused by the extracts derived from the samples of the more polluted areas, industry and urban, indicating a co-mitogenic activity of the $PM_{0.5}$-adsorbed chemical compounds (figure 8). Regarding the dose-dependent co-stimulatory effects of the extracts on proliferation the response of the PBMC of the six analyzed donors showed only minor variability.

Contrary to the stimulated vitality/proliferation, the production of interferon- γ cytokine was significantly suppressed by all of the highest extract concentrations. But in this, the samples differently affected production of interferon-γ (IFN-γ). The strongest suppression of IFN-γ production was caused by the $PM_{0.5}$-extract from the industrial area (figure 9). The suppressive effect of the extract derived from the urban area also caused strong IFN-γ suppression, whereas extracts of $PM_{0.5}$ derived from the control area was least suppressive.

Our data clearly indicate the dissimilar effects of the three samples on the cytokine production of human PBMC and furthermore showed that the $PM_{0.5}$-adsorbed chemicals pose a differential risk for the correct function of immune regulation.

This study, to the best of our knowledge, is the first study that shows differential impact of chemical compounds adsorbed to sub-micrometer particles ($PM_{0.5}$) on human peripheral blood cells and their cytokine production.

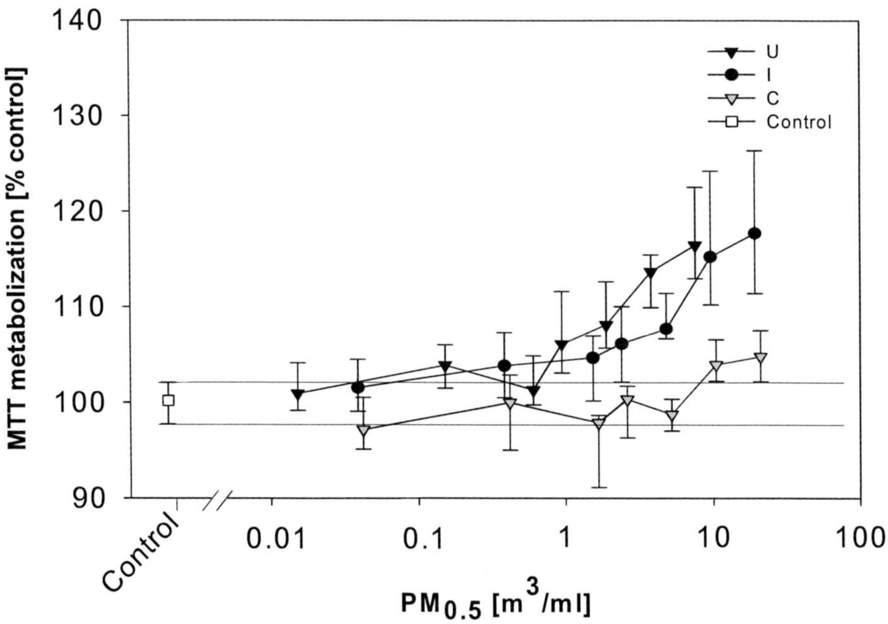

Figure 8. Dose-dependent effects of PM0.5-extracts on the MTT-metabolizing capacity of human PBMC as measure for their vitality/proliferation. The doses [m3/ml] are given as equivalent of m3 of original airborne particulate matter with an aerodynamic diameter <0.49 µm (PM0.5) in extract per ml cell culture or 1 x 106 PBMC, respectively. (I): industrial area; (U)traffic influenced urban area; (C): rural control area. Shown are the median and the interquartile range of the MTT metabolization normalized to solvent control.

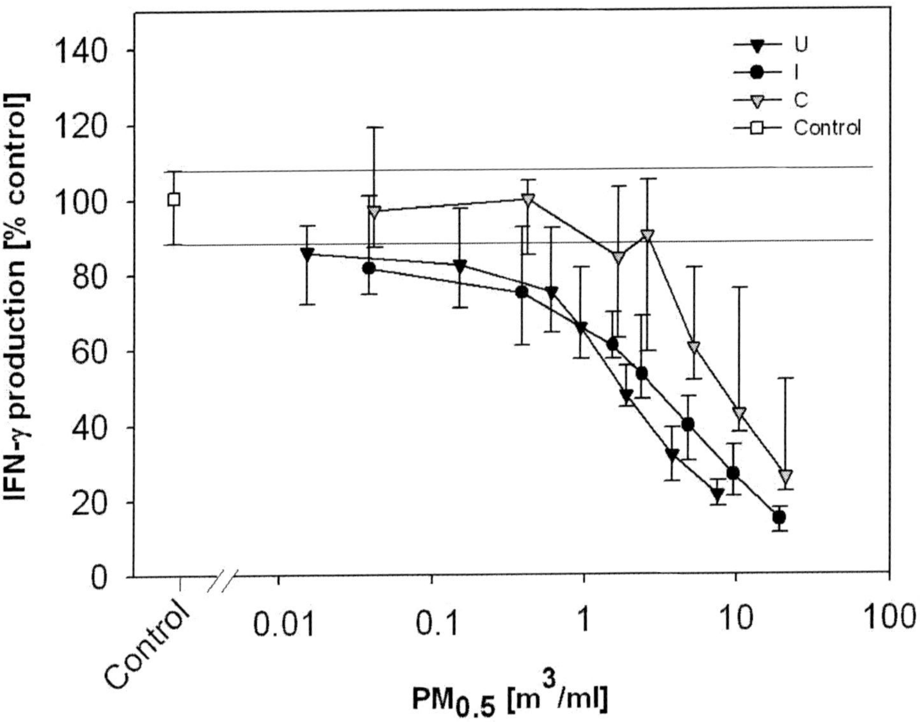

Figure 9. Dose-dependent effects of PM0.5-extracts on the IFN-γ production of human PBMC. The doses [m3/ml] are given as equivalent of m3 of original airborne particulate matter with an aerodynamic diameter <0.49 µm (PM0.5) in extract per ml cell culture or 1 x 106 PBMC, respectively. (I): industrial area; (U)traffic influenced urban area; (C): rural control area. Shown are the median and the interquartile range of the MTT metabolization normalized to solvent control.

As shown in this study, the production of the Th1 cytokine IFN-γ by human immune cells was suppressed much stronger by $PM_{0.5}$ of a traffic-influenced city or $PM_{0.5}$ of an industrial area than by those of an rural control area. This suggests that increasing $PM_{0.5}$ concentrations in ambient air and the PAHs and other sVOC adsorbed to $PM_{0.5}$ particles might strongly suppress the IFN-γ production of human blood leukocytes. This could trigger a dose-dependent ($PM_{0.5}$-dependent) bias in the Th1/Th2 balance toward dominance of Th2 responses. This is made likely by the obtained IFN-γ data. Thus, this thesis must be proofed in further investigations by analysis of IL-4 since interpretation of our data appears to be much clearer following a calculation of IFN-γ/IL-4-ratios.

Our results have implications for allergy development, since IL-4 (and IL-13) are the major factors in induction and maintenance of immediate type allergies promoting Ig switching to IgE (Brinkmann et al., 1997), and IFN-γ is the major regulatory factor in counter-regulation of these IL-4 effects (Romagnani 1991, 1996). The $PM_{0.5}$-adsorbed chemical compounds of the three regions suppress IFN-γ differentially and this uneven suppression might be the essential fact explaining a pollution-associated shift in the Th1/Th2 balance of exposed individuals and also the association of increased $PM_{0.5}$ and allergy development. Our observation that environmental pollution increases the number of $PM_{0.5}$ and the amount of PAHs adsorbed to these particles thus leads to an augmentation of these shifting effects. This is in agreement with observations in epidemiological studies (Takaki

1988) and the fact that animal experiments have shown an adjuvant effect of diesel exhaust particles (probably due to PAHs and other sVOC present in these particles) in triggering allergic diseases (Takafuji et al., 1987).

Moreover, the suppression of IFN-γ has also implications for cancer development since these cytokine is also the major factor in activation of cytotoxic T cells, NK cells and macrophages which are regulatory involved in the deletion of mutated or transformed cells. It is very likely that, additional to increased mutagenic and cancerogenic effects due to increased numbers of $PM_{0.5}$ particles and the increased amount of PAHs adsorbed to these particles, the suppression of IFN-γ leads to augmentation of these cancerogenic effects, since the effective deletion of mutated cells is suppressed by particles and the thereon adsorbed chemical compounds. Thus, the effective clearance of the body from somatic mutated cells might be disabled by suppression of IFN-γ production due to increased presence of particles. This effect might be involved in the mechanisms leading to the increased incidence of carcinogenesis in association with increased exposure to PAHs as well as to fine particles which was shown in epidemiological studies (see above).

In summary, our results have implications for allergy and cancer development. Therefore, the emission of particles and especially of sub-micrometer particles should be reduced to the maximum possible extent to avoid or at least reduce the negative effects on human health.

4.4. Risk Assessment of Life Time Cancer Risk (LCR)

Quantifying the health risk of air particles and thereon bound PAHs is a difficult task. The hazard potential of PAH can be determined in terms of benzo(a)pyrene (BaP), which is an indicator for carcinogenicity. This form to measure the hazard potential requires to ignore the rest of PAHs, some of them with a carcinogenic potency as high as BaP, and would therefore feature an inadequate evaluation of the risk of hydrocarbons altogether. To avoid this shortcoming, Toxic Equivalency Factors (TEFs) have been established as a useful tool for the estimation of the health risk of multi-component PAH exposure (Nisbet and LaGoy 1992). This approach assumes the toxicity of a mixture to be simply the sum of the expected effect of each mixture component. Synergistic or antagonistic interactions are negated. Several studies showed different results about effects of PAH mixtures.

4.4.1. Risk Assessment of Carcinogenic PAHs

To estimate the LCR of PAHs, the concentration of BaP and benzo(a)pyrene equivalents (BaP_{eq}) were used as a marker for carcinogenic PAHs. Table 10 shows the concentrations of analyzed PAH, calculated in terms of BaP_{eq}, and the total sum of BaP_{eq} of all studied areas using the TEFs of Nisbet and LaGoy (1992). BaP was used as a reference compound; a TEF of 1.0 was applied. In a first step, the BaP_{eq} was calculated by the multiplication of the individual PAH concentration by its TEF. In a second step, the carcinogenic potency of all considered PAHs as the sum of each individual BaP_{eq} was estimated. Statistical analyses and correlation factors were performed using STATISTICA 7.0. A *p*-value below 0.05 was regarded to be statistically significant.

The corresponding concentrations of BaP producing a life time cancer risk of 1×10^{-4}, 1×10^{-5} and 1×10^{-6} would be 1.2, 0.12 and 0.012 ng m^{-3}, respectively (WHO 2000). Considering this values, the lifetime cancer risk was calculated for the different areas in La Plata (Table 3).

Table 10. Benzo(a)pyrene equivalents (BaP_{eq}) calculated by their toxic equivalency factor (TEF)

PAH analyzed	TEF	Industry		Urban		Control	
Naphthalene	0.001	< 0.0001	*	< 0.0001		< 0.0001	
Acenaphthene	0.001	< 0.0001	*	< 0.0001	*	< 0.0001	
Fluorene	0.001	< 0.0001	**	< 0.0001		< 0.0001	
Phenanthrene	0.001	< 0.0001	**	< 0.0001		< 0.0001	
Anthracene	0.01	< 0.0001		< 0.0001		< 0.0001	
Fluoranthene	0.001	< 0.0001	**	< 0.0001		< 0.0001	
Pyrene	0.001	< 0.0001	*	< 0.0001		< 0.0001	
Benzo[*a*]anthracene	0.1	0.044	**	0.021	*	0.007	
Chrysene	0.01	0.006	**	0.003	*	0.001	
Benzo[*b*]fluoranthene	0.1	0.119	**	0.066		0.021	
Benzo[*k*]fluoranthene	0.1	0.043	**	0.024		0.008	
Benzo[*a*]pyrene	1	0.751	**	0.365	*	0.127	
Dibenz[*a,h*]anthracene	1	0.095	**	0.046		0.019	*
Benzo[*g,h,i*]perylene	0.01	0.013	**	0.010		0.003	
Indeno[*123*]cd-pyrene	0.1	0.097	**	0.061		0.023	
Sum BaP_{eq}		1.169	**	0.597	*	0.209	

Significant correlation with the results of the Ames-test * $p<0,05$, ** $p<0,001$.

Table 11. Lifetime cancer risk (LCR) calculated by BaP and the sum of BaP_{eq}

Study area	LCR to BaP	LCR to sum BaP_{eq}
Industry	6.2×10^{-5}	9.6×10^{-5}
Urban	3.04×10^{-5}	4.9×10^{-5}
Control	1.05×10^{-5}	1.7×10^{-5}

The results indicate the lifetime cancer risk in the industrial area to be two times higher than in the urban area and six times higher than in the control region. According to an acceptable risk value of $1 \times 10^{-6} - 1 \times 10^{-5}$ (WHO, 2000) a higher risk for the industrial as well as for the urban area in La Plata can be supposed.

However, these results are not congruently with those from the bioassays (chapter 4.1, 4.2). A significant correlation to the results of the mutagenicity test was found only for the BaP_{eq} of the industrial area. For the urban area, only the BaP_{eq} of a few compounds correlated with mutagenic effects (Table 10). Nevertheless, also for extracts from particles of the urban area strong mutagenic as well as cytotoxic effects were observed, nearly similar to those found for the industrial area (chapter 4.1). In addition to BaP, other substances can cause mutagenic and cytotoxic effects too. PAH mixtures in ambient air differ in concentration and in composition. BaP is an important indicator of such mixtures, but it cannot be an exclusive indicator of their carcinogenic potency. As demonstrated here, the possible risk probably is underestimated if only BaP is used for LCR assessment.

4.4.2. Risk Assessment of Benzene Exposure

Calculation of cancer risk requires the carcinogenic potency of the specific compound and the mean exposure of the target group. The lifetime cancer risks (LCR) associated to benzene in the different areas of study were calculated as a multiple of the chronic daily intake (CDI) and the potency factor (IPCS, 2000; Hoddinott and Lee, 2000; Muller at al., 2003; Guo et al., 2004). The CDI in mg/kg/day were calculated according to the following equation:

CDI= (CC x IR x ED x EF x LE) / (BW x AT x NY)

where CC is the contaminant concentration (mg m^{-3}),

IR the inhalation rate (m^3 h^{-1}),
ED the exposure duration (h $week^{-1}$),
EF the exposure frequency (weeks $year^{-1}$),
LE the length of exposure (years),
BW the body weight (kg),
AT the average time (period over which exposure is averaged),
NY the number of days per year (365 days).

The IRIS system benzene potency factor (0,029 (mg (kg day^{-1}))$^{-1}$) was used for the calculations (USEPA, 1998).

LCR was calculated for children exposure to benzene in indoor and outdoor environments in the different areas of study (Table 12). Children's risk was considered in order to be a vulnerable segment of the population to air pollutants exposure. For the calculations we assumed an average body weight of 17 kg and an IR average of 5 m^3 of air per day (USEPA, 1994; Hoddinott and Lee, 2000). The absorption factor of the VOCs for man was assumed to be 90% and a lifetime of 70 years was applied (Guo et al, 2004). An annual average of 20 hours per day in indoor environments and 4 hours per day outdoors were assumed. For the calculations the median concentrations of benzene was used.

Results in all cases showed an LCR one order of magnitude higher for indoor environment than outdoor. In this, most of the areas showed similar values of LCR indoors. The results from outdoor calculations showed LCR values two times higher in urban area compared to semi-rural and residential sites. Otherwise, an LCR one order of magnitude higher for children from industrial area in indoor and outdoor environments than the other areas was observed. The World Health Organization (WHO) considered a LCR in the range between 1×10^{-5} and 1×10^{-6} as "acceptable" while the USEPA below 1×10^{-6} (Miller et al, 1999; IPCS, 2000)

For all target groups of children, this study showed that the exposure to benzene may lead to a LCR greater 1×10^{-6} for indoor and outdoor exposure, higher than those considered as "acceptable" by the USEPA. For indoor exposure, all target groups showed also greater LCR than those considered by WHO for all areas while for outdoor exposure only the group from industrial area exceeded the WHO values.

Table 12. Children's lifetime cancer risk (LCR) for indoor and outdoor exposure in the different study areas

	Indoor LCR	Outdoor LCR
Industry	1.41×10^{-4}	2.62×10^{-5}
Urban	2.94×10^{-5}	5.03×10^{-6}
Control	2.90×10^{-5}	2.92×10^{-6}

For all target groups of children this study showed a LCR associated to benzene greater than the "acceptable" levels for indoor and outdoor exposure, the highest values corresponded to the industrial area. For outdoor exposure, the urban area showed a LCR higher than in the control areas. The health risks from benzene exposure may be minimized increasing the controls to reduce the industrial emissions as well as traffic pollution, taking also into consideration a low VOC indoor emission.

CONCLUSION

To evaluate the risk of industrial emissions besides the determination of PM, the analyses of semi-volatile and volatile compound, additionally extracts of PM were investigated in various bioassays. Samples were carried out from urban, industrial and control areas in La Plata, Argentina, a city influenced by traffic and the petrochemical industry because Argentina's main oil refinery is located nearby.

Sampling of airborne particulate matter in fractions ranging from 0.49 to 10 µm enabled a differentiated assessment of biological effects of particle size fractions. Organic extracts from particles less than 1.5 µm have the highest mutagenic potency, cytotoxicity and immunomodulating effects in agreement with the chemical analysis. These particles are more likely to reach the lung alveoli and the deep respiratory tract causing significant health problems.

Chemicals adsorbed to particulate matter are able to affect the energy providing metabolism and trigger a decreased efficiency of ATP production. These effects are associated with a simultaneously increased cytotoxicity. Especially smaller particles below 1 µm aerodynamic diameter from traffic influenced urban area and from an area with petrochemical industry caused the strongest effects. The observed imbalance between oxygen consumption and ATP concentration could be caused by oxidative stress, due to extractable chemical compounds bound on particulate matter.

The analysis VOC exposure has provided interesting insights on the characterization of volatile organic compounds in indoor and outdoor air in the region of La Plata. Ambient air in the vicinity of the industrial site is significantly different than the air quality found in the urban, and control area. Whereas traffic was identified as the main source of outdoor VOC exposure in the urban and control areas, in the industrial area, outdoor as well indoor air is strongly affected by emissions especially of hexane and the carcinogenic compound benzenes, originated from the petrochemical industry which documents a high risk for human health in the industrial area.

Due to the complexity of effects related to PM and their associated chemical compounds, various bioassays necessarily need to be investigated to obtain more information about their possible involvement in pathogenic processes. Our integrated approach of risk assessment involving epidemiological investigations combined with exposure measurements, the sampling of PM and the corresponding chemical analyses and furthermore also the performance of different types of bioassays to measure mutagenic, cytotoxic as well as immunomodulating effects offers the chance to estimate the risk for exposed human beings in a realistic manner. In general the individual investigations included in the unique approach described here, shows the greater risk of exposure to airborne particles and VOC from industrial areas. Our results in the various investigations clearly point to an enormous risk for adverse health effects associated especially to $PM_{0.5}$ and the chemicals adsorbed to these ultra-fine particles.

REFERENCES

Adgate, JL.; Church, TR; Ryan, AD; Ramachandran, G; Fredrickson, AL; Stock, TH; Morandi, MT; Sexton, K. Outdoor, Indoor, and Personal Exposure to VOCs in Children. *Environ. Health Persp.* 2004; 112 (14) 1386-1392.

APHA American Public Health Association, American Water Works Association, Water Environment Federation. 1998. Standard Methods for examination of water and wastewater. 20th Edition. Washington DC, USA.

Ataullahanov, FI ; Vitvitski, VM. What determined the intracellular ATP concentration. *Biosci. Rep.* 2002; 22: 501-511.

Begerow, J; Jermann, E; Kelles, T; Ranft, U; Dunemann, L. Passive sampling for Volatile Organic Compounds (VOCs) in air at environmentally relevant concentration levels. *Fresen J Anal Chems.* 1995; 351: 549-554.

Bernstein, L; Kaldor, J; McCann, J; Pike, MC. An empirical approach to the statistical analysis of mutagenesis data from the Salmonella test. *Mut. Res.* 1982; 97:267-281.

Brenner, JR; Bruce, A; Gorden, GE; Wise, SA. Mobile sources of atmospheric polycyclic aromatic hydrocarbons: a roadway tunnel study. *Environ. Sci. Technol.* 1989; 23: 1269-1278.

Brinkmann, V; Kinzel B. IL-4 and IL-13 in atopy: Regulation of production and effects. In: Vercelli D, editor. IgE-regulation: molecular mechanisms. John Wiley and Sons, Inc., Chichester, New York, Weinheim, Brisbane, Singapore, Toronto, 1997 p 75-101.

Brookes PS. Mitochondrial H+ leak and ROS generation: An odd couple. *Free Radical Biology and Medicine* 2005; 38: 12-23.

Brunnekreef, B; Holgate ST.. Air pollution and health. *Lancet* 2002; 360:1233-1242.

Butler, J; Jayson, GG; Swallow, AJ. The reaction between superoxide anion radical and cytochrome c. *Biochim. Biophys. Acta* 1975; 408:215– 222.

Černa, M; Pochmanova, D; Pastorkova, A; Beneš, I; Leniček, J; Topinka, J; Binkova, B.. Genotoxicity of urban pollutants in the Czech Republic Part I. Bacterial mutagenic potencies of organic compounds adsorbed on PM_{10} particulates. *Mut. Res.* 2000; 469:71-82.

Cetin, E; Odabasi, M; Seyfioglu, R. Ambient volatile organic compound (VOC) concentrations around a petrochemical complex and a petroleum refinery. *Sci Total Environ.* 2003; 312 (1-3): 103-112.

Chatzis, C; Evangelos, C; Alexopoulos, T; Linos, A. Indoor and outdoor personal exposure to benzene in Athens, Greece. *Sci Total Environ.* 2005; 349: 72– 80.

Dellinger, B; Pryor, WA; Cueto, R; Squadrito, GL; Hegde, V; Deutsch, WA.. Role of free radicals in the toxicity of airborne fine particulate matter, *Chem. Res. Toxicol.* 2001;14:1371–1377.

Dockery, DW; Schwarz, J; Spengler, JD. Air pollution and daily mortality: Association with particles and acid arerosols. *Environ. Res.* 1992; 59: 362-373.

Edgerton, SA; Holdren, MW; Smith, DL; Shah, JJ. Interurban comparison of ambient volatile organic compounds concentrations in U.S. cities. *Journal of the Air and Waste Management Association* 1989; 39: 729-732.

Edward, RD; Jurvelin, J; Koistinen, K; Saarela, K; Jantunen, M. VOC source identification from personal and residential indoor, outdoor and workplace microenvironment samples in EXPOLIS-Helsinki, Finland. *Atmos Environ.* 2001; 35: 4829–4841.

Fahrmeier, L; Hamerle, A; Tutz, G. Multivariate statistische Verfahren, Walter de Gruyter. 1996.

Ferin, J; Oberdörster, G; Penney, DP. Pulmonary retention of ultrafine and fine particles in rats. *Am. J. Respir. Cell Mol. Biol.* 1992; 6: 535-542.

Fernandes, M; Brickus, L; Moreira, J; Cardoso, J. Atmospheric BTX and polyaromatic hydrocarbons in Rio de Janeiro, Brazil. *Chemosphere* 2002; 47: 417-425.

Gee, I; Sollars, C. Ambient air levels of volatile organic compounds in Latin American and Asian cities. *Chemosphere* 1998: 6(11): 2497-2506.

Gosselin, RE; Smith, RP; Hodge, HC. Clinical Toxicology of Commercial Products.: 5th Edition. Williams and Wilkins, Baltimore. 1984.

Gräf, W; Gräf, H; Wenz, M. Tetrahymena pyriformis in the Ciliate Mobility Test. Validation and description of a testing procedure for registration of harmful substances in the air as well as the effect of cigarette smoke on the human respiratory ciliated epithelium. *Zbl Hyg Umweltm* 1998/1999; 201:451–472.

Guo, H; Lee, S; Li, W; Cao, J. Source characterization of BTEX in indoor microenviroments in Hong Kong. *Atmos. Environ.* 2003; 37: 73-82.

Guo, H; Lee, S; Chan, L; Li, W. Risk assessment of exposure to volatile organic compounds in different indoor environments. *Environ. Res.* 2004; 94: 57-6.

Harrison, RM; Jones, M. The chemical composition of airborne particles in the UK atmoshere. *Sci. Total Environ.* 1995;168:195-214.

Hathaway, GJ; Proctor, NH; Hughes, JP; Fischman, ML. Proctor and Hughes' chemical hazards of the workplace. 3rd ed. New York, NY (1991): Van Nostrand Reinhold.

Hauser, R; Godleski, J; Hatch, V; Christiani, DC. Ultrafine particles in human lung macrophages. *Arch Environ Health.* 2001; 56: 150-156.

Herbarth, O; Fritz, G; Krumbiegel, P; Diez, U; Franck, U; Richter, M. Effect of sulfur dioxide and particulate pollutants on bronchitis in children – a risk analysis. *Environ. Toxicol.* 2001; 16: 269-276.

Hetland, RB; Cassee, FR; Refnes, M; Schwarze, PE; Lag, M; Boere, AJF; Dybing, E.. Release of inflammatory cytokines, cell toxicology and apoptosis in epithelial lung cells after exposure to ambient air particles of different size fractions. *Toxicology in vitro*. 2004; 18: 203-212.

Hoddinott, K; Lee, A. The use of environmental risk assessment methodologies for an indoor air quality investigation. *Chemosphere* 2000; 41:77-84.

Hodgson, AT; Beal, DJ; McIlvaine, ER. Sources of formaldehyde, other aldehydes and terpenes in a new manufactured house. *Indoor Air* 2002; 12: 235-243.

Hopke, PK. Receptor Modelling in Environmental Chemistry. John Wiley and Sons, New York. 1985.

Hsiao, WLW; Mo, ZY; Fang, M; Shi, X; Wang, F.Cytotoxicity of $PM_{2.5}$ and $PM_{2.5\text{–}10}$ ambient air pollutants assessed by the MTT and the Comet assays. *Mutat. Res.* 2000; 471:45–55.

IAPG (Instituto Argentino del Petróleo y del Gas). La República Argentina y su Industria Petroquímica; 1999. Available on line: http://www.iapg.org.ar/iapg/.

IARC - International Agency for Research on Cancer. Monographs on the evaluation of carcinogenic risk to humans. 1983; Vol 32

IARC (International Agency for Research on Cancer). Monograph on the evaluation of carcinogenic risk to humans, WHO, Geneva, 1995 vol. 63: pp.75 and 159;

IARC - International Agency for Research on Cancer. *Monograph on the evaluation of carcinogenic risk to humans*. 2002; Vol 82: 367.

IPCS, International Programme on chemical safety (2000) Environmental Health criteria 214: Human Exposure Assessment. Geneva, World Health Organization.

Iuchi, S; Weiner, L.. Cellular and Molecular Physiology of *Escherichia coli* in the adaptation to aerobic environment. *J. Biochem.* 1996; 120: 1055-1063.

Junker, M; Kasper, M; Röösli, M; Camenzind, M; Künzli, N; Moon, Ch; Theis, G; Braun-Fahrländer, Ch. Airborne particles number profiles, particle mass distribution and particle-bound PAH concentrations within the city environment of Basel: an assessment as part of the BRISKA Project. *Atmos. Environ.* 2000; 34: 3171-3181.

Kawanaka, Y; Matsumoto, E; Sakamoto, K; Wang, N; Yun, SJ. Size distribution of mutagenic compounds and mutagenicity in atmospheric particulate matter collected with a low-pressure cascade impactor. *Atmos. Environ.* 2004; 38: 2125– 2132.

Kiss, G; Varga-Punchony, Z; Rohrbacher, G; Hlavay, J. Distribution of polycyclic aromatic hydrocarbons on atmospheric aerosol particles of different sizes. *Atmos. Res.* 1998; 46: 253-261.

Knaapen, AM; Borm, PJ; Albrecht, C; Schins, RP. Inhaled particles and lung cancer. Part A: mechanism, *Int. J. Cancer* 2004; 109: 799-809.

Lau, C; Fiedler, H; Hutzinger, O; Schwind, K; Hosseinpour, J. Levels of selected organic compounds in materials for candle production and human exposure to candle emissions. *Chemosphere* 1997; 4: 1623-1630.

Lee, DD; Seung, HS. Unsupervised learning by convex and conic coding. *Advances in Neural Information Processing* 1997; 9: 515-521.

Lee, DD; Seung, HS. Learning the parts of objects by non-negative matrix factorization. *Nature* 1999; 401: 788-791.

Lee, DD; Seung, HS. Algorithms for Non-negative Matrix Factorization. Advances in Neural Information *Processing Systems* 2001; 13: 556-562.

Lee, E; Chan, CK; Paatero, P. Application of positive matrix factorization in source apportionment of particulate pollutants in Hong Kong. *Atmospheric Environment* 1999; 33: 3201-3212.

Lewtas, J. Genotoxicity of complex mixtures: strategies for the identification and comparative assessment of airborne mutagens and carcinogen from combustion sources. *Fund Appl. Toxicol* 1988; 10: 571-589.

Lewtas, J. Experimental evidence for the carcinogenicity of air pollutants. In: *Tomatis L. Air pollution and human cancer.* Monographs, European School of Oncology. Springer-Verlag, 1993; 49-61

Lucarelli, F; Mando, PA; Nava, S; Prati, P; Zucchiatti, A. One-year study of the elemental composition and source apportionment of PM10 aerosols in Florence, Italy, *Journal of the Air and Waste Management Association* 2004; 54(11): 1372-1382.

Mailer, K. Superoxide radical as electron donor for oxidative phosphorylation of ADP. *Biochem. Biophys. Res. Commun.* 1990; 170: 59–64.

Maron, DM; Ames, BN. Revised methods for the Salmonella mutagenicity test. *Mut Res* 1983; 113:173-215.

Marple, VA; Liu, BYH. Characteristics of Laminar Jet Impactors. *Environ. Sci. Technol.* 1974; 8: 648-654.

Massolo, L; Müller, A; Tueros, M; Rehwagen, M; Franck, U; Ronco A. Assessment of mutagenicity and toxicity of different size fractions of air particulates from La Plata Argentina and Leipzig Germany. *Environ. Toxicol.* 2002; 17:219– 31.

McCubbin, AE; Ersing, N; Frank, ME. Mutagenicity of sediments from the Detroit River. *J. Great Lakes Res.* 1991; 17:314-321.

Miller, G; Fortney, P; Lantos, T; Thompson, M; Pelosi, N; Eshoo, A; Lee, B; Woolsey, L. Exposure to hazardous air pollutants in the San Francisco Bay area. 1999. Available on line: http://www.house.gov/geogemiller/airreport.html.

Monarca, S; Crebelli, R; Feretti, D; Zanardini, S; Fuselli, L; Filini, L; Resola, S; Bonardelli, PG; Nardi, G. Mutagens and carcinogens in size-classified air particulates of a Northern Italian town. *Sci. Total Environ.* 1997; 205:137-144.

Muller, E; Diab, R; Binedell, M; Hounsom, R. Health risk assessment of kerosene usage in an informal settlement in Durban, South Africa. *Atmospheric Environment* 2003; 37:2015–2022.

Müller, A; Alzuet, P; Herbarth, O; Ronco, AE. Assessment of Toxicity and Mutagenicity in air particulte matter from an urban industrial area in the cost of the Rio de la Plata. *Environ. Tox* 2001;16:151-157.

Müller, A; Wichmann, G; Massolo, L; Rehwagen, M; Gräbsch, C; Loffhagen, N; Herbarth, O; Ronco A. Cytotoxicity and oxidative stress caused by chemicals adsorbed on particulate matter. *Environmental Toxicology,* 2006; 21 (5), 457-463.

Na, K; Kim, Y; Moon, I; Moon, K. Chemical composition of major VOC emission sources in the Seoul atmosphere. *Chemosphere* 2004; 55: 585-594.

NIOSH Registry of toxic effects of chemical substances: Hexane. Cincinnati, OH: U.S. Department of Health and Human Services, Public Health Service, Centers for Disease Control, National Institute for Occupational Safety and Health, Division of Standards Development and Technology Transfer, Technical Information Branch. (1991).

Nisbet, ICT;.LaGoy, PK. Toxic Equivalency factors (TEFs) for polycyclic aromatic hydrocarbons (PAHs) Regulatory *Toxicology and Pharmacology* 1992; 16: 290-300.

OJ (Official Journal of the European Communities) L 313, 13/12/2000. Directive 2000/69/EC Relating to Limit Values for Benzene and Carbon Monoxide in Ambient Air, OJ L 313; 2000.

Pagano, P; De Zaiacomo, T; Scarcella, E; Calamosca, M. Mutagenic activity of total and particle-sized fractions of urban particulate matter. *Environ. Sci. Techno* 1996; 30: 3512-3516.

Pannwitz, K. Dräger Probenahmehandbuch, Dräger Aktiengesellschaft. Lübeck 1991: 49-54.

Panther, BC; Hooper, MA; Tapper, NA. Comparison of air particulate matter and associated polycyclic hydrocarbons in some tropical and temperate urban environments. *Atmos Environ.* 1999; 33: 4087-4099.

Peace, H; Owen, B; Raper, DW. Identifying the contribution of different urban highway air pollution sources, *Science of the Total Environment* 2004; 334: 347-357.

Pekey, H; Karakas, D; Bakoglu, M. Source apportionment of trace metals in surface waters of a polluted stream using multivariate statistical analysis, *Marine Pollution Bulletin* 2004 49(9-10), 809-818.

Pierce, RC; Katz, M. Dependency of polynuclear aromatic hydrocarbons content on size distribution of atmospheric aerosols. *Environ. Sci. Technol.* 1975; 9: 347-353.

Rappenglück, B; Fabian, P; Kalabokas, P; Viras, LG; Ziomas, IC. Quasi-continuous measurements of non-methane hydrocarbons (NMHC) in the greater Athens area during MEDCAPHOT-TRACE, *Atmospheric Environment* 1998; 32:(12), 2103-2121.

Rehwagen, M; Schlink, U; Herbarth, O. Seasonal cycle of VOCs in apartments. *Indoor Air* 2003; 13: 1-9.

Rehwagen, M; Müller, A; Massolo, L; Herbarth, O; Ronco, A. Polycyclic aromatic hydrocarbons associated with particles in ambient air from urban and industrial areas. *Science of Total Environ* 2005; 348: (1-3),199-210

Reyes, DR; Rosario, O; Rodriguez, JF; Jimenez, BD. Toxic evaluation of organic extracts from airborne particulate matter in Puerto Rico. *Environ. Health Persp.* 2000; 108:635-640.

Risom, L; Moller, P; Loft, St.. Oxidative Stress-induced DNA damage by particulate air pollution. *Mutation research* 2005; 592: (1-2), 119-137.

Romagnani S.. Human Th1 and Th2 subsets: doubt no more. *Immunol. Today* 1991; 12: 256-257.

Romagnani S. Development of Th 1- or Th 2- dominated immune responses: what about the polarizing signals? *Int. J. Clin. Lab. Res.* 1996; 26: 83-98.

Saarela, K; Tirkkonen, K; Laine-Ylijoki, J; Jurvelin, J; Nieuwenhuijsen, MJ; Jantunene, M. Exposure of population and microenvironmental distributions of volatile organic compound concentrations in the EXPOLIS study. *Atmos. Environ.* 2003: 37: 5563–5575.

Sack, T.M; Steele, DH; Hammerstrom, K; Remmers, J. A survey of household products for volatile organic compounds. *Atmos Environ.* 1992: 26A (6): 1063–70.

Sauvant, MP; Pepin, D; Piccinni, E. *Tetrahymena pyriformis*—a tool for toxicological studies. A review. *Chemosphere* 1999; 38:631–1669.

Schlink, U; Rehwagen, M; Damm, M; Richter, M; Borte, M; Herbarth O. Seasonal cycle of indoor-VOCs: comparison of apartments and cities. *Atmos. Environ.* 2004: 38: 1181-1190.

Schnelle-Kreis, J; Gebefügi, I; Welzl, G; Jaensch, T; Kettrup, A. Occurrence of particle-associated polyclic aromatic compounds in ambient air of the city of Munich. *Atmos Environ*. 2001; 35: 71– 81.

Schwartz, J; Neas, LM. Fine particles are more strongly associated than coarse particles with acute respiratory health effects in schoolchildren. *Epidemiology* 2000; 11: 6-10.

Schwartz J. Air pollution and daily mortality: a review and meta analysis. *Environ. Res*. 2004; 64, 36-52

Seigneur C. $PM_{2.5}$ modelling: Current status and research needs. In Proceedings, PM2.5: A Fine Particle Standard (Chow JC and Koutrakis P, eds.), pp. 713-724. Air and Waste Management Association, Pittsburgh, PA. (1998).

Sheu, HL; Lee, WJ; Lin, SJ; Fang, GC; Chang, HC; You, WC. Particle bound PAH content in ambient air. *Environ Pollut*. 1997; 96:369– 82.

Sies H. Strategies of antioxidant defense. *Eur. J. Biochem*. 1993; 215: 213–219.

Sirce, MA; Marty, JC; Saliot, A. Aliphatic and aromatic hydrocarbons in different size aerosols over the Mediterranean Sea: occurrence and origin. *Atmos. Environ*. 1987; 12: 2247– 2259.

Sokal, RR; Rohlf, FJ. Biometry. Chapters 13, 14. W.H. Freeman: New York; 1981.

Son, B; Breysse, P; Yang, W. Volatile organic compounds concentration in residential indoor and outdoor and its personal exposure in Korea. *Environ Int*. 2003; 29: 79-85.

Spinosa De Martinis, BR; Kado, NY; de Carvalho Lilian, RF; Okamoto, RA; Gundel, Lara A.. Genotoxicity of fractionated organic material in airborne particles from Sao Paulo, Brazil. *Mut Res*. 1999; 446: 83-94.

Srivastava, A. Source apportionment of ambient VOCs in Mumbai city, *Atmospheric Environment* 2004; 38: 6829-6843.

Takaki, M. Environmental and immunological survey of sensitization with Japanese cedar cryptomeria-japonica pollen in children. I. Environmental factors influencing sensitization with Japanese cedar cryptomeria-japonica pollen among asthmatic children. *Jpn J Allergol* 1988; 37:1065-1071.

Takafuji, S; Suzuki, S; Koizumi, K; Tadokoro, K; Miyamoto, T; Ikemori R.. Diesel exhaust particulates inoculated by the intranasal route have an adjuvant activity for IgE production in mice. *J Allergy Clin Immunol* 1987; 79: 639-643.

Tang, J; Chan, CY; Wang, X; Chan, LY; Sheng, G; Fu J. Volatile organic compounds in a multi-storey shopping mall in Guangzhou, South China. *Atmos. Environ*. 2005; 39: 7374–7383.

Tuominen, J; Salomaa, S; Pysalo, H; Skyttä, E; Tikkanen, L; Nurmale, T; Sorsa, M; Pohjola, V; Sauri, M; Himberg, K.. Polynuclear aromatic compounds and genotoxicity in particulate and vapor phases of ambient air: Effect of traffic, season and meteorological. *Environ. Sci. Techn*. 1988; 22:1228-1234.

Tzonou, A; Maragoudakis, G; Trichopoulos, D.. Urban living, tobacco smoking and chronic obstructive pulmonar disease: a study in Athens. *Epidemiology* 1992; 3: 57-60.

US EPA- United States Environmental Protection Agency. 1983. Interim Procedures for Conducting the Salmonella / Microsomal Mutagenicity Assay (Ames test). EPA/600/4-82/068, Las Vegas, NV.

US EPA. United States Environmental Protection Agency Methods for derivation of inhalation reference concentrations and application of inhalation dosimetry. 1994. Office of Health and Environmental Assessment, Environmental Criteria and Assessment Office, Research Triangle Park, NC, EPA/600/8- 90/066F.

US EPA (a). United States Environmental Protection Agency Air Quality Criteria for Particulate Matter EPA/600/P-95/00 1cF. US Environmental Protection Agency Office of Research and Development Washignton DC 1996; 3.

US EPA (b). United States Environmental Protection Agency Review of the National Ambient Air Quality Standards for Particulate Matter: Policy Assessment of Scientific and Technical Information OAQPS Staff Paper EPA-452/R-96-013 Office of Air Quality Planning and Standards. US Environmantal Protection Agency Research Triangle Park NC.1996.

USEPA. United States Environmental Protection Agency 1997. Health and Environmental Effects of Particulate Matter. United States Environmental Protection Agency, Office of Air and Radiation, Office of Air Quality Planning and Standards.

US EPA. Integrated risk information system—benzene. 1998. URL:http://www.epa.gov/ iris/subst.o276.htm

US EPA (2001) CMB8 User Manual, EPA-454/R-01-XXX, USEPA, QAQPS Research Triangle Park, NC 27711.

Vasconcellos, PC; Zacarias, D; Pires, MAF; Pool, CS; Carvalho, LRF. Measurements of polycyclic aromatic hydrocarbons in airborne particles from the metropolitan area of São Paulo City, Brazil. *Atmos. Environ.* 2003; 37: 3009-3018.

Venkataraman, C; Lyons, JM; Fiedlander, S. Size distribution of aromatic hydrocarbons and elemental carbon, Sampling measurement methods and source characterization. *Environ Sci Technol.* 1994; 28: 535-562.

Viras, LG; Siskos, PA; Samara, C; Kouimtzis, T; Athanasious, K; Vavatzandis, A.. Polycyclic aromatic hydrocarbons and mutagens in ambient air particles sampled in Thessalonik Greece. *Atmos. Environ.* 1991; 10: 999-1007.

Wang, S; Ang, HM; Tade, MO. Volatile organic compounds in indoor environment and photocatalytic oxidation: State of the art. *Environment International* 2007; 33: 694–705

Watson, JG; Chow, JC; Fujita, E. Review of volatile organic compound source apportionment by chemical mass balance. *Atmos. Environ.* 2001; 35: 1567-1584.

Watson, JG; Chow, JC Particulate Pattern Recognition. In: *An Introduction to Environmental Forensics* (R. Morrison and B. Murphy, eds.), Academic Press, San Diego, 2002 pp. 429-460.

Whitby, RA; Altwicker, ER. Acetylene in the atmosphere: sources, representative ambient concentrations, and ratios to other hydrocarbons, *Atmospheric Environment* 1978; 12: 1289-1296.

WHO (World Health Organization). Air Quality Guidelines for Europe. WHO Regional Publications, European Series, No.91, Second Edition, Copenhagen; 2000.

Xue, L; Stahura, FL; Bajorath, E. Similarity search profiling reveals effects of fingerprint scaling in virtual screening. *Journal of Chemical Information and Computer Sciences* 2004; 44(6): 2032-2039.

Ye, SH; Zhou, W; Song, J; Peng, B-C; Yuan, D; Lu Y-M; Qi P-P. Toxicity and health effects of vehicle emissions in Shanghai. *Atmosph. Environ.* 1999; 34: 419-429.

Zar, J.H. Bioestadistical Analysis; Chapters 17, 18. ISBN 0-13-081542-X. Prentice Hall: New Jersey, 4ta ed, Ed Ryu T; 1998.

Zhao, Y; Wang, ZB; Xu, JX.. Effect of cytochrome c on the generation and elimination of O2• − and H_2O_2 in mitochondria. *J. Biol. Chem.* 2003; 278:2356– 2360.
Zou, S; Lee, S; Chan, C; Ho, K; Wang, X; Chan, L; Zhang, Z. Characterization of ambient volatile organic compounds at a landfill site in Guangzhou, South China. *Chemosphere* 2003; 51: 1015-1022.
Zwoździak, J; Jadzyk, P; Kucharczyk, J.. Seasonal variability of the mutagenic of airborne particles in the town center. *Aerosol science* 2001; 32:409-423.

Recommended reviewer:

We would like to recommend Bo Strandberg, PhD, who is working on the topic of indoor and outdoor air pollution in the European Commission-Joint Research Centre, Ispra

Bo Strandberg, PhD
Associate Professor
Department of Occupational and Environmental Medicine
Sahlgrenska Academy at Göteborg University
Box 414, Medicinaregatan 16
SE 405 30 Göteborg
Sweden
Phone: +46 31 773 2892
Fax: +46 31 825004
E-mail: bo.strandberg@amm.gu.se

In: Industrial Pollution Including Oil
Editor: Harry Newbury and William De Lorne
ISBN 978-1-60456-917-9

Chapter 5

ADVERSE HEALTH EFFECTS OF HEAVY INDUSTRIAL AND AGRICULTURAL POLLUTION IN EASTERN SLOVAKIA

Pavel Langer[1], Anton Kočan[2], Mária Tajtáková[3], Åke Bergman[4], Tomáš Trnovec[2], Daniela Gašperíková[1] and Iwar Klimeš[1]
[1]Institute of Experimental Endocrinology, Slovak Academy of Sciences, Bratislava, Slovakia
[2]Department of Toxic Organic Pollutants, Slovak Medical University, Bratislava, Slovakia
[3]Clinic of Internal Medicine, Faculty of Medicine, P.J. Šafárik University, Košice, Slovakia
[4]Department of Environmental Chemistry, tockholm University, Stockholm, Sweden

ABSTRACT

Few years after the war three large chemical factories were built in eastern Slovakia which were designed to produce artificial fibers, wood products and various large scale chemical products including polychlorinated biphenyls (PCBs). At the same time the agricultural production in this area also increased which was accompanied by nearly unlimited use of pesticides and fertilizers. Resulting from environmental negligence heavy waterborn and airborn pollution developed, but either any survey or any public or scientific information about the actual status has been strictly banned. From such reason scientific surveys on possible health effects were started as late as after the previous regime has been changed in 1990. These were based on comparing large numbers of adults and children from polluted area (POLL) with these from the area of background pollution located about 60 km upwind and upstream (BCGR). Thus, in POLL very high levels of PCBs, DDE and HCB in environmental as well as in human samples were found. Recently also the data on high level of dioxins, furans as well as selected OH-PCBs., $MeSO_2$-PCBs and $MeSO_2$-DDE in human serum were obtained. Although such levels in BCGR were much lower, the background pollution did not appear quite negligible. Medical examinations showed increased thyroid volume, increased prevalence of thyroid antibodies in blood and thyroid disorders namely of overt and subclinical autoimmune thyroiditis as well as changes in thyroid hormone and thyrotropin level. Moreover, increasing prevalence of impaired fasting glucose level and other dysglycemias was found thus possibly supporting the increasing prevalence of diabetes. In addition, the examinations of cognitive and psychomotoric functions as well as these of hearing apparatus and dental status were conducted in children. All such signs of adverse health status were significantly associated with increasing level of PCBs which were used as a marker of other organochlorines cocktail. Although the manufacturing of PCBs has been terminated in 1984, heavy environmental pollution is still persisting and several investigations are still in progress. Among others, it appears that high organochlorine levels in blood of adult women are influencing their newborns due to transplacental transmission and some adverse effects already appeared in the third generation.

ABBREVIATIONS

BCGR – control area of backhround pollution
DDE – dichlorodiphenyl dichloroethylene
DDT –
FT3 - free triidoothyronine
FT4 – free thyroxine
HCB – hexachlorobenzene
OCs - organochlorines
PCB – polychlorinated biéhenyls
PCDDs – polychlorinated dibenzo-dioxins
PCDFs – polychlorinated dibenzo-furans
POLL – polluted area
PCBRISK – acronyme of the project
SIR – Standard Incidence Rate
SPEECD – acronyme of the project
TG - thyroglobulin
TGab – thyroglobulin antibodies
TPOab – thyroperoxidase antibodies
TRab – thyrotropin receptor antibodies
ThCa – thyroid cancer
ThV – thyroid volume
TSH – thyroid stimulating hormone (thyrotropin)
TT3 – total triiodothyronine
TT4 – total thyroxine

HISTORICAL INTRODUCTION

In addition to global warming, the mankind is also facing a global pollution by persistent toxic and mostly organohalogenated chemicals. One of dramatic chapters of this story is going on in eastern Slovakia. It started after the war when the Iron Curtain separated Eastern

Europe from other world not only politically, but also economically. In order to overcome the consequences of war events such as heavy devastation of the country and its infrastructure there was an urgent need to increase industrial production as rapidly as possible and thus huge factories were rapidly built up. However, under the lack of know-how and financial resources such rapid industrialization certainly was not any easy task. Thus, the only possible way was to limit the importation radically and, at the same time, to develop the home production rapidly by using exclusively domestic intellectual and material resources.

In eastern Slovakia three of those factories were designed to large scale chemical production and were built in close vicinity of the rivers (Figure 1). Thus, Chemlon (synthetic fibers) and Chemko factory (various large scale chemical products including polychlorinated biphenyls and explosives) were both located on the river Laborec which is supplying highly polluted water to the large artificial Sirava lake, while Bukoza factory (wood treatment plant) was located on the river Topla which flows to the river Ondava.

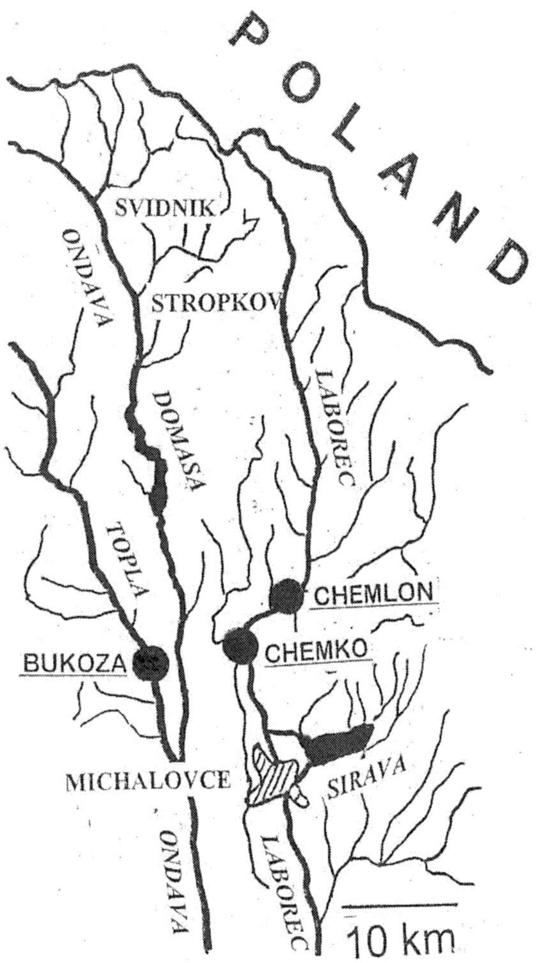

Figure 1. Schematic map of northeastern Slovakia. UNDERLINED – chemical factories, VERTICAL LETTERS – district towns; ITALICS – rivers and lakes.

Large scale production in those factories was started in early fifties. Since the main necessity was to keep the production in such state owned factories as cheap as possible, all preventive ecological and health measures were underestimated and simply neglected, while all information about any possible risk has been strictly banned to keep the pollution out of public concern as long as possible. From such reasons all chemical waste has been simply and silently dumped to those rivers. Thus, in addition to PCBs, the waste from Chemko contained also benzene, nitrosamine, formaldehyde etc., while that from Bukoza included mercaptanes, fural, methanol, methylacetate, acetone and presumably also pentachlorophenol (which all together resulted in "life free" lower segment of Ondava river) and that from Chemlon consisted mainly from caprolactame. After few years the reality on the waterborn and considerable airborne pollution (sulfur and nitrogen oxides, ammonia etc.) was known only to very few people who were put to silence or, when asking for some preventive measures, were claimed as "enemies of the regime".

However, in spite of that there were some findings which turned the attention to potential risk and also few individuals appeared who simply could not keep silence. Among them was Dr. Michal Šak from the Clinic of Dermatology, Faculty of Medicine in Košice who treated 56 male employees from Chemko with severe dermatological illnesses including severe chloracne (Figure 2) as referred in his Thesis (Šak, 1978). Few years later the pathologist from Regional Hospital in Michalovce Dr. Milan Botka turned the attention to high stillbirths and newborn mortality rate mainly due to a considerable clustering of lethal Potter's syndrome. Such syndrome has been described by Potter (1946) as total absence or severe malformation of kidneys, facial malformations (Potter's face), pulmonary hypoplasia, skeletomuscular anomalies etc. Among 143 subsequent autopsies of newborns in Michalovce Hospital during 1975-79 Botka found 11 cases of Potter's syndrome who were either stillborn or died within few hours after birth. However, in 1980 he observed other 7 cases of Potter's syndrome among 28 subsequent autopsies of newborns who died within 10 days after birth which prompted him to write private report to be submitted to local health authorities (Botka, 1980). Based on this report, in 1998 we attempted to collect retrospective data on the prevalence of Potter's syndrome in Slovakia. However, since the State Institute of Health Information and Statistics is including such syndrome in the total sum of all inborn errors, we were only able to obtain data from the autopsy protocols of 19 willing hospitals covering the period of previous 10-20 years.

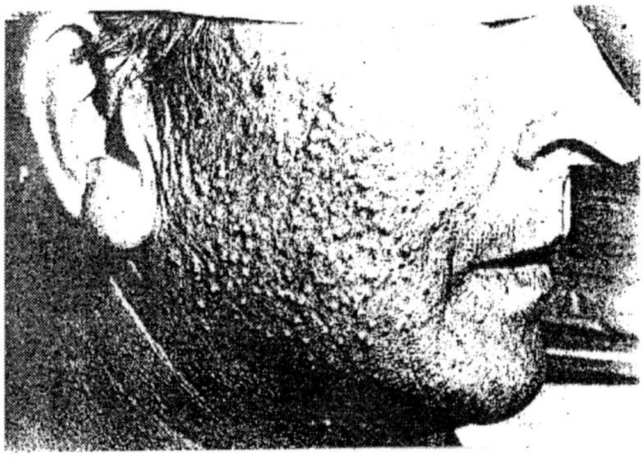

Figure 2. Employee of Chemko with severe chloracne after working in PCB manufacture for 17 years.

Thus, such incomplete data showed that in three districts where Chemko, Chemlon and Bukoza were located, a total of 83 cases of Potter's syndrome were found and, if one neighboring district was included, the total increased to 105. However, 38 cases were reported from 15 hospitals located in Western and Central Slovakia (Kočan et al., 1998). In contrast, several reports from various countries showed very low general incidence of Potter's syndrome. Thus, for instance, 92 cases in 625,000 consecutive births in British Columbia (Wilson and Baird, 1985) or 0.1- 0.3 cases/1000 births in Spain (Vinals et al., 1992) were found, but unexplained clustering which exceeded usual incidence of 1/6250 births was reported from northern Tennessee (Prouty and Myers, 1987). Although the interrelation between this syndrome and OCs pollution in Michalovce was not elucidated, there was namely a highly suspicious time coincidence of this syndrome with the period of highest PCB production and environmental pollution in that district. In fact, Botka's report became one of most important arguments used by the fist of those courageous enough who started to argue against PCB manufacturing which was finally terminated in 1984. However, since no other effective preventive measures were undertaken, such considerable pollution is still persisting.

Between 1959 and 1984 Chemko factory manufactured 21,500 tons of PCBs. However, it was not permitted to start any scientific survey either of the actual status of pollution or of possible adverse health effects on the population before so called "velvet revolution" in November 1989 which was stimulated by Gorbatchow's "perestroika" and resulted in the fundamental change of regime in this country.

PILOT SURVEYS

The first survey on serum organochlorine (OCs) levels in the population in five selected areas of western, central and eastern Slovakia was started in 1992 (Kočan et al., 1994 a,b; 2001). In the district of Michalovce the average level of 15 PCB congeners sum (IUPAC No. 28, 52, 101, 105, 114, 118, 123, 138^{+163}, 153, 156^{+171}, 157, 167, 170, 180 and 189) was 2-3 times higher (average of 4150 ng/lipid) than that in four western and central districts (average district values were from 1260 to 1880 ng/g lipid). Such findings were apparently related to heavy environmental PCBs pollution around Michalovce. However, even the PCBs level found in those control districts were relatively high at least as compared to those reported from some other countries (Table 1). Surprisingly, in all districts we also found considerably high levels of DDE (2,2'-*bis*(4-chlorophenyl)-1,1-dichloroethylene; average value from 5 districts being 6060 ng/g lipid) and HCB (hexachlobenzene; average value from 5 districts being 5380 ng/g lipid). Such levels of DDE and HCB were even higher than these of PCBs and apparently resulted from nearly unlimited use of those OCs in agriculture during several previous decades.

Later, extensive surveys of OCs level in Slovak population and mainly in polluted eastern Slovakia were conducted (Kočan et al., 2001). Table 2 shows that in environmental and food samples obtained in 1998 the level of PCBs from polluted area (POLL) was considerably higher than that from the area of background pollution in Stropkov and Svidnik districts (BCGR) located about 60 km upstream and upwind from POLL (Figure 1).

Table 1. The level of PCBs in blood serum (ng/g lipid) in general population of several countries

Reference	Country	Number. (M= males; F= females)	Age (years)	PCB analyzed	Mean (ng/g lip)	Median (ng/g lip)
Apostoli et al. (2005)	Italy	311 M	~ 50	24 congeners	897	705
Masuda et al. (2005)	Japan	151 M	> 46	35 congeners	386	
Kiviranta et al. (2005)	Finland	420 M	> 46	36 congeners	502	437
Jönsson et al. (2005)	Sweden-west	97 M	not shown	CB-153[A)]	170	---------
------ " ------	Sweden-east	92 M	not shown	CB-153[A)]	210	---------
------ " ------	Greenland-W	286 M	not shown	CB-153[A)]	220	---------
------ " ------	Greenland-E/N	29 M	not shown	CB-153[A)]	600	---------
Hunter et al. (1997)	USA	230 F	~ 50	not shown	774[C)]	---------
Turyk et al. (2006a)	USA	26 M	46 - 59	89 congeners	224	---------
Moysich et al. (2002)	USA	192 F	63 ± 7.2	69 congeners	618[C)]	---------
------- " -------	USA	78 MF	~ 50	----- " -----	682[C)]	---------
Turyk et al. (2006b)	USA	29 M	not shown	89 congeners	806[B)]	---------
Schell et al. (2004).	USA	115 MF	10 - 16	44 congeners	256[C)]	---------
Takser et al. (2005)	Canada	146 F	not shown	17 congeners	-------	~50[C)]
Meeker et al. (2007)	USA	341 M	20 - 54	57 congeners	222[B)]	---------

[A)] = CB-153 is one of the main *di-ortho*-PCB congeners which represents cca 10-30 % of total PCB (depends mainly from how many of most abundant congeners are analyzed); [B)] = geometric mean; [C)] = the authors indicated PCB concentration as ppb (pars per billion) which, for instance, means such as ng/g serum wet weight; such concentration was converted to ng/g serum lipid by multiplying (ng/g x 150) assuming the approximate concentration of one gram lipid per 150 ml (or 150 g) serum.

However, since the data shown in Table 2 were obtained in 1998 which was 14 years after the termination of PCB manufacturing in Chemko, they also show the status of continuing pollution, the main source being the sediments in the former Chemko industrial canal to the Laborec river. Although this opened canal is no more in use since mid eighties, after each rain or melting snow the water still flows to Laborec (see the line "Laborec down of Chemko" in Table 2) and from Laborec to Sirava lake (see the line "Laborec-Sirava channel" in Table 2). Thus, from the data shown in Table 2 including serum levels of OCs obtained in 1998 clearly show that heavy pollution is still persisting. However, as stimulated by Stockholm Convention the project of removing the pollution has been recently developed under United Nations Development (UND) in cooperation with UNIDO (United Nations Industrial Development Organization) and the government of Slovak Republic with the use of non-combustion destruction technology to be supported by Global Environment Facility (GEF). Such project is aimed to remove about 1000 tons of Chemko wastes, 1000 tons of PCB equipment and capacitators, 1500 tons of various other wastes and 40,000 tons of contaminated soil and sediments.

Table 2. The level of PCBs in human samples, selected "home-made" foods, wildlife and environment in polluted area of Michalovce and background pollution area of Svidnik/Stropkov

	Unit	POLLUTED AREA			BACKGROUND AREA		
		Range	Mean	N	Range	Mean	N
Human samples							
Human milk (in 1994) A)	ng/g lip	610 - 2790	1370	12	610 - 1230	822	5
Adipose tissue (in 1994) A)	-- " --	2560 – 16,930	6870	12	363 – 2190	1180	9
Human serum (in 2001)	-- " --	336 – 101,413	3107	1008	148 – 16,390	871	1038
Foods (home-made)							
Cow milk	ng/liter	9.9 – 341.7	52.7	23	7.9 – 61.6	19.2	7
Lard	ng/g lip	3.8 – 274.4	49.3	25	3.0 – 55.5	16.5	10
Chicks	-- " --	5.3 – 1703.7	374.3	6	6.7 – 51.9	24.6	6
Egg yolk	-- " --	36.1 – 55,204	2764.2	29	16.4 – 351.6	128.9	12
Wildlife							
Deer, roe, wild pig	ng/g lip	34.2 – 331.3	76.6	17	9.6 – 48.4	32.3	16
Fish (planktivores)	-- " --	94.9 – 587,410	223,550	11	731 – 2820	654	7
Fish (predators)	-- " --	4490 – 933,770	375,430	9	1200 – 3750	2270	4
Environment							
Air	pg/m^3	---------------	≤ 200	3	-------------	$\leq 1500^{B)}$	3
Soil	ng/g dry	4.9 – 2.0	6.9	4	4.1 – 28.0	17.5	11
Laborec up of Chemko	ng/liter	9 – 16	3	2	-------------	-------	-----
Laborec down of Chemko	-- "--	49 – 92	69	3	-------------	-------	-----
Channel Chemko – Laborec	-- "--	1070 – 2830	2120	6	-------------	-------	-----
Sirava lake	-- "--	12 – 26	21	6	-------------	-------	-----
Ondava river	-- "--	---------------	----------	-----	7 - 11	9	5
Domasa lake	-- "--	---------------	----------	-----	6 - 8	7	3
Sediments							
Laborec up of Chemko	ng/g dry	51 – 54	------	2	-------------	-------	-----
Laborec down of Chemko	-- "--	830 – 9700	6850	3	-------------	-------	-----
Chemko – Laborec channel	-- "--	3.000,100 C)	------	2	-------------	-------	-----
Laborec – Sirava channel	-- "--	2600 – 9400	7250	3	-------------	-------	-----
Sirava – Laborec channel	-- "--	220 – 600	490	3	-------------	-------	-----
Sirava lake	-- "--	1200 – 3900	2800	6	-------------	-------	-----
Ondava river	-- "--	--------------	---------	-----	5 - 64	24	4
Domasa lake	-- "--	--------------	---------	-----	4 - 12	7	4

Explanatory comments: Chemical waste was flowing from Chemko via an opened channel (about 5 km long) to Laborec river (see also Fig. 1). Artificial Sirava lake was fed by polluted water from Laborec river upstream of Michalovce city via opened channel Laborec-Sirava. Some water from Sirava was flowing back to Laborec downstream of Michalovce via opened channel Sirava-Laborec.

A) = sum of 9 PCB congeners (28, 52, 101, 138, 153, 180, 105, 118, 156); all other data show a sum of 15 PCB congeners: 28, 52, 101, 105, 114, 118, 123, 138^{+163}, 153, 156^{+171}, 157, 167, 170, 180, 189;

B) = as estimated in the area of Chemko factory.

C) = this shows a tremendous high concentration of PCBs in the sediments of about 5 km long channel.

Our attention has been focused to strikingly high PCBs level in home-produced chicks and egg yolk, but namely to that in fish from polluted Laborec river and Sirava lake which we later showed as one of the main sources of extremely high PCBs level in human serum (Langer, 2007a). Similarly as shown in Table 2 for PCBs, Chovancová et al. (2005) recently found strikingly high values of PCDDs, PCDFs and dioxin-like non-*ortho* and mono-*ortho*-PCBs in home-produced eggs and chicks raised with access to soil as compared to these from commercial poultry farms which was in agreement with the observations by Goldman et al. (1999) in California. From the same reasons much higher OCs level was found in milk of small private farmers compared to that from commercial farms. Serum levels of 15 PCB congeners, DDE, HCB, DDT and α-. β-, γ-HCH (hexachlorocyclohexane), as obtained in 2001 from 2046 adults within the PCBRISK survey to be discussed bellow, were significantly higher in POLL than in BCGR (Petrík et al., 2006; Langer et al., 2007b). Similar significant difference between those two areas was later also found for 4 non-*ortho*-, 8 mono-*ortho*- and 45 other *ortho*-substituted PCB congeners (Jursa et al., 2006) as well as for selected OH-PCBs and $MeSO_2$-PCBs metabolites (Hovander et al., 2006). Overview of those findings is shown in Table 3.

With respect to the above findings on high serum OCs levels in POLL and reports by others on increased goiter frequency in subjects consuming rice oil poisoned by PCBs and furans such as Yusho patients in Japan (Murai et al., 1987) and Yucheng patients in Taiwan (Guo et al., 1999), we decided to use a long-term experience of one of us (M.T.) in thyroid ultrasound diagnostics and thus to conduct the survey of thyroid status in the Chemko employees and inhabitants of Michalovce. Actually, since for several foregoing years we conducted thyroid surveys by ultrasound in adolescents and adults in several districts of eastern Slovakia which were aimed to other purposes (Tajtáková et al., 1988, 1992, 1998; Langer et al., 1994), we used the findings obtained in such districts with background pollution as control data. Hence, we measured thyroid volume by ultrasound for the first time within such PCBs related thyroid survey, since this appeared much more precise method than the palpation by hand used by the above mentioned authors (Murai et al., 1987; Guo et al., 1999). In addition, since Bahn et al. (1980) found increased frequency of microsomal antibodies (identical with present thyroperoxidase antibodies) in workers exposed to polybrominated biphenyls, we also attempted to measure several thyroid antibodies.

Table 3. Overview on serum levels of major organochlorines and selected metabolites in two areas

SUBSTANCE	UNITS	BACKGROUND AREA			POLLUTED AREA		
		N	Mean ± S.E.	5 % - 95 %	N	Mean ± S.E.	5 % - 95 %
PCBs	ng/g lip	1038	871 ± 23	354 - 1698	1008	354 - 1698	3107 ± 170
DDE	ng/g lip	1038	1797 ± 45	350 - 4493	1008	350 - 4493	3160 ± 83
HCB	ng/g lip	1038	828 ± 25	95 - 2237	1008	95 - 2237	1015 ± 40
PCDDs [A)]	pg/g lip	174	22.1 ± 0.7	10.9 - 38.2	133	41.1 ± 3.8	9.3 - 120.6
OH-PCB[B)]	ng/dl	174	87.4 ± 3.8	23.5 - 343.7	133	341.9 ± 39.5	61.9 - 1036.4
MeSO2-PCB [B)]	ng/dl	174	1236 ± 81	126 - 10,649	133	6,798 ± 1023	462 – 27,107
MeSO2-DDE [B)]	ng/dl	174	4627 ± 396	186 - 27,859	133	9,412 ± 965	690 - 33,690

A) = sum of PCDDs, PCDFs, *non-ortho-* and *mono-ortho*-PCBs;

B) = data obtained within PCBRISK project by Hovander et al. (2006).

Within such first survey (Langer et al., 1998) we found significantly higher thyroid volume (ThV) in 238 Chemko employees (median of 17.3 ml, 75 % of 22.8 ml) than that in 486 adults from several much less polluted districts as examined previously (median of 11.3 ml, 75 % of 15.5 ml; $p<0.001$). In addition, we also found significantly higher ThV in 454 17-yr old adolescents from the polluted city of Michalovce (190 boys and 264 girls) than that in 965 referents of the same age (434 boys and 531 girls) from much less polluted districts (median of 9.0 ml, 75 % of 11.0 ml vs. median of 7.7 ml, 75 % of 9.8 ml, respectively; $p<0.001$). Although in this survey the level of OCs in blood has not been estimated, very high level in Chemko employees and citizens of Michalovce has been assumed which followed from previous survey by Kočan et al. (1994a) and was later supported by our second survey (Langer et al., 2003).

In addition, within the same first survey we found increased prevalence of TPOab in 190 women employed in Chemko as compared to 482 referent women (28.4 % vs. 20.5 %; $p<0.05$) and similar difference (21.3 % vs. 14.6 %, $p<0.05$) in the prevalence of thyroglobulin antibodies (TGab) in similar groups of women. In addition, increased frequency of thyrotropin receptor antibodies (TRab) was found in 238 male and female Chemko employees compared to 238 age and sex matched referent subjects (10.5 % vs. 2.5 % – $p<0.001$).

Within our second field survey in 1998 conducted by joined chemical and medical staff our previous findings on increased ThV and frequency of TPOab were supported (Langer et al., 2003). In cooperation with several district physicians we examined a total of 101 Chemko employees and few subjects living in a close vicinity of the factory (59 men and 42 women; median age of 51 years) and 360 referent adults (180 men and 180 women; median age of 46 years) from BCGR as well as from four villages south of Sirava lake (Figure 1). As shown in Table 4, the average PCBs level in Chemko employees was nearly six times higher than that in BCGR area of Stropkov and about three times higher than that in the villages south of Sirava lake. However, the level of HCB was about the same in all areas, while the average DDE level was about the same only in three southern areas and that in each area was significantly higher compared to northern less agricultural area of Stropkov. Such nearly equally high levels of HCB and DDE were similar to these found in 1992 in the whole Slovakia (see above) and apparently resulted from almost unlimited use of pesticides in the agriculture. It may be supposed that similar levels could be found even in surrounding countries (e.g. Poland, Hungary) in which similar regime was in action until 1989.

Table 4. Serum organochlorine levels in four areas examined within 1998 survey

Group	Area*)	Number of subjects	Serum organochlorine level (means ± SE; ng/g lipid)		
			PCB	DDE	HCB
1	Chemko and vicinity	101	$6493 \pm 705^{2,3,4}$	3636 ± 274^{4}	1859 ± 157
2	Michalovce	91	$4571 \pm 684^{3,4}$	4222 ± 337^{4}	1799 ± 142
3	South of Sirava lake	58	2233 ± 170^{4}	3763 ± 316^{4}	1024 ± 246
4	Stropkov	205	1230 ± 62	2317 ± 156	1631 ± 136

*) – for the location of area see Figure 1.

2,3,4 – statistical significance ($p<0.05$) versus the group with the appropriate number.

Since possible role of PCBs and other OCs in the etiology of cancer has been the topic of numerous surveys, we also conducted an epidemiological cross sectional study of cancer incidence between 1985 and 1994 in the districts of Michalovce and Svidník. We calculated standardized incidence ratios (SIRs) and 95 % confidence intervals for each district using indirect standardization and rates from eastern Slovakia (Pavuk et al., 2004a). Among males from Michalovce there was an excess of cancer of the tongue, stomach, lung, testis and kidney, while in females an excess of lip, stomach and lung cancer was found. However, lower than expected was the incidence of prostate cancer in Michalovce. In contrast, in Svidnik an excess of peritoneal and laryngeal cancer in males and that of kidney and thyroid cancer in females was found which was not observed in Michalovce. Moreover, in 24 breast cancer patients from Michalovce district positive association of such cancer with high levels of DDE was found (Pavuk et al., 2004b).

As shown above, in eastern Slovakia multiple OCs pollution has been found. Thus, in addition to PCBs there were also high levels namely of DDE and HCB, but also of chlorinated dibenzo-dioxins and dibenzo-furans (Table 4). Actually, several components may participate in the adverse health effects, but the question of such possible "cocktail effect" still remains to be elucidated. Although the interactive effects of all pollutants cannot be excluded, the appropriate toxic action of each of them cannot be satisfactorily quantified. Thus, it is difficult to specify to which extent each substance is responsible for certain biological effect.

Since PCBs were the main target of this survey and also showed the largest range of values, they were used as a marker of the whole spectrum of OCs and thus all other data to be presented bellow were also stratified in terms of PCBs level. However, from this follows that simultaneously changing levels of all other OCs should be always kept in mind. Although several statistical approaches are currently using sophisticated software, all findings thus obtained must not be necessarily compatible with biological interrelations and thus they still seem to show rather superficial associations, while several black boxes still remain between the input, for instance such as blood level of certain OCs and final endpoint such as some hormone or antibody level.

Extensive Surveys in Adults

As based on the data obtained within above described pilot surveys, an international study PCBRISK funded by European Commission was conducted in 2001-04. Thus, in polluted area of Michalovce district (POLL) and area of background pollution (BCGR) of Svidnik and Stropkov districts (Figure 1) a total of 2046 adults (834 men and 1212 women aged 21-75 years) and 433 children 8-9 year-old (220 boys and 213 girls) were examined which were recruited by 28 district practitioners as described in more detail by Langer et al. (2007a). The methods used and the main findings will be quoted and described bellow. In addition, after PCBRISK study the next international study SPEECD (Slovakia PCB Exposure and Early Childhood Development) funded by National Institute of Health (USA) was started under the coordination by professor Irva Hetz-Piccioto (Center of Children Environmental Health, University of California at Davis, CA), the first published findings obtained within such study will be also refered below.

Thyroid Volume and Structure

As shown above, within pilot surveys we repeatedly found the association of increased ThV as estimated by ultrasound with increased level of PCBs used as a marker of OCs cocktail. Since the level of OCs depends on the duration of exposure, it is obviously increasing with age. In spite of that, we showed that there is some additional effect of PCBs (or of OCs cocktail) on ThV other than that of age (Langer et al., 2007b), as also demonstrated in Figure 3. However, it should be noted that ThV in general, although being higher in POLL, was relatively low in both POLL and BCGR areas. From this follows that such subtle differences in ThV related to different OCs level may be detected only by ultrasound, while they can hardly be observed by such crude method as the palpation.

Actually, due to well monitored mandatory consumption of iodized salt since 50 years ago, Slovakia belongs to iodine replete countries and thus any interfering effect of iodine deficiency on the increase of ThV may be excluded. This was also supported by the European study (Delange et al., 1997) in which Slovak schoolchildren showed second lowest ThV and second highest urinary iodine among several European countries. Within our previous thyroid surveys we always found a low ThV and normal urinary iodine in schoolchildren and, similarly, also within PCBRISK study normal urinary iodine was found in adults (Langer et al., 2007b). So far, the effect of OCs on increased prevalence of goiter by palpation has been found in exposed humans consuming OCs poisoned rice oil in Japan (Murai et al., 1987; Aoki, 2001) and in Taiwan (Guo et al., 1999). Since these two countries are iodine replete namely due to high consumption of sea food, any interfering effect of iodine deficiency should be also excluded. In contrast, however, in some other countries only mild iodine deficiency itself could result in increased ThV and thus in false positive findings when evaluating the role of OCs on the thyroid.

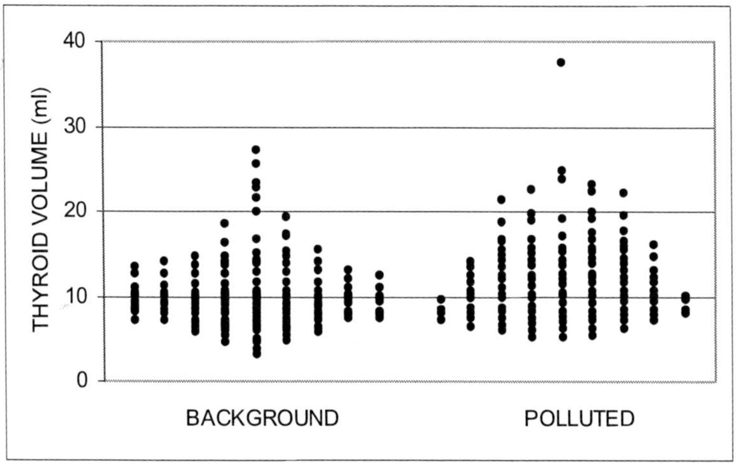

Figure 3. Thyroid volume in males aged 51-60 years from the area of background pollution (left panel, n = 156; mean ± SD = 10.27 ± 4.03 ml) and from polluted area (right panel, n = 133; mean ± SD = 12.59 ± 5.18 ml; p<0.0001) as observed within PCBRISK survey.

Nevertheless, the mechanism of such OCs effect on ThV still remains to be elucidated. Actually, in rats fed various OCs several thyroid morphological changes suggesting thyroid activation were found such as increased epithelium height, decrease of colloid area, increased numbers of intracellular colloid droplets and lysosomes (Akoso et al., 1982; Capen and Martin, 1989; Ness et al., 1993; Saeed and Hansen, 1997; Wade et al., 2002; Nishimura et al., 2003). However, using electron microscopy Collins and Capen (1980) observed a considerable difference between the effect of thyroid stimulation (e.g. by injection of exogenous thyrotropin) or thyroid inhibition (e.g. by administration of exogenous thyroxine which inhibits endogenous TSH release) on one side and the effect of PCBs feeding on the other. Thus, they concluded that antithyroid effect of PCBs is direct and specific. Such view was later supported by Capen and Martin (1987) who considered some alterations of thyroid ultrastructure in rats fed PCBs to result from a direct damage to the thyroid tissue.

Since one of possible outcomes of impaired growth mechanism of thyroid cells could be the thyroid cancer (ThCa), several studies evaluated such cancer prevalence in OCs exposed population, but only few of them possibly found some relevant data. Thus, Grimalt et al. (1994) reported three cases of ThCa which appeared within 10 years among 5003 inhabitants of a community located nearby of OCs producing factory. In 13,482 workers or sprayers exposed to herbicides and chlorophenols – among others - increased risk for thyroid cancer was reported (Saracci et al., 1991). Furthermore, increased mortality from ThCa among 2885 US workers employed in capacitator producing plant was reported by Mallin et al. (2004) or in agricultural region of Minnesota with extensively used herbicides by Schreinemachers et al. (1999). At 15 years after the dioxin "Seveso accident" Bertazzi et al. (1997) observed one thyroid cancer vs. 0.2 expected. Using Norwegian excellent Cancer Registry, Frich et al. (1997) found significantly elevated risk of ThCa in 40,849 women (174 cases per 1.210,683 person/years; SIR 1.91, CI 1.65-2.21) whose spouses were employed in fishing, whaling or sealing work. Although the authors did not analyze the reasons for such high incidence, it may be speculated that this may be well related to high OCs level in fish from northern Atlantic. Although there are some other reports on increased prevalence of ThCa, the interrelation with OCs is lacking. Among 2046 examined adults we found a total of 98 cases with thyroid nodules, among those 10 nodules with the diameter of 21-50 mm were found in POLL versus only 4 such cases in BCGR (Figure 4). Among those 98 subjects only 33 agreed with further examination by gammagraphy and, moreover, among those 33 only 8 agreed with thin needle biopsy. Finally, independently of such further examinations only 5 of those with large nodules agreed with surgical thyroidectomy. Among them three had benign follicular adenoma, one had benign colloid goiter and one had benign multinodular goiter.

Thus, we did not find any ThCa among the examined subjects, but there were three cases of ThCa already diagnosed and treated by others among their first degree relatives (2 from POLL and one from BCGR) as reported by questionnaires.

One of plausible possibilities of thyroid growth promoting effect may be possible estrogenic activity of some OCs. Since estrogen receptors were repeatedly found in normal and disordered thyroid (Kawabata et al., 2003; Tavangar et al., 2007) and the binding of several OCs (e.g. PCBs and OH-PCBs) to such receptors *in vitro* has been demonstrated (Shiraishi et al., 2002; Kitamura et al., 2005), it may be assumed that the thyroid growth could be also stimulated by estrogen-like effects namely of DDE or lower chlorinated PCB congeners (Arnold et al., 1996; Wolff et al., 1997; Hansen, 1998). However, the activation of aryl hydrocarbon receptor (AhR – specific receptor for several OCs), for instance by dioxin or coplanar PCBs, may further activate also the function of intracellular hormone signaling pathways including these of estrogens (Matthews and Gustafson, 2006).

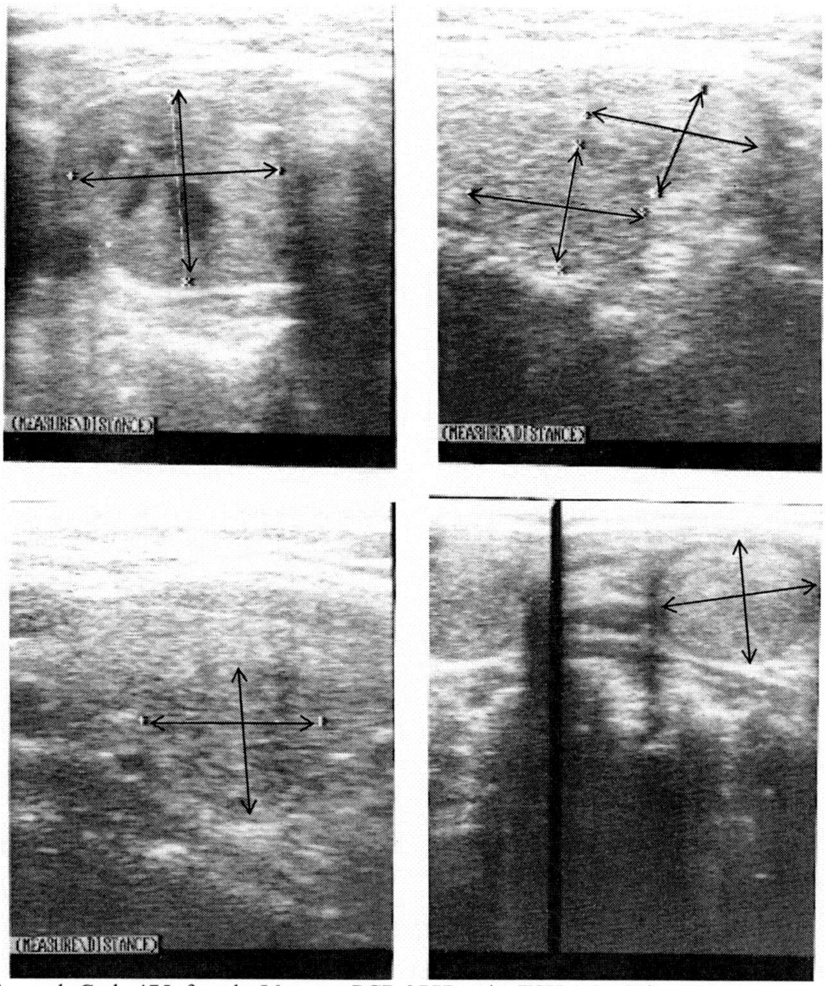

Upper left panel: Code 475, female 56 years; PCB 2777 ng/g; TSH 1.5 mU/L; FT4 17.6 pmol/L; TG 56 ng/mL; thyroid volume 37.5 ml; nodule size: 29 x 24 ml; gammagraphy: cold nodule; thin needle biopsy: no signs of malignity; histological finding after thyroidectomy: follicular adenoma.

Upper right panel: Code 269, female 50 years; PCB 1564 ng/g; TSH 4.2 mU/L; FT4 13.6 pmol/L; TG 67 ng/mL; thyroid volume 40.9 ml; size of individual nodules: 24x16 and 26x15 ml; gammagraphy: refused; thin needle biopsy: refused; histological finding after thyroidectomy: multinodular goiter, autoimmune (Hashimoto's) thyroiditis-

Lower left panel: Code 374, male 58 years; PCB 1972 ng/g; TSH 3.1 mU/L; FT4 15.6 pmol/L; TG 154 ng/mL; thyroid volume 41.2 ml; nodule size: 15x12 and 15x14 mm; gammagraphy: unhomogenous radioidine uptake (!); thin needle biopsy: refused; thyroidectomy: refused.

Lower right panel: Code 176, male 22 years; PCB 2094 ng/g; TSH 1.6 mU/L; FT4 21.7 pmol/L; TG 574 ng/ml; thyroid volume 25.5 ml; nodule size: 34x24 mm; gammagraphy: cold nodule; thin needle biopsy: suspected papillary cancer; thyroidectomy: refused. Note: request by phone in 2007 (6 years after the examination): no urgent problems with thyroid.

Figure 4. Selected cases of thyroid nodules from polluted area as obtained by ultrasound.

Estradiol was also found to stimulate the growth of FRTL-5 rat thyroid cell line containing estrogen receptors (Furlanetto et al. 1999). Rajapakse et al. (2002) observed the activation of human estrogen receptor alpha in response to combining effect of suboptimal concentrations of several xenoestrogens. Since PCBs were found to inhibit estrogen sulfotransferase – intracellular enzyme converting estrogens to their metabolites (Kester et al. 2000) - such mechanism could also influence estrogen availability in thyroid tissue.

Estrogens and phytoestrogens are well known to enhance the process of thyroid cell growth and tumorigenesis (Manole et al., 2001; Kung et al., 2002; Madej et al., 2002) and human thyroid was found to exhibit the potential for estradiol synthesis and autocrine or paracrine estrogen responsiveness which may contribute to such process (Valle et al., 1999; Vivacqua et al. 2006). Such effects may be supported by higher incidence of ThCa either in women during reproductive years compared to men or in postmenopausal women treated by estrogens for a long time compared to those not treated, except for just a low dose estrogen replacement (Persson et al., 1996). Manole et al. (2001) found estradiol a potent mitogen for benign and malignant thyroid tumor cells. Recently, Araujo et al. (2006) found considerably higher thyroid weight and follicular cell height after treating ovariectomized rats with conjugated equine estrogens (200 μg/kg) and tamoxifen (1 mg/kg) daily for 50 days as compared to untreated ovariectomized controls. All such findings support the view that estrogens could contribute to the growth of malignant and possibly also normal thyroid which might be potentially valid also for OCs displaying estrogenic activity. Finally, close interrelations between female gender and thyroid may be supported by very well known high prevalence of all thyroid disorders which, in each ethnic, is at least about four times higher than that in males.

Among other possibilities of thyroid growth stimulating effect might be the disruption of cellular membranes fluidity by strongly lipophilic OCs (Tan et al., 2003) thus possibly influencing the cross talk between thyroid membrane antigens and circulating lymphocytes. Such mechanism may further facilitate the development of autoimmune thyroiditis at least in hereditary predisposed individuals which also could contribute to the thyroid growth. Nevertheless, any possible other specific effect of OCs on thyroid growth cannot be excluded.

Thyroid Hormones in Blood and Pituitary-Thyroid Interrelations

Several authors found the binding of PCBs and their hydroxylated metabolites to transthyretin (TTR) – one of two major plasma thyroxine carrier proteins in humans and also in rodents (Van den Berg et al., 1991; Van Raaij et al., 1991; Lans et al., 1993; Cheek et al., 1999; Chauhan et al., 2000). Since that time such phenomenon has been considered to play a crucial role in frequently observed acute decrease of total thyroxine (TT4) level in rats which was assumed to result from the displacement of that hormone from carrier protein (Byrne et al., 1987; Van Raaij et al., 1993; Seo et al., 1995; Fowles et al., 1997; Kuriyama et al., 2003). However, since the doses of PCBs used in such experiments (e.g. mg/kg body weight) were perhaps by an order of magnitude higher than the assumed intake in humans, the question remains whether the displacement itself might influence the thyroid hormone level in humans. In addition, so far no binding of PCBs was found to TBG (thyroxine binding globulin) which is the major thyroxine binding protein in humans (Lans et al., 1994).

Moreover, the decrease of blood TT4 level in animals was further enhanced by the increase of T4 hepatic conjugation and biliary excretion due to induction of hepatic UDP-glucuronyltransferase after few weeks of high dose PCBs feeding as summarized by Liu J. et al. (1995), while later Hood and Klaasen (2000) specified such inducers of thyroxine (T4) and triiodothyronine (T3) glucuronidation in more detail. Schuur et al. (1999) demonstrated a 5-fold increase in glucuronidation activity in rats after a single high dose of 2,3,7,8-TCDD (10 μg/kg) and concluded that decreased thyroid hormone level in blood was predominantly due to increased activity of that hepatic enzyme. However, it still remains to be elucidated whether such glucuronidation plays any role in humans. In addition to a decrease of T4 and T3, after long-term feeding rats with PCBs also the increase of pituitary thyrotropin (TSH) was found which was apparently the response to a prolonged T4 level decrease (Byrne et al. 1987; Wade et al., 2002). Short-term effects of non-persistent *ortho*-PCBs 95 and 101 were studied by Khan et al. (2002) who fed a single dose of such OCs to twenty-one-day-old weanling female rat pups for two days. They found significantly decreased serum thyroxine and hypothalamic dopamine, while no changes in serum triiodothyronine, thyrotropin and prolactin appeared. However, increased thyroid epithelial cell height and decreased colloid area was found only after PCB-95.

However, so far several controversial data on possible effects of OCs on blood thyroid hormone level in humans were published and this question still remains to be definitely clarified. On the other hand, available data do not exclude such association (Hagmar, 2003). It should be noted that a great majority of such human studies faced the cohorts exposed to the cocktail of OCs consisting of the same major substances (PCBs, DDE and HCB). Several human studies were possibly influenced by those findings of decreased TT4 level in rats after high doses of PCBs (see above) and from such reason a similar decrease was expected also in humans. Thus, negative association of TT4 with PCBs and/or DDE was observed in Great Lakes fish consumers (Persky et al., 2001; Turyk et al., 2006a, 2007), that with PCBs, DDE and HCB in Akwesasne Mohawk youth (Schell et al., 2004) and that with HCB in professionally exposed workers (Sala et al., 2001), while no change of TT4 level was observed in New York anglers (Bloom et al., 2003). No difference in TT4 level was found between referents and dioxin exposed US workers, but the latter group showed significantly increased free thyroxine index (Calvert et al., 1999). In our pilot study we did not find any difference in TT4 level between highly exposed Chemko employees and referents (Langer et al., 1998). So far, at the only occasion to measure also the level of PCDDs, Turyk et al. (2007) found negative association with TT4.

Possible effects of major OCs on free T4 (FT4) and total T3 (TT3) were studied by several authors (Osius et al., 1999; Sala et al., 2001; Hagmar et al., 2001a, b; Pelletier et al., 2002; Schell et al., 2004; Takser et al., 2005; Rylander et al., 2006a) who reported either no association or negative association, while positive association with both FT4 and TT3 was found only for DDE by Meeker et al. (2007) who also observed negative association between PCB-153 and TT3. The effect of PCBs on free T3 level was perhaps studied only by Osius et al. (1999) who found negative association.

Within PCBRISK survey we observed predominantly negative correlation of low categories of PCBs, DDE and HCB level with FT4 and TT3, while positive correlation with FT4 (Figure 5) and TT3 was found in the range of higher categories of such OCs level (Langer et al., 2007c).

The association observed between various OCs and TSH level was found mostly positive such as that with mono-*ortho* PCB 118 in schoolchildren (Osius et al., 1999), that with PCBs, but not with DDE and HCB in young Mohawks (Schell et al., 2004), that with DDE but not with PCB-153 in Baltic fishermen (Kiviranta et al., 2006a) and that with TCDD in Vietnam veterans (Pavuk et al., 2003).

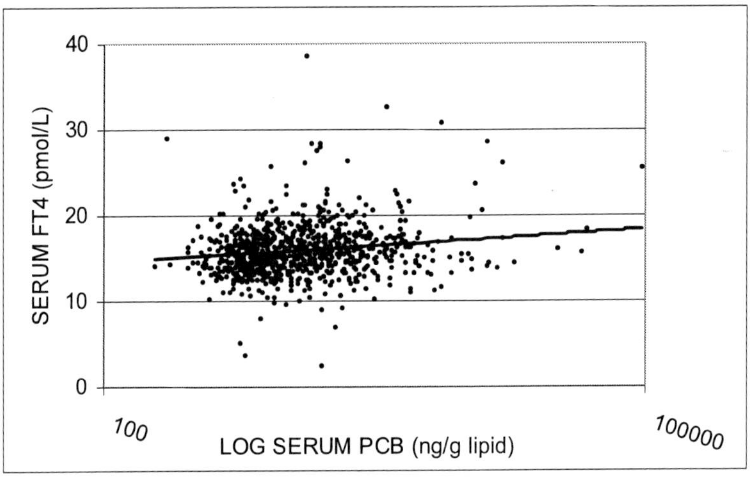

Figure 5. Positive correlation between PCBs (log scale) and free thyroxine (FT4) in women aged >40 years (n = 881, Pearson's r for logarithmically transformed values = 0.137; $p<0.001$) as observed within PCBRISK survey.

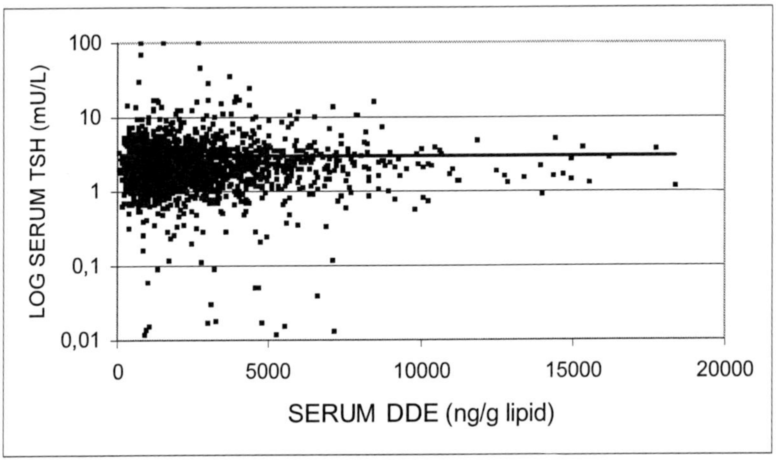

Figure 6. No correlation between the level of DDE and TSH (log scale) in pooled males and females from both areas (n = 2046; Pearson's r for log values = -0.0013) as observed within PCBRISK survey.

The interrelations of TSH with those major OCs observed by Persky et al. (2001) in Great Lakes fish consumers were inconsistent, while recently Turyk et al. (2007) reported positive association of TSH with PCBs and also with toxic equivalents of PCDDs, PCDFs and coplanar-PCBs in similar fish consumers. In breast fed babies of mothers exposed to dioxin a transient increase of serum TSH level was reported by Pluim et al. (1993) and Koopman-Esseboom et al. (1994).

Perhaps the only report published so far on negative association between DDE and TSH in adults appeared recently (Meeker et al., 2007). However, in spite of very large range of OCs level in 2046 adults, we could not find any correlation of TSH either with PCBs (5% to 95% range of 417 – 5480 ng/serum lipid; Rádiková et al.. 2007) or with DDE (5% to 95% range of 403 – 6851 ng/serum lipid; Figure 6).However, it should be noted that several authors (Osius et al., 1999; Bloom et al., 2001; Hagmar et al., 2001b; Schell et al., 2004; Takser et al., 2005; Meeker et al., 2007) reported relatively low median (cca 50 – 200 ng/g lipid) and also maximum serum PCBs levels (cca 200 - 900 ng/g lipid) in their cohorts. Thus, a majority of their cases apparently had PCBs level <1000 ng/g which, in our PCBRISK survey, resulted either in negative or no correlation with FT4 and TT3 (Langer et al., 2007c) thus resembling the data reported by the above authors. From such reason there still remains a question whether namely the negative associations of such relatively low PCBs level with FT4 and TT3 result from the true effect of PCBs or rather from some effects other than these of PCBs. However, considerably higher PCBs level (i.e. cca 350 – 1100 ng/g lipid for medians and cca 1100 – 3600 ng/g lipid for maximum) as found by the authors who examined either fishermen and fish consumers (Persky et al., 2001; Hagmar et al., 2001a; Rylander et al., 2006a; Turyk et al., 2006a,b) or industrially exposed workers (Sala et al., 2001) were more or less in agreement with predominantly positive associations of OCs with FT4 and TT3 as also found in the range of upper three quintiles of PCBs level in our PCBRISK study (Langer et al., 2007c).

However, although we did not find any negative association between PCBs and TSH in our total cohort, among the population from polluted area we detected a group of 13 subjects (4 males and 9 females aged 44-60 years) with very high level of PCBs (median of 7438 ng/g lipid) and, at the same time, with hormonal levels that showed hyperthyroid status. Thus, their mean FT4 level of 23.3 pmol/L (upper limit of reference range 22.0 pmol/L) was increased and, vice versa, their TSH level was decreased, being <0.21 mU/L in eleven among those 13 subjects (lower limit of reference range = 0.27 mU/L). Their average body mass index (BMI) of 26.2 was slightly above the normal value of 25.0 which shows that such hyperthyroid status was subclinical in reality (Manji et al., 2006). Thus, we classified such cases as "high PCB related subclinical hyperthyroidism" which apparently corresponds to the definition recently published by Monaco (2003) as "subclinical hyperthyroidism is an asymptomatic status, in which circulating levels of FT4 and TT3 are increased, but supersensitive TSH is decreased. It is not any disease entity, but a mild stage of thyroid hyperfunction."

Thyroid Autoimmune Disorders

With the exception of thyroid tumors, virtually all major thyroid diseases (i.e. both the hypothyroidism and hyperthyroidism) are of autoimmune origin. Thus, progressing destruction of thyroid follicular cells by anti-thyroperoxidase (TPOab) and anti-thyroglobulin (TGab) autoantibodies is the main cause of hypothyroidism, while the stimulation of thyrotropin receptor by thyrotropin receptor antibodies (TRab) results in high increase of thyroid hormone biosynthesis and release thus producing hyperthyroidism (thyrotoxicosis or Graves disease). Since several OCs were found to possess immunotoxic effects, it may be suggested that they could aggravate the development of such thyroid autoimmune disorders namely in hereditary predisposed individuals.

Numerous reports on general immunotoxic effects of OCs were published such as, for instance, impaired leucocyte function in workers formerly exposed to dioxins (Jung et al., 1998), increased frequency of otitis media and respiratory tract infections in Inuit children prenatally exposed to PCBs (Dewailly et al., 2004; Dallaire et al., 2006) and, similarly, decreased antidiphtheria vaccine response in children from Faroe Islands exposed to high

PCBs level in maternal milk (Heilmann et al., 2003). In addition, Baccarelli et al. (2002) found decreased IgG level in the population exposed to dioxin after industrial accident in Seveso (Italy) and Cooper et al. (2004) reported similar finding in American farmers with increased levels of DDE. Decreased plasma IgG was also found in OCs exposed polar bears (Bernhoft et al., 2000) and multiple lesions of immune system were observed after accidental human poisoning by HCB in Turkey (Michielsen et al., 1999). Decreased level of interferon-gamma, interleukin-4 and DR+ blood cell count in subjects occupationally exposed to PCBs and other organochlorines was found by Volker et al. (2001a,b). In subjects living near a pesticide dump site Vine et al. (2000) found decreased lymphoproliferative activity. In Flamish adolescents exposed to PCBs and dioxin-like compounds increased IgA was detected, while IgG and certain specific IgEs (e.g. against cat dander, house dust and grass pollen) were decreased (van den Heuvel et al. 2002). Recently the pathogenesis of autoimmune thyroid diseases that are caused by an interaction between susceptibility genes and environmental triggers has been extensively reviewed by Tomer and Davies (2003).

As mentioned above, anti-microsomal antibodies (which is the obsolete term for TPOab) were found in few workers employed in polybrominated biphenyl producing factory by Bahn et al. (1980). As far as we are aware, since that time no findings on increased thyroid antibody prevalence in humans exposed to OCs were published. However, considerably increased frequency of TPOab positively associated with increased PCBs level was found in all our surveys conducted so far (Langer et al., 1998, 2003, 2007b; Rádiková et al., 2007). As shown in Figure 7, such increase was much more striking in males than in females, namely in the 4th, 5th and 6th decade of age. Although positive TPOab level in serum is the humoral marker of autoimmune process and, at the same time, also a predictor of possible overt hypothyroidism in the future, each subject with actual positive value must not yet be clinically hypothyroid.

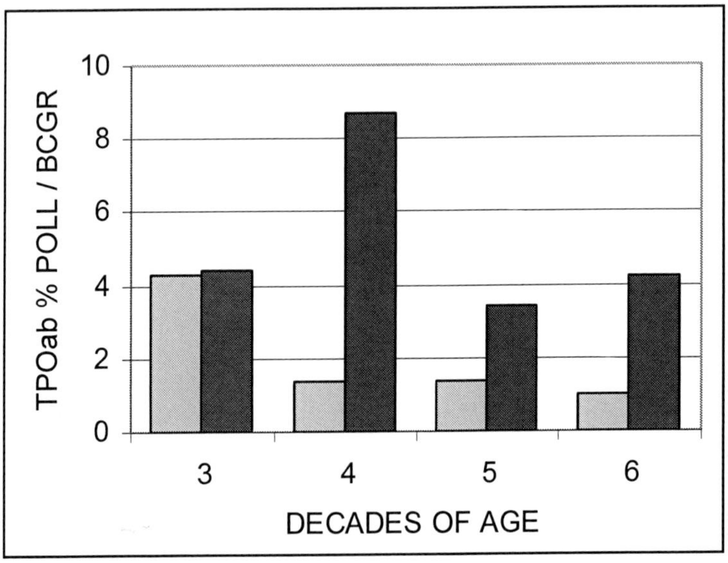

Figure 7. Ratio of percent cases with positive level of thyroperoxidase antibodies (TPOab) between polluted and background area in four decades of life (21-30, 31-40, 41-50 and 51-60 years of life) for males (black columns) and females (gray columns) as observed within PCBRISK survey. *** = $p<0.001$ or **** = $p<0.0001$.

However, the stage of progressing hypothyroidism resulting from the action of TPOab may be seen by ultrasound which reveals the extent of thyroid injury by hypoechogenicity of thyroid ultrasound image. Actually, we found significantly increased frequency of TPOab and hypoechogenicity coincidence between 5th and 1st quintile of serum PCB level in males (16/223 = 7.2 % vs. 1/101 = 1.0 %; $p<0.025$), while that in females was not significant (27/184 = 14.7 % vs. 35/308 = 11.3 %), although in terms of absolute numbers it was higher than that in males (Langer et al. 2007b). Thus, it appeared that thyroid autoimmunity should be included in the cluster of immune system injuries by organochlorines.

Impairments of Glucose Metabolism and Diabetes

Among several adverse effects of OCs on human health, the attention was also focused on impairment of glucose metabolism and prevalence of diabetes mellitus. Among the first epidemiological studies with positive findings was that of increased fasting and 2-hr postprandial glucose level in 989 Vietnam veterans spraying TCDD containing Agent Orange as compared to 1276 non involved veterans (Henriksen et al., 1997). Later Michalek et al. (1999) reported increased prevalence of diabetes together with serum glucose and insulin abnormalities in a cohort of veterans with the highest level of TCDD. Similarly, Cranmer et al. (2000) showed that plasma insulin levels at fasting and after glucose load were significantly increased in the subjects with high blood TCDD level thus possibly showing insulin resistance and, finally, Kern et al. (2004) supported such observation by finding high insulin resistance in 29 veterans with high TCDD level (median of 45.3 ppt) versus age matched veterans with low TCDD level (median 3.6 ppt).

The association of hexachlorodibenzo-*p*-dioxin, PCB-126 and DDT with diabetes in about 2100 participants of NHANES III study was reported by Everett et al. (2007). More detailed study by Lee et al. (2006) included 2016 NHANES III participants and their level of six organochlorines (heptachloro- and octachloro-dibenzo-*p*-dioxin, PCB 153, oxychlordane, DDE and transnonachlor). Hence, after adjusting for age, sex, race and ethnicity, poverty income ratio, BMI and waist circumference, diabetes prevalence was strongly positively associated with lipid adjusted serum concentration of all six substances.

Increased prevalence of self-reported diabetes with serum level of PCB-153 and DDE was also found in 196 Swedish women (median age of 60 years) and 184 Swedish men (median age of 64 years) by Rylander et al. (2005) as well as in 544 Swedish women (median age of 50 years) by Rignell-Hydbom et al. (2007). Among 11,273 women self-reporting pesticide related activities during first trimester of pregnancy within 25 years of their enrollment in Agricultural Health Study a total of 506 women (4.5 %) had gestational diabetes mellitus (Saldana et al., 2007).

The data obtained within our PCBRISK study showed significantly increased prevalence of impaired fasting glucose, impaired glucose tolerance and diabetes mellitus in the examined population associated with increasing PCBs level in serum (Klimeš et al., 2003; Rádiková et al., 2004). Figure 8 shows significantly higher prevalence of impaired fasting glucose level (5.6 to 6.9 mmol/L; according to American Diabetes Association guidelines; ADA, 2005) in the upper quartile of PCBs level (1989 – 101,413 ng/g) as compared to each of three lower quartiles. In addition, such prevalence in the second (697 - 1084 ng/g) and third quartile

(1085 – 1985 ng/g) of PCBs level was still significantly higher than that in the lower quartile (148 - 696 ng/g).

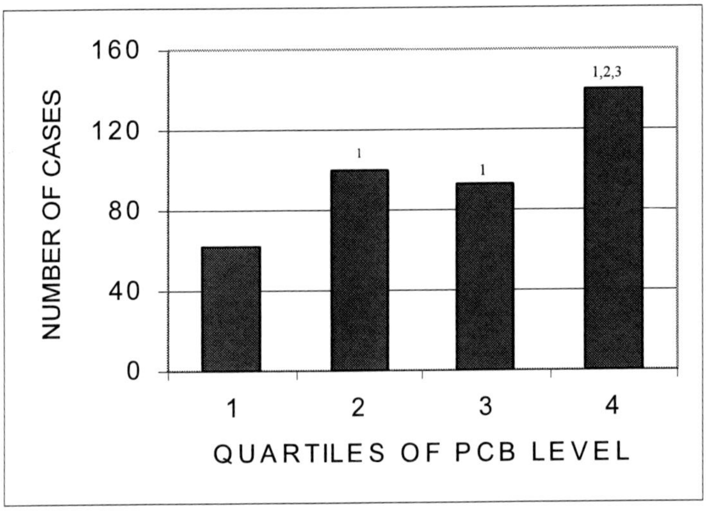

Figure 8. Number of cases with impaired fasting glucose level in quartiles of the sum of 15 PCBs congeners level in females aged 41-60 years (number of all females = 868, number per quartile = 217); the columns labeled by 1, 2, 3 show a significant difference ($p<0.001$ by chi-square test) from the quartile of that number.

In spite of those epidemiological studies showing the association of high OCs exposure with increased risk of diabetes, experimental links explaining such associations are still lacking. Liu P.C. and Matsumura (1995) found reduced glucose transport activity in adipocytes of dioxin-treated guinea pig and mice which also showed reduced copy numbers of glucose transporter GLUT4 and its mRNA. Such observation raised the possibility that dioxin might gradually reduce the production of GLUT4 and other glucose transporters and thereby cause progressive insulin resistance. Later Remillard and Bunce (2002) suggested that dioxin specific Ah-receptor may promote diabetogenic effect by antagonizing the function of peroxisome proliferator activated receptors (PPARs) which were recently found to be linked to cellular proliferation, differentiation and apoptosis as well as to obesity, diabetes, atherosclerosis, inflammation, cancer and aging. Hankinson (2005) stressed that the responsive genes required for TCDD toxicity still have to be defined. Recently, Fujiyoshi et al. (2006) concluded that the ratio of mRNA of GLUT 4 and nuclear transcription factor kappa B (NFκB) is the most sensitive and reliable indicator of dioxin induced diabetes. They suggested that exposure to dioxin-containing Agent Orange resulted in a diabetogenic shift occurring in the biochemistry of adipose tissue.

ADVERSE OCs EFFECTS IN CHILDREN

Human fetuses are exposed to OCs through the placenta, while infants are further exposed through maternal milk. However, recently has been strongly suggested to include also an epigenetic component into the considerations about such transgenerational transmission of endocrine disruptors effect (Anway and Skinner, 2006; Zoeller, 2007). Several authors recently showed high concentrations of PCBs and their hydroxylated (OH-PCBs) or methylsulfonated metabolites ($MeSO_2$-PCBs) as well as these of DDE, HCB, polychlorinated-dibenzo-*p*-dioxins (PCDDs), -furans (PCDFs) and several other OCs in umbilical cord plasma or umbilical cord tissue (Covaci et al., 2002; Sandau et al., 2002; Guvenius et al., 2003; Soechitram et al., 2004; Fukata et al., 2005; Wang S.-L. et al., 2005; Otake et al., 2007; Linderholm et al., 2007; Park, J-S. et al., 2007, 2008; Maervoet et al., 2007). Because of their low body mass, fetuses and infants are considered the most exposed fraction of human population. Such exposure apparently results in several adverse outcomes in children and adolescents either due to numerous toxic effects of OCs on the organism (Waring and Harris, 2005) resulting in specific alterations of thyroid function including blood thyroid hormone levels and pituitary-thyroid feed-back loop (Rogan and Ragan, 2003; Wang et al., 2005a; Otake et al., 2007). Such alterations may further specifically influence namely the development of central nervous system (Waring and Harris, 2005). Numerous new findings in this respect are continuously appearing and hardly deserve to be neglected (Tusscher and Koppe, 2004).

Impaired Immune Functions

One of such adverse effects in children appears the disruption of immune system as reviewed by Holladay (1999) and by Rao and Richardson (1999) who suggested that a fundamental step of such immunotoxic activity could be the thymic atrophy as repeatedly observed in rodents (Landreth, 2002), while Michielsen et al. (1999) as well as Tryphonas and Feeley (2001) pointed out also the role of thymus-independent immunopathology. Detailed description of human immune system development during intrauterine life has been recently provided by West (2002) and Hertz-Picciotto et al. (2008) who also discussed numerous studies on the adverse influence of environmental disturbances on impairment of immune system. Among them Park H-Y. et al. (2008) recently reported a significant association of high prenatal PCB exposure with smaller thymic volume in newborns from Michalovce district which provides the first evidence suggesting impaired immunological development in fetuses and infants which presumably will be followed by immunological disorders during adolescence and adulthood.

There are several outcomes of impaired immune system such as, for instance, increased frequency of otitis media and respiratory tract infections as observed in children from Inuit community exposed to PCBs via breast feeding (Dewailly et al., 2000; Dallaire et al. 2006) and similar findings such as an increased count of some lymphocyte subsets in Dutch children (Tusscher et al., 2003; Weisglas-Kuperus et al., 2000, 2004) or in children from Taiwan (Guo et al., 2003) prenataly exposed to the rice oil poisoned by PCBs and dibenzofurans (Yu-cheng disease). In such children Chang et al. (1981) described also a decreased IgA and IgM level as well as decreased total T cells, cytotoxic and suppressor T cells count and Chao et al. (1997) found increased frequency of middle-ear infections. In Flamish adolescents with increased serum level of PCBs and dioxin-like activity a decrease of eosinophils, natural killer cells, IgE and IgG was observed, while IgA was elevated (Van den Heuvel et al., 2002).

Moreover, several interrelations between OCs and thyroid related hormones in cord blood were reported. Although Longnecker et al. (2000) suggested that the range of background level exposure in the US may be only slightly related to thyroid hormone and TSH concentration, Wang S.-L. et al. (2005) found a negative association of PCDDs and coplanar PCBs with TSH and also positive correlation of PCDFs with total T3 and thyroxine binding globulin. In Thailand, Asawasinsopon et al. (2006) found negative association of high DDE level with total T4 in cord serum.

Onset of Puberty

Possible role of OCs on the attainment of puberty was reviewed by Wang R.Y. et al. (2005) who concluded that in US between 1988-1994 the onset of breast development for girls was 1 to 2 years earlier and the onset of genital development in boys was about 1 year earlier than in other countries. Blanck et al. (2000) observed premature thelarche and risk of premature menarche in girls who were breast-fed by mothers with serum polybrominated biphenyls (PBB) levels about 1000 ng/g lipid resulting from their exposure to milk from contaminated cows. Also Colon et al. (2000) found association of phytoestrogens with earlier onset of thelarche in girls.

In contrast, several studies showed somewhat less pronounced effects of OCs on the onset of puberty. Thus, *in utero* exposure to PCBs and DDE was not associated with the Tanner puberty stages in boys and girls (Gladen et al., 2000). When the effects of postnatal exposure to environmental chemical has been studied, Den Hond et al. (2002) after dioxin like compounds and PCBs as well as Sayied et al. (2003) after endosulphan observed either no effects or delayed effects on timing of pubertal milestones in boys and girls. Also Gladen et al. (2004) did not find any differences between high and low levels of DDT and DDE in stored maternal serum and pubertal measures (testosterone a dehydro-epinandrosterone level) obtained in 304 young males who were periodically examined up to 20 years of age. In 138 Akwesasne Mohawk girls exposed to PCBs, DDE, HCB, lead and mercury (Denham et al., 2005) the median predicted age of menarche was comparable to that obtained within NHANES III study, but was affected by a group of potentionally estrogenic PCB congeners (CB-52, -70, -187). Nevertheless, the question still remains whether OCs level in each population studied was subthreshold or not.

Impaired Psychoneuromotoric and Cognitive Functions

Several studies in mother-child pairs as conducted in the US as well as in Europe (namely Dutch studies) and other countries (Taiwan, Faroe Islands, arctic Quebec) showed predominantly negative association of prenatal and perinatal PCB exposure with measures of cognitive and psychoneuromotoric functions and neuropsychological disorders during infancy and childhood, as pioneered by Jacobson and Jacobson (1997) and thoroughly reviewed by Schantz et al. (2003) and by Tusscher and Koppe (2004). However, since several positive findings appeared also in cohorts subjected to relatively low exposure, it appeared important to consider the validity of individual neuropsychological tests and also to pay attention to possible effects of all individual components of "organochlorine cocktail", namely to dioxins.

In comparison to the cohorts reported by others the cross-sectional examination of neurobehavioral performance in 435 children 8-9 year-old within PCBRISK study (Šovčíková et al., 2008) showed some specific features such as a direct assessment of unusually wide range of organochlorine exposure (predominantly by PCBs, but also by DDE and HCB) by individual serum analyses as well as simultaneous dental, audiological and thyroid examinations. The role of some possible confounders has been assessed using the data obtained from questionnaires. Serum level of PCBs in breast fed children was significantly higher than in these fed by formula. In addition, such level was increasing with maternal age at delivery and with the length of breast feeding.

In agreement with the data by others (Chen et al., 1994; Winneke et al., 1998) a negative association between the serum PCBs level and performance in neuropsychological tests (Simple reaction time, Vienna discrimination test, Benton memory test) was found. Considering the experience obtained by Grandjean et al. (2001), Walkowiak et al. (2001) and Vreugdenhil et al. (2002, 2004), an important role of age specific confounders was anticipated and thus educational level of the mother and age of the child appeared as important confounders counteracting the adverse effects of prenatal OCs exposure on such cognitive and motor abilities. However, significant associations were found between the performances in higher mental tasks and variables including education of mother, maternal results in Raven test and the school from which the child has been recruited, with no relationship to serum PCBs level. However, Lee at al. (2008) recently reported the association of serum concentrations of OCs (PCB-126, dioxins, furans, DDE and trans-nonachlor) with learning disability and attention deficit disorder in 278 children aged 12-15 years participating in NHANES III study.

It was also found that the sound pressure level of transient evoked otoacoustic emissions was significantly associated with the results of all applied neurobehavioral tests targeted at complex mental processes with the exception of Raven intelligence test. Thus, for the first time it was shown that a decreased neurobehavioral performance may be associated with hearing impairment in children environmentally exposed to PCBs. Furthermore, the examination of possible impairments of hearing apparatus as related to PCBs exposure was conducted. Although the ototoxicity of PCBs has been evidenced in numerous animal studies (reviewed by Powers et al., 2006), human studies either did not confirm such type of adverse PCBs effect or were rather inconclusive at the given exposure levels (Rylander and Hagmar, 2000; Grandjean et al., 2001; Longnecker at al., 2004). Within the examination in eastern Slovakia we found a significant increase of hearing thresholds at the frequencies of 250–2000 Hz (left ear) and 250-1000 (right ear) as well as a decrease of the amplitude of transient otoacoustic emissions with increasing PCBs level (Trnovec et al., 2008). However, the prognosis of such findings of impaired hearing apparatus as obtained in these children is unknown. The outcome may be clinically significant in combination with noise exposure, with effects of other ototoxicants or with increasing age.

Developmental Dental Defects

Since the developing tooth is sensitive to a wide range of local and systemic disturbances, it appeared that some developmental dental effects should be expected in children from polluted area namely because of known transplacental and lactational transfer of toxic OCs from the mother to fetus and newborn (for OCs level in adults from POLL see Table 3). Such defects were first described in children whose mothers suffered from Yusho and Yucheng disease due to the intake of rice oil poisoned by PCBs and furans (Rogan et al, 1988; Wang et al. 2003). Later Alaluusua et al. (1999) described developmental enamel defects in children exposed to dioxins and furans via mother's milk, while similar defects after PCBs were weaker. As about the mechanism of dental defects resulting from OCs exposure, Miettinen et al. (2002) found in rats that teeth are mostly vulnerable during the early morphogenesis.

Within PCBRISK survey including 435 children aged 8-9 years the proportion of different types of dental defects (e.g. developmental enamel defects, demarcated opacities and/or hypoplasia, diffuse opacities – Figure 9) significantly correlated with serum PCB level (Jan et al., 2007). However, no association between PCB exposure and carries susceptibility, gingival health or number of teeth was observed. Among 3523 deciduous teeth evaluated. 4.5 % were affected with developmental dental effects, while among 6340 permanent teeth evaluated such proportion was 65.0 %.

Increased prevalence of dental caries has been later found in children from Faroe Islands exposed to high PCBs level from sea food (Jan and Reinert, 2008) where 114 children of the same age as those from Slovakia (i.e. 8-9 years) were examined. The number of carious surfaces was found significantly correlated ($r=0.26$, $p<0.01$) with PCBs level in maternal milk (average of 2205, range 70 – 11,150 ng/g lipid).

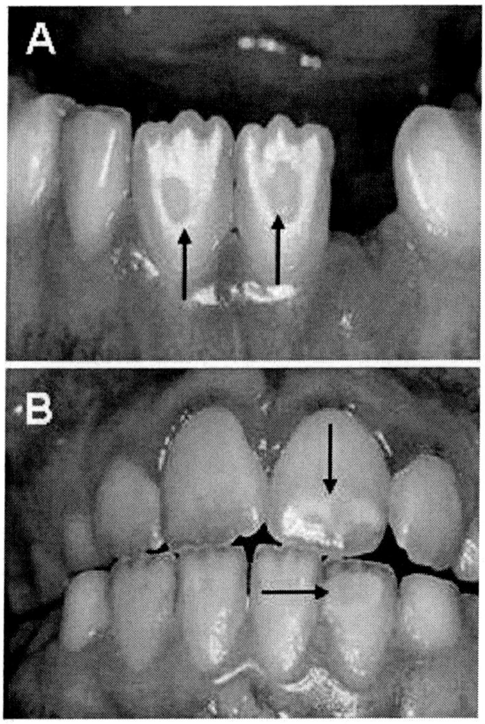

Figure 9. Some of developmental dental defects found in children from Michalovce.Upper panel – hypoplastic defects on permanent lower incisors, Lower panel – demarcated opacity (seen as demarcated white lesions) on deciduous upper canine. This figure was kindly supplied by Janja Jan, M.D., Department of Dental Diseases, University of Ljubljana, Ljubljana, Slovenia.

The latter authors explained some discrepancy between the data obtained by them in Slovakia (Jan et al., 2007) and in Faroes (Jan and Reinert, 2008) by different methods used to evaluate the prevalence of caries. Thus, in Faroes the outcome was the number of carious surfaces (i.e. more than one surface per tooth) while in Slovakia it was the number of teeth without respect to the number of individual carious surfaces which possibly resulted in underreporting.

EPILOGUE

Recently much effort is being paid to environment-gene axis and growing body of information appeared that environmental influences may extend beyond the DNA sequences of human genome. It appeared that, in addition to nucleotide sequence of DNA containing information for RNA and protein, eucaryotic cells contain an additional level of information superimposed on the DNA double helix. Such new field related to this topic is called "epigenetics" which is focused to the study of gene expression changes that are not accompanied by alterations in DNA sequences. The major message emerging from this new form of biologic regulation is that DNA sequence is static, but epigenome – embedded in the protein constituents of chromatin - is a dynamic entity that changes during the cell cycle being influenced, among others, by environmental changes.

Some evidence recently appeared which showed that one of those environmental influences may be endocrine disruptors. Such chemicals actually behave as environmental signals that mimic the signals of endogenous hormones and thus may disrupt their effects on target cells. It is well known that some disruptors mimic estrogens or act as antiestrogens or antiandrogens. At the same time the question appeared whether an epigenetic record of previous generations may be carried via transgenerational transmission to the next generations (Zoeller, 2006). Perhaps the first observation of such event was that of reproductive tract cancers in adult daughters whose pregnant mothers were treated by diethylstilbestrol as reported several decades ago and reevaluated by Newbold (2004). Recently Anway and Skinner (2006) found a number of prostate, testis, kidney and immune system disorders in adult offsprings of pregnant rats treated with antiandrogenic fungicide vinclozin on 8th to 14th day of pregnancy. So far, similar effects were observed also in next three generations (F2 - F4).

Since the story of endocrine disruptors started less about one century ago, the question remains whether and how such transgenerational passage of endocrine disruptors (e.g. as described above) will continue to influence the human health and disease in the future.

ACKNOWLEDGEMENTS

Great part of this study was supported by 5th Framework Programme of European Commision (PCBRISK study - QLK4-2000-00488) and by Ministry of Environment, Slovak Republic (Contract No. 65/97/2.4). The support from Children Fund (Geneva), Henning Berlin and Merck Slovakia is greatly acknowledged. The authors wish to express their appreciation also to 28 district practitioners from Michalovce, Svidník and Stropkov districts, to public health authorities of these districts as well as to several colleagues and other members of field survey staff for their invaluable help in the planning, arrangement and carrying out the field surveys and endless laboratory estimations.

REFERENCES

Akoso, B.T., Sleight, S.D., Nachreiner, R.F. and Aust, S.D. (1982). Effects of purified polybrominated biphenyl congeners on the thyroid and pituitary gland in rats. *J. Am. Coll. Toxicol. 1,* 23-36.

Alaluusua, S., Lukinmaa, P.L., Torpa, J., Tuomisto, J. and Vartiainen, T. (1999). Developing teeth as biomarker of dioxin exposure. *Lancet 353* (9148), 206.

American Diabetes Association. (2005). Standards of medical care in diabetes. *Diabetes Care 28, suppl. 1,* S4-S35.

Anway, M.D. and Skinner, M.K. (2006). Epigenetic transgenerational actions of endocrine disruptors. *Endocrinology 147,* S43-S49.

Aoki, Y. (2001). Polychlorinated biphenyls, polychlorinated dibenzo-*p*-dioxins, and polychlorinated dibenzofurans as endocrine disrupters – what we have learned from Yusho disease. *Environ. Res., Section A 86*: 2-11.

Apostoli, P., Magoni, M., Bergonzi, R., Carasi, S., Indelicato, A., Scarcella, C. and Donato, F. (2005). Assessment of reference values for polychlorinated biphenyl concentration in human blood. *Chemosphere 61,* 413-421.

Araujo, L.B.F., Soares, Jr., J.M., Simoes, J.S., Calio, P.L., Oliveira-Filho, R.M., Simoes, M. de J., Haidar, M.A. and Baract, E.C. (2006). Effect of conjugated equine estrogens and tamoxifen administration on thyroid gland histomorphology of the rat. *Clinics 61,* 321-326.

Arnold, S., Klotz, D., Collins, B., Vonier, P., Guillette, L., Jr. and McLachlan, J. (1996). Synergistic activation of estrogen receptor with combinations of environmental chemicals. *Science 272,* 1489-1492.

Asawasinsopon, R., Prapamontol, T., Prakobvitavakit, O., Vaneesorn, Y., Mangklabruks, A. and Hock, B. (2006). The association between organochlorine and thyroid hormone levels in cord serum: a study from northern Thailand. *Environ. Int. 32,* 554-559.

Baccarelli, A., Mocarelli, P., Patterson, D.G., Bonzini, M., Pesatori, A.G., Caporaso, N. and Landi, T. (2002). Immunologic effects of dioxin: New results from Seveso and comparison with other studies. *Environ. Health Perspect. 110,* 1169-1173.

Bahn, A.K., Mills, J.L., Snyder, P.J, Gamm, P.H., Houten, L., Bialik, O., Hollmann, M.P.H. and Utiger, R.D. (1980). Hypothyroidism in workers exposed to polybrominated biphenyls. *NE J. Med. 302,* 31-33.

Bernhoft, A., Skaare, J.U., Woog, O., Derocher, A.E. and Larsen, H.J.S. (2000). Possible immunotoxic effects of organochlorines in polar bears (*ursus maritimus*) at Svalbard. *J. Toxicol. Environ. Health. 59,* 561-574.

Bertazzi, P.A., Zochetti, C., Guercilena, S., Consonni, D., Tironi, A., Landi, M.T. and Pesatori, A.C. (1997). Dioxin exposure and cancer risk: A 15-year mortality study after the "Seveso accident". *Epidemiology 8,* 646-652.

Blanck, H.M., Marcus, M., Tolbert, P.E., Rubin, C., Henderson, A.K., Hertzberg, V.S., Zhang R.H. and Cameron L. (2000). Age at menarche and Tanner stage in girls exposed in utero and postnatally to polybrominated biphenyl. *Epidemiology 11,* 641-647.

Bloom, M.S., Weiner, J.M., Vena, J.E. and Beehler, G.P. (2003). Exploring associations between serum level of select organochlorines and thyroxine in a sample of New York state sportsmen: the New York State Cohort Study. *Environ. Res. 93,* 52-66.

Botka, M. (1980). Private report. Regional Hospital, Michalovce, Slovakia.

Byrne, J.J., Carborne, J.P. and Hanson, E.A. (1987). Hypothyroidism and abnormalities in the kinetics of thyroid hormone metabolism in rats treated chronically with polychlorinated biphenyl and polybrominated biphenyl. *Endocrinology 21,* 520-527.

Calvert, G.M., Sweeney, M.H., Deddens, J. and Wall, D.K. (1999). Evaluation of diabetes mellitus, serum glucose, and thyroid function among United States workers exposed to 2,3,7,8,-tetrachlorodibenzo-p-dioxin. *Occup. Environ. Med. 56,* 270-276.

Capen , C.C. and Martin, S.L. (1989). The effects of xenobiotics on the structure and function of thyroid follicular and C-cells. *Toxicol. Pathol. 17,* 266-293.

Chang, K.J., Hsieh, K.H., Lee, T.P., Tang, S.Y. and, Tung, T.C. (1981) Immunologic evaluation of patients with polychlorinated biphenyl poisoning: Determination of lymphocyte subpopulations. *Toxicol. Appl. Pharmacol. 61,* 58-63.

Chao, W.Y., Hsu, C.C. and Guo, Y.L.L. (1997). Middle-ear diseases in children exposed prenatally to polychlorinated biphenyls and polychlorinated dibenzofurans. *Arch. Environ. Health 52,* 257-262.

Chauhan, K.R., Kodavanti, P.R.S. and McKinney, J.D. (2000). Assessing the role of *ortho*-substitution on polychlorinated biphenyl binding to transthyretin, a thyroxine transport protein. *Toxicol. Appl. Pharmacol. 162,* 10-21.

Cheek, A.O., Kow, K., Chen, J., Chen, J. and McLachlan, J.A. (1999). Potential mechanisms of thyroid disruption in humans: interaction of organochlorine compounds with thyroid receptor, transthyretin, and thyroid binding globulin. *Environ. Health Perspect . 107,* 273-278.

Chen, YJ, Yu, M-L. M., Rogan, W.J., Gladen, B.C. and Hsu, C.C. (1994). A 6-year follow-up of behavior and activity disorders in the Taiwan Yu-cheng children. *Am. J. Public Health, 84,* 415-421.

Chovancová, J., Kočan, A. and Jursa, S. (2005). PCDDs, PCDFs and dioxin-like PCBs in food of animal origin (Slovakia). *Chemosphere 61,* 1305-1311.

Collins, W.T. and Capen, C.C. (1980). Ultrastructural and functional alterations of the rat thyroid gland produced by polychlorinated biphenyls compared with iodide excess and deficiency, and thyrotropin and thyroxine administration. *Virchow Arch. B Cell. Path. 33,* 213-231. 180.

Colon, I., Caro, D., Bourdony, C.J. and Rosario, O. (2000). Identification of phthalate esters in the serum of young Puerto Rican girls with premature breast development. *Environ. Health Perspect. 108,* 895-900.

Cooper, G.S., Martin, S.A., Longnecker, M.P., Sandler, D.P. and Germolec, D.R. (2004). Associations between plasma DDE levels and immnunologic measures in African-American farmers in North Carolina. *Environ. Health Perspect. 112,* 1080-1084.

Covaci, A., Jorens, P., Jacquemyn, Y. and Schepens, R. (2002). Distribution of PCB and organochlorine pesticides in umbilical cord and maternal serum. *Sci. Total Environ. 298,* 45-53.

Cranmer, M., Louie, S., Kennedy, R.H., Kern, P.A. and Fonseca, V.A. (2000) Exposure to 2,3,7,8-tetrachlorodibenzo-*p*-dioxion (TCDD) in associated with hyperinsulinism and insulin resistance. *Toxicol. Sci. 56,* 431-436.

Dallaire, F., Dewailly, E., Vézina, C., Muckle, G., Weber, G., Bruneau, S. and Ayotte, P. (2006). Effect of prenatal exposure to polychlorinated biphenyls on incidence of acute respiratory infections in preschool Inuit children. *Environ. Health Perspect. 114,* 1301-1305.

Delange, F., Benker, G., Caron, Ph., Eber, O., Ott, W., Peter, F., Podoba, J., Simescu, M., Szybinski, Z., Vertongen, E., Vitti, P., Wiersinga, W. and Zamrazil, V. (1997). Thyroid volume and urinary iodine in European schoolchildren: standardization of values for assessment of iodine deficiency. *Eur. J. Endocrinol. 136,* 180-187.

Denham, M., Schell, L.M., Deane, G., Gallo, M.V., Ravenscroft, J. and DeCaprio, A.P. (2005). Relationship of lead, mercury, mirex, dichlorodiphenyldichloroethylene, hexachlorobenzene, and polychlorinated biphenyls to timing of menarche among Akwesasne Mohawk girls. *Pediatrics 115,* 127-134.

Den Hond, E., Roels, H.A., Hoppenbrouwers, K., Nawrot, T., Vandermeulen, C., Winnekke, G., Vanderschuren, D. and Staessen, A. (2003). Sexual maturation in relation to polychlorinated aromatic hydrocarbons: Sharpe and Skakebaek's hypothesis revisited. *Environ. Health Perspect. 110,* 774-776.

Dewailly, E., Ayotte, P., Bruneau, S., Gingras, S., Belles-Isles, M. and Roy, R. (2004). Susceptibility to infections and immune status of Inuit infants exposed to organochlorines. *Environ. Health Perspect. 108,* 206-211.

Everett, C.J., Frithsen, I.L., Diaz, V.A., Koopman, R.J., Simpson, W.M., Jr. and Mainous, A.G. (2007). Association of a polychlorinated dibenzo-*p*-dioxin, a polychlorinated biphenyl, and DDT with diabetes in the 1999-2002 National Health and Nutrition Examination Survey. *Environ. Res. 103,* 413-418.

Fowles, J.R., Fairbrother, A., Trust, K.A. and Kerkvliet, N.T. (1997). Effect of Aroclor 1254 on the thyroid gland, immune functions, and hepatic cytochrome P450 activity in mallards. Environ. Res. 75, 119-129.

Frich, L., Akslen, L.A. and Glattre, E. (1997). Increased risk of thyroid cancer among Norwegian women married to fishery workers - a retrospective cohort study. *Brit. J. Cancer 75,* 385-389.

Fujiyoshi, P.T., Michalek, J.E. and Matsumura, F. (2006). Molecular epidemiologic evidence for diabetogenic effects of dioxin exposure in U.S. Air Force veterans of the Vietnam war. *Environ. Health Perpect. 114,* 1677-1683.

Fukata, H., Omori, M., Osada, H., Todaka, E. and Mori, C. (2005). Necessity to measure PCBs and organochlorine pesticide concentrations in human umbilical cords for fetal exposure assessment. *Environ. Health Perspect. 113,* 297-303.

Furlanetto, T.V., Nguyen, L.Q., and Jameson, J.L. (1999). Estradiol increases proliferation and down-regulates the sodium/iodide symporter gene in FRTL-5 cells. *Endocrinology 140,* 5705-5711.

Gladen, B.C., Klebanoff, M.A., Hediger, M.L., Katz, S.H., Barr, D.B., Davis, M.D. and Longnecker, M. (2004). Prenatal DDT exposure in relation to anthropometric and pubertal measures in adolescent males. *Environ. Health Perspect. 112,* 1761-1767.

Goldman, L.R., Harnly, M., Flattery, J., Patterson, D.G., Jr. and Needham, L.L. (2000). Serum polychlorinated dibenzo-*p*-dioxins and polychlorinated dibenzofurans among people eating contaminated home-produced eggs and beef. *Environ. Health Perspect. 108,* 13-19.

Grandjean, P, Weihe, P., Burse, V.W., Needham, L.L., Storr-Hansen, E., Henzow, B., Nawrot, T., Thus, L., Vandermeulen, C., Winneke, G., Vanderschuren, D. and Staessen, A. (2001). Neurobehavioral deficits associated with PCBs in 7 year old children prenatally exposed to seafood neurotoxicants. *Neurotoxicol. Teratol. 12,* 305-317.

Grimalt, J.O., Sunyer, J., Moreno, V., Amaral, O.C., Sala, M., Rosell, A., Anto, J.M. and Albaiges, J. (1994). Risk excess of soft-tissue sarcoma and thyroid cancer in community exposed to airborne organochlorinated compound mixtures with a high hexachlorobenzene content. *Int. J. Cancer, 56*: 200-203.

Guo, Y.L., Yu, M.-L., Lin, L.-Y., Hsu, C.C. and Rogan, W.J. (1999). Chloracne, goiter, arthritis and anemia after polychlorinated biphenyl poisoning: 14-year follow-up of the Taiwan Yucheng cohort. *Environ. Health Perspect. 107,* 715-719.

Guvenius D.M., Aronsson A., Ekman-Ordeberg, G., Bergman, A. and Norén, K. (2003). Human prenatal exposure to polybrominated diphenyl ethers, polychlorinated biphenyls, polychlorobiphenylols, and pentachlorophenol. *Environ. Health Perspect. 111,* 1235-1241.

Hagmar, L., Rylander, L., Dyremark, E., Klasson-Wehler, E. and Erfurth, E.M. (2001a). Plasma concentrations of persistent organochlorines in relation to thyrotropin and thyroid hormone levels in women. *Int. Arch. Occup. Environ. Health 74,* 184-188.

Hagmar, L., Björk, J., Sjödin, A., Bergman, A. and Erfurth, E.M. (2001b). Plasma levels of persistent organohalogens and hormone levels in adult male humans. *Arch. Environ. Health 56,* 138-143.

Hagmar, L. (2003). Polychlorinated biphenyls and thyroid status in humans: A review. *Thyroid 13,* 1021-1028.

Hankinson, O. (2005). Role of coactivators in transcriptional activation by the aryl hydrocarbon receptor. Arch. Biochem. Biophys. 433, 379-386.

Hansen, L. (1998). Stepping backward to improve assessment of PCB congener toxicities. *Environ. Health Perspect. 106, suppl. 1,* 171-189.

Heilmann, C., Grandjean, P., Weihe, P. (2003). Decreased childhood vaccine response in children exposed to PCBs from maternal seafood diet. *Organohalogen Compounds 60-65,* 1-4.

Henriksen, G.L., Ketchun, N.S., Michalek, J.E. and Swaby, J.A. (1997). Serum dioxin and diabetes mellitus in veterans of Operation Ranch Hand. *Epidemiology 8,* 252-258.

Hertz-Picciotto, I., Park, H.-Y., Dostal, M., Kocan, A., Trnovec, T. and Sram, R. (2008). Prenatal exposures to persistent and non-persistent organic compounds and effects on immune system development. *Basic and Clin. Pharmacol. Toxicol. 102,* 146-154.

Holladay, S.D. (1999). Prenatal immunotoxicant exposure and postnatal autoimmune disease. *Environ. Health Perspect. 102,* 684-691.

Hood, A. and Klaasen, C.D. (2000). Differential effects of microsomal enzyme inducers on in vitro thyroxine (T4) and triiodothyronine (T3) glucuronidation. *Toxicol Sci 55,* 78-84, 2000.

Hovander, L., Linderholm, L., Athanasiadou, M., Athanassiadis, I., Bignert, A., Fangström, B., Kočan, A., Petrík, J., Trnovec, T. and Bergman, Å. (2006). Levels of PCB and their metabolites in the serum of residents of a highly contaminated area in eastern Slovakia. *Environ. Sci. Technol. 40,* 3696-3703.

Hunter, D.J., Hankinson, S.E., Laden, F., Colditz, G.A., Manson, J.E., Willett, W.C., Speizer, F.E. and Wolff, M.S. (1997). Plasma organochlorine levels and the risk of breast cancer. *N. Eng. J. Med. 137,* 1253-1258.

Jacobson, J.L. and Jacobson, S.W. (1997). Evidence for PCBs as neurodevelopmental toxicants in humans. Neurotoxicology *18,* 415-424.

Jan, J., Sovcikova, E., Kocan, A., Wsolova, L. and Trnovec, T. (2007). Developmental dental defects in children exposed to PCBs in eastern Slovakia. *Chemosphere 67,* S350-S353.

Jan, J. and Reinert, K. (2008). Dental caries in Faroese children exposed to polychlorinated biphenyls. *Environ. Toxicol. Pharmacol. 25,* 188-191.

Jönsson, A.G., Rylander, L., Lindh, C., Rignell-Hydbom, A., Giwercman, A., Toft, G., Pedersen, H.S., Ludwicki, J.K., Goralczyk, K., Zvyezday, V., Spano, M., Bizzaro, D., Bonefeld-Jörgensen, E.C., Manicardi, G.C., Bonde, J.P. and Hagmar, L. (2005). Inter-population variations in concentrations, determinants of and correlations between 2,2',4,4',5,5'-hexachlorobiphenyl (CB-153) and 1,1-dichloro-2,2-bis (*p*-chlorophenyl)-ethylene (p,p'-DDE): a cross-sectional study of 3161 men and women from Inuit and European populations. *Environ. Health 4,* 27-40.

Jung, D., Berg, P.A., Edler, L., Ehrenthal, W., Fenner, D., Flesch-Janys, D., Huber, C., Klein, R., Koitka, C., Lucier, G., Manz, A., Muttray, A., Needham, L., Päpke, O., Pietsch, M., Portier, C., Patterson, D., Prellwitz, W., Rose, D.M., Thews, A., Konietzko, J. (1998). Immunologic findings in workers formerly exposed to 2,3,7,8-tetrachlorodibenzo-*p*-dioxin and its congeners. *Environ. Health Perspect. 106,* suppl. 2, 689-694.

Jursa, S., Chovancová, J., Petrík, J. and Lokša, J. (2006). Dioxin-like and non-dioxin-like PCBs in human serum of Slovak population. *Chemosphere 64,* 686-691.

Kawabata, W., Suzuki, T., Moriya, T., Fujimori, K., Naganuma, H., Inoue, S., Kinouchi, Y., Kameyama, K., Takami, H., Shimosegawa, T. and Sasano, H. (2003). Estrogen receptors (alpha and beta) and 17beta-hydroxysteroid dehydrogenase type 1 and 2 in thyroid disorders: possible in situ estrogen synthesis and actions. *Mol. Pathol. 16,* 437-444.

Kern, P.A., Said, S., Jackson, W.G. and Michalek, J.E. (2004). Insulin sensitivity following Agent Orange exposure in Vietnam veterans with high blood levels of 2,3,7,8-tetrachlorodibenzo-*p*-dioxin. *J. Clin. Endocrinol. Metab. 89,* 4665-4672.

Kester, M.H.A., Bulduk, S., Tiboel, D., Meinl, W., Glatt, H., Falany, C.N., Coughtrie, M.W.H., Bergman, A., Safe, S.H., Kuiper, G.G.J.M., Schuur, A.G., Brouwer, A. and Visser, T.J. (2000). Potent inhibition of estrogen sulfotransferase by hydroxylated PCB metabolites: A novel pathway explaining the estrogenic activity of PCBs. *Endocrinology 141,* 1897-1900.

Khan, M.A., Lichtensteiger, C.A., Faroon, O., Mumtaz, M., Schaeffer, D.J. and Hansen, L.G. (2002). The hypothalamo-pituitary-thyroid (HPT) axis: A target of nonpersistent *ortho*-substituted PCB congeners. *Toxicol. Sci. 65,* 51-61.

Kitamura, S., Jinno, N., Suzuki, T., Siguhara, K., Ohta, S., Kuroki, H. and Fujimoto, N. (2005). Thyroid hormone-like and estrogenic activity of hydroxylated PCBs in cell culture. *Toxicology 208,* 377-387.

Kiviranta, H., Tuomisto, J.T., Tuomisto, J., Tukiainen, E. and Vartiainen, T. (2005). Polychlorinated dibenzo-*p*-dioxins, dibenzofurans, and biphenyls in the general population in Finland. *Chemosphere 60,* 854-859.

Klimeš, I., Koška, J., Kšinantová, L., Bučková, K., Červenáková, Ž., Imrich, R.., Petrík, J., Jursa, S., Tajtáková, M., Suchánek, P., Vigaš, M., Langer, P., Kočan, A., Trnovec, T.,

Šeböková, E. (2003). Increased frequency of glucose intolerance in the population of specific areas of eastern Slovakia chronically exposed to contamination with polychlorinated biphenyls (PCB). 63rd ADA Scientific Sessions. *Diabetes, Abstract Book 52,* Suppl. 1, 955-P.

Kočan, A., Petrík, J., Drobná, B. and Chovancová, J. (1994a). Levels of PCBs and some organochlorine pesticides in the human population of selected areas of the Slovak Republic. *Chemosphere 29,* 2315-2325.

Kočan, A., Petrík, J., Chovancová, J. and Drobná, B. (1994b). Method for the group separation of non-*ortho*, mono-*ortho* and multi-*ortho*-substituted polychlorinated biphenyls and polychlorinated dibenzo-*p*-dioxins/polychlorinated dibenzofurans using activated carbon chromatography. *J. Chromatogr. A 665,* 139-153.

Kočan, A., Drobná, B., Chovancová, J., Kočan, J., Petrík, J. and Szabová, E. (1998). Research report. Institute of Preventive and Clinical Medicine. Bratislava, pp. 80-82.

Kočan, A., Petrík, J., Jursa, S., Chovancová, J. and Drobná, B. (2001). Environmental contamination with polychlorinated biphenyls in the area of their former manufacture in Slovakia. *Chemosphere 93,* 595-560.

Koopman-Esseboom, C., Morse, D.C., Weisglas-Kuperus, N., Lutkeschipholt, I.J., van der Paauw, C.G., Tuinstra, L.G.M.T., Brouwer, A. and Sauer, P.J.J. (1994). Effects of dioxins and polychlorinated biphenyls on thyroid hormone status of pregnant women and their infants. *Pediatr. Res. 36,* 468-476.

Kung, A.W., Chau, M.T., Lao, T.T., Tam, S.C. and Low, L.C. (2002). The effect of pregnancy on thyroid hormone formation. *J. Clin. Endocrinol. Metab. 87,* 1010-1014.

Kuriyama, S., Fidalgo-Neto, A., Mathar, W., Palavinskas, R., Friedrich, K. and Chahoud, I. (2003). Effect of low dose mono-ortho 2,3',4,4',5-pentachlorobiphenyl on thyroid hormone status and EROD activity in rat offspring: consequences for risk assessment. *Toxicology 15,* 11-20.

Landreth, K.S. (2002). Critical windows in development of the rodent immune system. *Hum. Exp. Toxicol. 21,* 493-498.

Langer, P., Tajtaková, M., Podoba, J., Košťálová, L. and Gutekunst, R. (1994). Thyroid volume and urinary iodine in school children and adolescents in Slovakia after 40 years of iodine prophylaxis. *Exp. Clin. Endocrinol. 102,* 394-398.

Langer. P., Tajtáková, M., Fodor, G., Kočan, A., Bohov, P., Michálek, J. and Kreze, A. (1998). Increased thyroid volume and prevalence of thyroid disorders in an area heavily polluted by polychlorinated biphenyls. *Eur. J. Endocrinol. 139,* 402-409.

Langer, P., Tajtáková, M., Bohov, P. and Klimeš, I. (1999). Possible role of genetic factors in thyroid growth rate and in the assessment of upper limit of normal thyroid volume in iodine-replete adolescents. *Thyroid 9,* 552-562.

Langer, P., Kočan, A, Tajtáková, M., Petrík, J., Chovancová, J., Drobná, B., Jursa, S., Pavúk, M., Koška, J. Trnovec, T., Šeböková, E. and Klimeš, I. (2003). Possible effects of polychlorinated biphenyls and organochlorinated pesticides on the thyroid after long-term exposure to heavy environmental pollution. *J. Occup. Environ. Med. 45,* 526-532.

Langer, P., Tajtáková, M., Podoba, J., Košťalová, L. and Gutekunst, R. (1994). Thyroid volume and urinary iodine in school children and adolescents in Slovakia after 40 years of iodine prophylaxis. *Exp. Clin. Endocrinol. 102,* 394-398.

Langer, P., Kočan, A., Tajtáková, M., Petrík, J., Chovancová, J., Drobná, B., Jursa, S., Rádiková, Ž, Koška, J., Kšinantová, L., Hučková, M., Imrich, R., Wimmerová, S.,

Gašperíková, D., Shishiba, Y., Trnovec, T., Šebökova, E. and Klimeš, I. (2007a). Fish from industrially polluted freshwater as main source of organochlorinated pollutants and increased frequency of thyroid disorders and dysglycemia. *Chemosphere 67,* S379-S385.

Langer, P., Tajtáková, M., Kočan, A., Petrík, J., Koška, J., Kšinantová, L., Rádiková, Ž., Ukropec, J., Imrich, R., Hučková, M., Chovancová, J., Drobná, B., Jursa, S., Vlček, M., Bergman, A., Athanasiadou, M., Hovander, L., Shishiba, Y., Trnovec, T., Šebökova, E. and Klimeš, I. (2007b). Thyroid ultrasound volume, structure and function after long-term high exposure of large population to polychlorinated biphenyls, pesticides and dioxin. *Chemosphere 69,* 118-127.

Langer, P., Kočan, A., Tajtáková, M., Rádiková, Ž., Petrík, J., Koška, J., Kšinantová, L., Imrich, R., Hučková, M., Chovancová, J., Drobná, B., Jursa, S., Bergman, A., Athanasiadou, M., Hovander, L., Vlček, M., Trnovec, T., Šebökova, E.. and Klimeš, I. (2007c). Possible effects of persistent organochlorinated pollutants cocktail on thyroid hormone levels and pituitary-thyroid interrelations. *Chemosphere 70,* 110-118.

Lans, M.C.V., Klasson-Wehler, E., Willemsen, M., Meussen, E. and Safe, S. (1993). Structure-dependent, competitive interaction of hydroxy-polychlorobiphenyls, -dibenzo-*p*-dioxins and –dibenzofurans with human transthyretin. *Chem-Biol. Interactions 88,* 7-21.

Lans, M.C.V., Spiertz, C., Brouwer, A. and Koeman, J.H. (1994). Different competition of thyroxine binding to transthyretin and thyroxine-binding globulin by hydroxy-PCBs, PCDDs and PCDFs. *Eur. J. Pharmacol. 270,* 129-136.

Lee, D.H., Lee, I.K., Song, K., Steffes, M., Toscano, W., Baker, B.A. and Jacobs, D.R. (2006). A strong dose-response relation between serum concentrations of persistent organic pollutants and diabetes. *Diabetes Care 29,* 1638-1644.

Linderholm, L., Park, J-S., Kočan, A., Trnovec, T., Athanasiadou, M., Bergman, A. and Hertz-Piciotto, I. (2007). Maternal and cord serum exposure to PCB and DDE methyl sulfone metabolites in eastern Slovakia. *Chemosphere 69,* 403-410.

Liu, J., Liu, Y., Barter, R. and Klaasen, C.D. (1995). Alteration of thyroid homeostasis by UDP-glucoronosyltransferase inducers in rats: A dose-response study. *J. Pharmacol. Exper. Therap. 273,* 977-985.

Liu, P.C. and Matsumura, F. (1995). Differential effects of 2,3,7,8-tetrachloro-dibenzo-*p*-dioxin on the "adipose-type" and "brain-type" glucose transporters in mice. *Mol. Pharmacol. 47,* 65-73.

Longnecker, M.P., Hoffman, H.J., Klebanoff, M.A., Brock, J.W., Zhou, H., Needham, I.., Adera, T., Guo, X. and Gray, K.A. (2004). In utero exposure to polychlorinated biphenyls and sensorineural hearing loss in 8-year old children. *Neurotoxicol. Teratol. 26,* 629-637.

Longnecker, M.P., Gladen, B.C., Patterson, D.G. and Rogan, W.J. (2000). Polychlorinated biphenyl (PCB) exposure in relation to thyroid hormone levels in neonates. *Epidemiology 11,* 239-241.

Madej, A., Persson, E., Lundh, T. and Ridderstrale, Y. (2002). Thyroid gland function in ovariectomized ewes exposed to phytoestrogens. *J. Chromatogr. B 777,* 281-287.

Maervoet, J., Vermeir, G., Covaci, A., Van Larebeke, N., Koppen, G., Schoeters, G., Nelen, V., Baeyens, W., Schepens, P., Viaene, M.K. (2007). Association of thyroid hormone concentration with levels of organochlorine compounds in cord blood of neonates. *Environ. Health Perspect, 115,* 1780-1786.

Mallin, K., McCann, K. and D'Aloisio, A. (2004). Cohort mortality study of capacitator manufacturing workers, 1944-2000. *J. Occup. Environ. Med. 46,* 565-576.

Manji, N., Boelaert, K., Sheppard, M.C., Holdert, R.L., Gough, S.C. and Franklyn, J.A. (2006). Lack of association between serum TSH or free T4 and body mass index in euthyroid subjects. *Clin. Endocrinol. 64,* 125-128.

Manole, D., Schildknecht, B., Gosnell, B., Adams, E. and Derwahl, M. (2001). Estrogen promote growth of human thyroid tumor cells by different molecular mechanisms. *J. Clin. Endocrinol. Metab. 86,* 1072-1077.

Matthews, J. and Gustafsson, J-A. (2006). Estrogen receptor and aryl hydrocarbon receptor signaling pathways. *NRS 4,* 1-4.

Masuda, Y., Haraguchi, K., Kono, S., Tsuji, H. and Päpke, O. (2005). Concentrations of dioxins and related compounds in the blood of Fukuoka residents. *Chemosphere 58,* 329-344.

Meeker, J.D., Altshul, L. and Hauser, R. (2007). Serum PCBs, *p,p'*-DDE and HCB predict thyroid hormone levels in men. *Environ. Res. 104,* 296-301.

Michalek, J.E., Akhtar, F.Z. and Kiel, J.L. (1999). Serum dioxin, insulin, fasting glucose, and sex hormone-binding globulin in veterans of Operation Ranch Hand. *J. Clin. Endocrinol. Metab. 84,* 1540-1543.

Michielsen, C.C.P.P.C., van Loveren, H. and Vos, J.G. (1999). The role of the immune system in hexachlorobenzene-induced toxicity. *Environ. Health Persp. 107, suppl. 5,* 783-792.

Miettinen, H.M., Alaluusua, S., Tuomisto, J. and Viluksela, M. (2002). Effect of in utero and lactational 1,2,7,8-tetrachlorodibenzo-*p*-dioxin exposure on rat molar development: the role of exposure time. *Toxicol. Appl. Pharmacol. 184,* 57.

Monaco, F. (2003). Classification of thyroid diseases: Suggestions for a revision. *J. Clin. Endocrinol. Metab. 88,* 1428-1432.

Moysich, K.B., Ambrosone, C.B., Mendola, P., Kostyniak, P.J., Greizerstein, H.B., Vena, J.E., Menezes, R.J., Swede, H., Shields, P.G. and Freudenheim, J.L. (2002). Exposure associated with serum organochlorine levels among postmenopausal women from western New York State. *Am. J. Ind. Med. 41,* 102-110.

Murai, K., Okamura, K., Tsuji, H., Kajiwara, E., Watanabe, H., Akahi, K. and Fujishima, M. (1987). Thyroid function in "Yusho" patients exposed to polychlorinated biphenyls (PCB). *Environ. Res. 44,* 179-187.

Ness, D.K., Schantz, S.L., Moshtaghian, J. and Hansen, L. (1993). Effects of perinatal exposure to specific PCB congeners on thyroid hormone concentrations and thyroid histology in the rat. *Toxicology Letters 68,* 311-323.

Newbold, R.R. (2004). Lessons learned from perinatal exposure to diethylstilbestrol. (2004). *Toxicol. Appl. Pharmacol. 199,* 142-150.

Nishimura, N., Yonemoto, J., Miyabara, Y., Sato, M. and Tohyama, C. (2003). Rat thyroid hyperplasia induced by gestational and lactational exposure to 2,3,7,8-tetrachlorodibenzo-*p*-dioxin. *Endocrinology 144,* 2075-2083.

Osius, N., Karmaus, W., Kruse, H., and Witten, J. (1999). Exposure to polychloinated biphenyls and levels of thyroid hormones. *Environ. Health Perspect. 107,* 845-849.

Otake, T., Yoshinaga, J., Enomoto, T., Matsuda, M., Wakimoto, T., Ikegami, M., Suzuki, E., Naruse, H., Yamanaka, T., Shibuya, N., Yasumizu, T. and Kato, N. (2007). Thyroid hormone status of newborns in relation to *in utero* exposure to PCBs and hydroxylated PCB metabolites. *Environ. Res. 105,* 375-381.

Park, J-S., Bergman, A., Linderholm, L., Athanasiadou, M., Kočan, A., Petrík, J., Drobná, B., Trnovec, T., Charles, M.J. and Hertz-Picciotto, I. (2008). Placental transfer of polychlorinated biphenyls, their hydroxylated metabolites and pentachlorophenol in pregnant women from eastern Slovakia. *Chemosphere 70,* 1676-1684.

Park, J-S., Linderholm, L., Charles, J.M., Athanasiadou, M., Petrík, J., Kočan, A., Drobná, B., Trnovec, T., Bergman, A. and Hertz-Picciotto, I. (2007) Polychlorinated biphenyls and their hydroxylated metabolites (OH-PCBs) in pregnant women from eastern Slovakia. *Environ. Health Perspect. 115,* 20-27.

Park, H-Y., Hertz-Picciotto, I., Petrík, J., Palkovičová, L., Kočan, A. and Trnovec, T. (2008). Prenatal PCB exposure and thymus size at birth in neonates in Eastern Slovakia. *Environ. Health Perspect. 116,* 104-109.

Pavuk, M., Schecter, A., Akhtar, F., Michalek, J. (2003). Serum 2,3,7,8-tetrachlorodibenzo-*p*-dioxin (TCDD) levels and thyroid system effects in the US Air Force veterans. *Ann.. Epidemiol. 13,* 335-345.

Pavuk, M., Cerhan, J.R., Lynch, C.F., Schecter, A., Petrik, J., Chovancova, J. and Kocan, A. (2004a). Environmental exposure to PCBs and cancer incidence in eastern Slovakia. *Chemosphere 54,* 1509-1528.

Pavuk, M., Cerhan, J.R., Lynch, C.F., Kocan, A., Petrik, J. and Chovancova, J. (2004b). Case-control study of PCBs, other organochlorines and breast cancer in eastern Slovakia. *CJ. Expo. Anal. Environ. Epidemiol. 13,* 267-275.

Pelletier, C., Doucet, E., Imbeault, P. and Tremblay, A. 2002. Associations between weight loss-induced changes in plasma organochlorine concentrations, serum T3 concentration, and resting metabolic rate. *Toxicol. Sci. 67,* 46-51.

Persky, V., Turyk, M., Anderson, A., Hanrahan, L.P., Falk, C., Steenport, D.N., Chatterton, R. and Freels, S. (2001). The effects of PCB exposure and fish consumption on endogenous hormones. *Environ. Health Perspect. 109,* 1275-1283.

Persson, I., Yuen, J., Bergkvist, L. and Schairer, C. (1996). Cancer incidence and mortality in women receiving estrogen and estrogen-progestin replacement therapy-long-term follow-up of Swedish cohort. *Int J Cancer 67,* 327-332.

Petrík, J., Drobná, B., Pavúk, M., Jursa, S., Wimmerová, S. and Chovancová, J. (2006). Serum PCBs and organochlorine pesticides in Slovakia: Age, gender, and residence as determinants of organochlorine concentrations. *Chemosphere 65,* 410-418.

Pluim, H.J., de Viljder J.J., Olie, K., Kok, J.H., Vulsma, T., van Tijn, D.A., van der Slikke, J.V. and Koppe, J.G. (1993). Effects of pre- and postnatal exposure to chlorinated dioxins and furans on human neonatal thyroid hormone concentration. *Environ. Health Perspect. 101,* 504-508.

Potter, E.L. (1946). Facial characteristics of infants with bilateral renal agenesis. *Am. J. Obstetr. Gynecol. 51,* 885-888.

Powers, B.E., Widholm, J.J., Lasky, R.E. and Schantz, S.L. (2006). Auditory deficits in rats exposed to an environmental PCBs mixture during development. *Toxicol. Sci. 89,* 415-422.

Prouty, L.A. and Myers, T.L. (1987). Oligohydramnions sequence (Potter's syndrome): case clustering in northeastern Tennessee. *South Med. J. 80,* 585-592.

Radikova, Z., Koska, J., Ksinantova, L., Imrich, R., Kocan, A., Petrik, J., Huckova, M., Wsolova, L., Langer, P., Trnovec, T., Sebokova, E. and Klimes. I. (2005). Increased frequency of diabetes and other forms of dysglycemia in the population of specific areas of eastern Slovakia chronically exposed to contamination with polychlorinated biphenyls (PCB). *Organohalogen Compounds 66,* 3547-3550.

Rádiková, Z., Tajtáková, M., Kočan, A., Trnovec, T., Šeböková, E., Klimeš I.. and Langer, P. (2007). Possible effects of environmental nitrates and organochlorines on human thyroid in highly polluted areas in Slovakia. *Thyroid 18,* 353-362.

Rajapakse, N., Silva, E. and Kortenkamp, A. (2002). Combining xenoestrogens at levels below individual no-observe-effect concentrations dramatically enhances steroid hormone action. *Environ. Health Perspect. 110,* 917-921.

Rao, T. and Richardson, B. (1999). Environmentally induced autoimmune diseases: Potential mechanisms. *Environ. Health Perspect. 107,* 737-742.

Remillard, R.B.J. and Bunce, N.J. (2002). Linking dioxins to diabetes: Epidemiology and biological plausibility. *Environ. Health Perspect. 110*, 853-857.

Rignell-Hydbom, A., Rylander, L. and Hagmar, L. (2007). Exposure to persistent organochlorines pollutants and type 2 diabetes mellitus. *Hum. Exp. Toxicol. 26,* 447-452.

Rogan, W.J., Gladen, B.C., Hong, K.L., Koong, S.L., Shih, L.I., Taylor, J.C., Wu, Y.C., Young, D., Ragan, N.B. and Hsu, C.C. (1988). Congenital poisoning by polychlorinated biphenyls and their contaminants in Taiwan. *Science 241,* 334-336.

Rogan, W.J. and Ragan, N.B. (2003). Evidence of effects of environmental chemicals on the endocrine system in children. *Pediatrics 112,* 247-252.

Rylander, L. and Hagmar, L. (2000). Medical and psychometric examinations of conscripts born to mothers with a high intake of fish contaminated with persistent organochlorines. *Scand. J. Work. Environ. Health 26,* 207-212.

Rylander, L., Rignell-Hydbom, A. and Hagmar, L. (2005). A cross-sectional study of the association between persistent organochlorine pollution and diabetes. *Environ. Health 4,* 28-33.

Rylander, L., Wallin, E., Jonsson, B.A.G., Stridsberg, M., Erfurth, E.M. and Hagmar, L. (2006). Associations between CB-153 and *p,p'*-DDE and hormone levels in serum in middle-aged and elderly men. *Chemosphere 65,* 675-681.

Saeed, A. and Hansen, L. (1997). Morphometric changes in the prepubertal female rat thyroid gland following acute exposure to 2,2',4,4'-tetrachlorobiphenyl and Aroclor 1242. *J. Toxicol. Environ. Health. 51,* 503–513.

Šak, M. (1978). Thesis. Faculty of Medicine, P.J. Šafárik University, Košice, Slovakia.

Sala, M., Sunyer, J., Herrero, C., To-Figueras, J. and Grimalt, J. (2001). Associations between serum concentrations of hexachlorobenzene and polychlorinated biphenyls with thyroid hormone and liver enzymes in a sample of general population. *Occup. Environ. Med. 58,* 172-177.

Saldana, T.M., Basso, O., Hoppin, J.A., Baird, D.D., Knott, C., Blair, A., Alavanja, M.C. and Sandler, D.P. (2007). Pesticide exposure and self-reported gestational diabetes mellitus in the Agricultural Health Study. *Diabetes Care 30,* 529-534.

Sandau, C.D., Ayotte, P., Dewailly, E., Duffe, J., and Norstrom, R. (2002). Pentachlorophenol and hydroxylated polychlorinated biphenyl metabolites in umbilical cord plasma of neonates from coastal populations in Quebec. *Environ. Health Perspect. 110,* 411-417.

Saracci, R., Kogevinas, M., Bertazzi, P.A., De Mesquita, B.H., Coggon, D., Green L.M., Kauppinen, T., L'Abbé, K.A., Littorin, M., Lynge, E., Mathews, J.D., Neuberger, M.,

Osman, J., Pearce, N. and Winkelmann, R. (1991). Cancer mortality in workers exposed to phenoxy herbicides and chlorophenols. *Lancet 338,* 1027-1032.

Schantz, S.L., Widholm J.J. and Rice, D.C. (2003). Effects of PCB expoasure on neuropsychological functions in children. *Environ. Health Perspect, 11,* 357-376.

Schell, L.M., Gallo, M., DeCaprio, A.P., Hubicki, L., Denham, M. and Ravenscroft, J. (2004). Thyroid function in relation to burden of PCBs, *p,p'*-DDE, HCB, mirex and lead among Akwesasne Mohawk youth: a preliminary study. *Environ. Toxicol. Pharmacol. 18,* 91-99.

Schreinemachers, D.M., Creason, J.P. and Garry, V.F. (1999). Cancer mortality in agricultural regions of Minnesota. *Environ. Health Perspect. 107,* 205-211.

Schuur, A.G., Brouwer, A., Bergman, A., Goughtrie, M.W.H. and Visser, T.J. (1998). Inhibition of thyroid hormone sulfation by hydroxylated metabolites of polychlorinated biphenyls. *Chem-Biol. Interactions 109,* 293-297.

Seo, B.W., Li, M.H., Hansen, L.G., Moore, R.W., Peterson, R.E. and Schantz, S.L. (1995). Effects of gestational and lactational exposure to coplanar polychlorinated biphenyl (PCB) congeners or 2,3,7,8,-tetrachloro-*p*-dibenzo-dioxin (TCDD) on thyroid hormone concentrations in weanling rats. *Toxicol. Lett. 78,* 253-262.

Shiraishi, F., Okumura, T., Nomachi, M., Serizawa, S., Nishikawa, J., Edmonds, J.S., Shiraishi, H. and Morita, M. (2003). Estrogenic and thyroid hormone activity of a series of hydroxy-polychlorinated biphenyls. *Chemosphere 52,* 33-42.

Soechitram, S.D., Athanasiadou, M., Hovander, L., Bergman, A. and Sauer P.J.J. (2004). Fetal exposure to PCBs and their hydroxylated metabolites in a Dutch cohort. *Environ. Health Perspect. 112,* 1208-1212.

Šovčíková, E., Trnovec, T., Wsolová, L., Wimmeroví, S., Drobná, B., Hušťák, M., Petrík, J., Kočan, A. and Langer, P. PCB exposure, neurobehavioral performance and hearing impairment in children. In: Hansen L, Robertson L, editors. PCBs – Human and Environmental Disposition and Toxicology, University of Illinois Press, 2008 (in press).

Tajtáková, M., Hančinová, D., Langer, P., Tajták, J., Malinovský, E. and Varga, J. (1988). Thyroid volume in East Slovakian adolescents determined by ultrasound 40 years after the introduction of iodized salt. *Klin. Wochenschr. 66,* 749-751.

Tajtáková, M. and Langer, P. (1992) Retrospectively evaluated finding of larger thyroid volume in east Slovakian adolescents in 1987 than in 1992: effect of Chernobyl? *Horm. Metab. Res. 25,* 644-645.

Tajtáková, M., Langer, P., Bohov, P., Gonsorčíková, V. and Hančinová, D. (1998). Recognition of a group of adolescents with rapidly growing thyroids under iodine-replete conditions: seven year follow up. *Eur. J. Endocrinol. 138,* 674-680.

Tajtáková, M., Semanová, Z., Tomková, Z., Szökeová, E., Majoroš, J., Rádiková, Z., Šeboková, E. and Klimeš, I. (2006). Increased thyroid volume and frequency of thyroid disorders signs in Takser, L., Mergler, D., Baldwin, M., de Grosbois, S., Smargiassi, A. and Lafond, J. (2005). Thyroid hormones in pregnancy in relation to environmental exposure to organochlorine compounds and mercury. *Environ. Health Perspect. 113,* 1039-1045.

Takser, L., Mergler, D., Baldwin, M., de Grosbois, S., Smargiassi, A. and Lafond, J. (2005). Thyroid hormones in pregnancy in relation to environmental exposure to organochlorine compounds and mercury. *Environ. Health Perspect. 113,* 1039-1045.

Tan, Y., Li, D., Song, R., Lawrence, D. and Carpenter, D.O. (2003). Ortho-substituted PCBs kill thymocytes. *Toxicol. Sci. 76,* 328-337.

Tavangar, S.M., Monajemzadeh, M., Larijani, B. and Haghpanah, V. (2007). Immunochemical study of oestrogen receptors in 351 human thyroid glands. *Singapore Med 48,* 744-747.

Tomer, Y. and Davies, T. (2003). Searching for the autoimmune thyroid disease susceptibility genes: from gene maping to gene function. *Endocrine Reviews 24,* 694-717.

Trnovec, T., Sovcikova, E., Hustak, M., Wimmerova, S., Kocan, A., Jureckova, D., Langer, P., Palkovicova, L. and Drobna, B. (2007). Exposure to polychlorinated biphenyls and hearing impairment in children. Environ. *Toxicol. Pharmacol. 25,* 183-187, 2008.

Tryphonas, H. and Feeley, M. (2001). Polychlorinated biphenyl-induced immunomodulation and human health effects. In: Robertson L, Hansen L, editors. PCBs. University Press of Kentucky, 2001.

Turyk, M.E., Anderson, H.A., Hanrahan, L..P., Falk, C., Steenport, D.N., Needham, L.L., Patterson, D.G., Freels, S., Persky, V.W. and the Great Lakes Consortium (2006a). Relationship of serum levels of individual PCB, dioxin, and furan congeners and DDE with Great Lakes sport-caught fish consumption. *Environ. Res. 100,* 173-183.

Turyk, M.E., Anderson, H.A., Freels, S., Chatterton Jr., R., Needham, L.L., Patterson, D.G., Steenport, D.N., Knobeloch, L., Imm, P., Persky, V.W. and the Great Lakes Consortium (2006b). Associations of organochlorines with androgenous hormones in male great lakes fish consumers and nonconsumers. *Environ. Res. 102,* 299-307.

Turyk, M.E., Anderson, H.A. and Persky, V.W. (2007). Relationship of thyroid hormones with polychlorinated biphenyls, dioxins, furans and DDE in adults. *Environ. Health Perspect. 115,* 1197-1203.

Tusscher ten, G.V., Steerenberg, P.A., Loveren van, H., Vos, J.G., Borne von dem, A.E.G.K., Westra, M., Slikke van der, J.W., Olie, K., Pluim, H.J. and Koppe, J.G. (2003). Persistent hematologic and immunologic disturbances in 8-year-old Dutch children associated with perinatal dioxin exposure. *Environ. Health Perspect. 111,* 1519-1523.

Tusscher ten, G.W. and Koppe, J.G. (2004). Perinatal dioxin exposure and later effects – a review. *Chemosphere 54,* 1329-1336.

Valle, L.D., Ramina, A., Vianello, S., Fassina, A., Belvedere, P. and Colombo, L. (1998). Potential for estrogen synthesis and action in human normal and neoplastic thyroid tissues. *J. Clin. Endocrinol. Metab. 83,* 3702-3709.

Van den Berg, K.J., van Raaij, J.A.G.M., Bragt, P.C. and Notten, W.R.F. (1991) Interactions of halogenated industrial chemicals with transthyretin and effects on thyroid hormone levels in vivo. *Arch. Toxicol. 65,* 15-19, 1991.

Van Den Heuvel, R.L., Koppen, G., Staessen, J.A., Den Hond, E., Verheyen, G., Nawrot, S., Roels, H.A., Vlieting, R. and Schoeters, G.E.R. (2002). Immunologic biomarkers in relation to exposure markers of PCBs and dioxins in Flemish adolescents (Belgium). *Environ. Health Perspect. 110,* 595-600.

Van Raaij, J.A.G.M., van den Berg, K.J. and Notten WRF. (1991). Hexachlorobenzene and its metabolites pentachlorophenol and tetrachlorohydroquinone interaction with thyroxine binding sites of rat thyroid hormone carriers ex vivo and in vitro. *Toxicol. Lett. 59,* 101-107, 1991.

Van Raaij, J.A.G.M., Kaptein, E., Visser, T.J. and van den Berg, K.J. (1993). Increased glucuronidation of thyroid hormone in hexachlorobenzene treated rats. *Biochem. Pharmacol. 45,* 627-631.

Vinals, F., Sepulveda, W.H., Klaasen, R., Ciuffardi, I. and Doneth, G. (1992). Bilateral renal agenesis: a 10-year experience. *Rev. Chil. Obstetr. Ginecol. 57,* 171-175.

Vine, M.F., Stein, L., Weigle, K., Schroeder, J., Degnan, D., Tse, C-K.J., Hanchette, C. and Backer, L. (2000). Effects on the immune system associated with living near a pesticide dump site. *Environ. Health Persp. 108,* 1113-1124.

Vivacqua, A., Bonofiglio, D., Albanito, L., Madeo, A., Rago, V., Carpino, A., Musti, A.M., Picard, D., Ando, S. and Maggiolini, M. (2006). 17β-estradiol, genistein, and 4-hydroxytamoxifen induce the proliferation of thyroid cancer cells through the G protein-coupled receptor GPR30. *Mol. Pharmacol, 70,* 1414-1423.

Volker, D., Huber, W., Bauer, K. Suesal, C., Conradt, C. and Opelz, G. (2001a). Associations of blood levels of PCB, HCHs, and HCB with numbers of lymphocyte subpopulations, in vitro lymphocyte response, plasma cytokine levels. and immunoglobulin autoantibodies. *Environ. Health Perspect. 109,* 173-178.

Volker, D., Huber, W., Bauer, K. Suesal, C., Mytilineos, J., Melk, C., Conradt, C. and Opelz, G. (2001b). Associations of elevated blood levels of pentachlorophenol (PCP) with cellular and humoral immunodeficiencies. *Arch. Environ. Health 56,* 77-83.

Vreugdenhil, H.J., Lanting, C.I., Mulder, P.G., Boersma, E.R. and Weisglas-Kuperus, N. (2002). Effects of prenatal PCB and dioxin background exposure on cognitive and motor abilities in Dutch children at school age. *J. Pediatr. 140,* 48-56.

Vreugdenhil, H.J., Mulder, P.G., Emmen, H.H. and Weisglas-Kuperus, N. (2004) Effects of perinatal exposure to PCBs on neuropsychological functions in the Rotterdam cohort at 9 years of age. *Neuropsychology 18,* 185-193.

Wade, M.G., Parent S., Finnson, K.W., Foster, W., Younglai, E., McMahon, A., Cyr, D.G. and Hughes, C. (2002). Thyroid toxicity due to subchronic exposure to a complex mixture of 16 organochlorines, lead, and cadmium. *Toxicology 67,* 207-218.

Walkoviak, J., Wiener, J.A., Fastabend, A., Heinzow, B., Schmidt, E. and Steingruber, H.J. (2001). Environmental exposure to polychlorinated biphenyls and quality of the home environment: effects on psychodevelopment in early childhood. *Lancet 359,* 1602/1607.

Wang, S-L., Su, P-H., Guo, Y.L., Chou, W-L., Päpke, O. (2005). In utero exposure to dioxins and polychlorinated biphenyls and its relations to thyroid function and growth hormone in newborns. *Environ. Health Perspect. 113,* 1645-1650.

Wang, R.Y., Needham, L. and Barr, D.B. (2005b). Effects of environmental agents on the attainment of puberty: Considerations when assessing exposure to environmental chemicals in the national children's study. *Environ. Health Perspect. 113,* 1100-1107.

Warring, R.H. and Harris, R.M. (2005). Endocrine disrupters: A human risk? *Mol. Cell. Endocrinol. 244,* 2-9.

Weisglas-Kuperus, N., Patandin, S., Berbers, G.A.M., Sas, T.C.J., Mulder, P.G.H., Sauer, P.J.J. and Hooijkaas, H. (2000). Immunological effects of background exposure to polychlorinated biphenyls and dioxins in Dutch preschool children. *Environ. Health Perspect. 108,* 1203-1207.

Weisglas-Kuperus, N., Vreugdenhil, H.J.I. and Mulder, P.G.H. (2004). Immunological effects of environmental exposure to polychlorinated biphenyls and dioxins in Dutch school children. *Toxicol. Lett.* 149, 281-285.

Wilson, R.D. and Baird, P.A. (1985). Renal agenesis in British Columbia. *Am. J. Med. Genet. 21,* 153-169.

Winneke, G., Bucholski, A., Heinzow, B., Kramer, U., Schmidt, E. and Walkowiak, J. (1998). Developmental neurotoxicity of polychlorinated biphenyls (PCBS): cognitive and psychomotor functions in 7-month old children. *Toxicol. Letters 102-103,* 423-428.

Winneke, G., Walkowiak, J. and Lilienthal H. (2002). PCB-induced neurodevelopmental toxicity in human infants and its potential mediation by endocrine dysfunction. *Toxicology 27,* 161-165.

Wolff, M.S., Camman, D., Gammon, M. and Stellman, S.D. (1997). Proposed PCB congener groupings for epidemiological studies. *Environ Health Perspect. 105,* 13-14, 1997.

Zoeller, R.T. (2006). Endocrine disruptors: Do family lines carry an epigenetic record of previous generations' exposures? *Endocrinology 147,* 5513-5514.

In: Industrial Pollution Including Oil
Editor: Harry Newbury and William De Lorne ISBN 978-1-60456-917-9

Chapter 6

ENERGY RECOVERY SYSTEMS FROM INDUSTRIAL PLANT WASTE: PLANNING OF AN INDUSTRIAL PARK LOCATED IN THE SOUTH OF ITALY

Silvana Kühtz*, Francesca Intini, Sara Bellini and Giovanna Matarrese
Università della Basilicata, Facoltà di Ingegneria DIFA – via Lazzazzera, 75100 Matera, Italy

ABSTRACT

In this chapter, we compare environmental and technological aspects of some innovative energy recovery systems from industrial waste. We present the results of a research study that we are conducting on an Italian firm that produces polyethylene terephthalate (PET) supports for waterproof membranes from plastic bottles. For this firm (and all of the firms in the same industrial park), the waste represents only an undesired cost rather than a potential energy source. We first compared a traditional thermal waste treatment with a molecular dissociator and then with a specific gasifier. All three technologies can be fed with basically every type of waste, and can produce electric and/or thermal energy. In particular, the latter two produce syngas that can be burned after depuration to produce energy. A cost-benefit analysis is then carried out to plan how the whole industrial park can use the industrial waste to produce the energy it needs, with economic and environmental benefits for all.

INTRODUCTION

The gradual depletion of environmental resources is one of the most complex and pressing problems facing the world today. Too often in the past people have failed to take into account that the environment is not an inexhaustible resource and that its indiscriminate use would have led sooner or later to a highly critical situation. Waste management in industrial parks is a particularly relevant issue for the purpose of minimizing the global impact on the environment and introducing an advanced model of industrial ecology.

The plans to set up a dense network of relations within an industrial park, aiming to achieve shared goals of environmental and economic performance, have been carried out through the creation of eco-industrial parks. In this chapter, we are going to analyze various waste disposal methods by identifying their specific skills in the ecological exploitation of waste products and the respective plant/environment interactions.

To this purpose, it is necessary to introduce innovative methods and tools of analysis, design and planning, which take into account the complex relationships between the environmental variables.

Life cycle assessment, combined with cost-benefit analysis, is part of the methodological apparatus developed for an objective environmental assessment aimed at comparing alternative scenarios, and it can be used as a decision support tool for the framing of industrial and environmental policies.

In order to make our analysis more realistic, our comparative assessment of technological and management aspects of the various systems present in the waste to energy sector has referred to the results of a research study we have conducted on an Italian firm that produces polyethylene terephthalate (PETP) support membranes from plastic bottles, and which regards (like all of the firms in the same industrial park) industrial waste only as an undesired cost, whereas it could represent an energy source.

1. A Strategy for Sustainable Management of Industrial Parks

The environmental management of industrial parks was a new topic until some time ago. The situation has evolved rapidly, thanks to the introduction of regulations and directives concerning this sector, and it has increasingly drawn the attention of public and private institutions which are becoming increasingly aware of the need to promote sustainable development, especially in terms of energy saving and of improving life and work conditions. These notions, that only a few years ago were considered too innovative and as potential obstacles to industrial development, are today topics of wide interest.

1.1. Environmental Qualification of Industrial Parks

The issue of the environmental qualification of industrial sites is relatively new, even on an international level: it is only since the early 1990s that the U.S., together with Asia and Europe, has witnessed the spread of voluntary experimentations aimed at creating industrial areas equipped with tools for minimizing the impact on the environment [1].

Eco-industrial parks (EIP), as theorized by Lowe [2], are *networks of manufacturing companies and service firms which are linked by a joint management and commit themselves to improving their own environmental, economic and social performances, by collaborating*

* silvana.kuhtz@unibas.it

in dealing with environmental issues and resource use (including energy, water and minerals). This integrated approach pursues collective benefits that are greater than the sum of individual benefits that each company would obtain individually from the optimization of its own performances. The strategies to achieve such a goal require a new design or a revitalisation of infrastructures and of industrial park planning, cleaner production, protection from pollution, energy efficiency and business cooperation.

We distinguish three basic categories of eco-industrial projects:

1) Eco-industrial park or estate (EIP) – an industrial park developed and managed as a real estate development enterprise and seeking high environmental, economic, and social benefits as well as business excellence.
2) By-product exchange (BPX) – a set of companies seeking to utilize each other's by-products (energy, water, and materials) rather than disposing of them as waste.
3) Eco-industrial network (EIN) – a set of companies collaborating to improve their environmental, social, and economic performance in a region (see Figure 1).

We believe these distinctions are important to maintain, although there are various ways projects can overlap. EIPs and EINs may include by-product exchange programs. One or more EIPs may participate in either a BPX or an EIN [3].

An eco-industrial network may include stand-alone companies, companies in industrial parks, and the park management organizations. EIN members collaborate to enhance their performance and to create shared services and facilities. One form of collaboration is to exchange by-product materials, energy, or water among companies, when feasible.

Hence, the strategic objectives of environmental performance pursued by the EIPs are:

- efficient resource use
- environmental impact reduction
- management of interactions between the environment and surrounding communities.

Some of the shared services may be: environmental management systems of a single production cycle, logistics, by-products exchange, recruitment of other businesses, external promotion, management of green areas, etc.

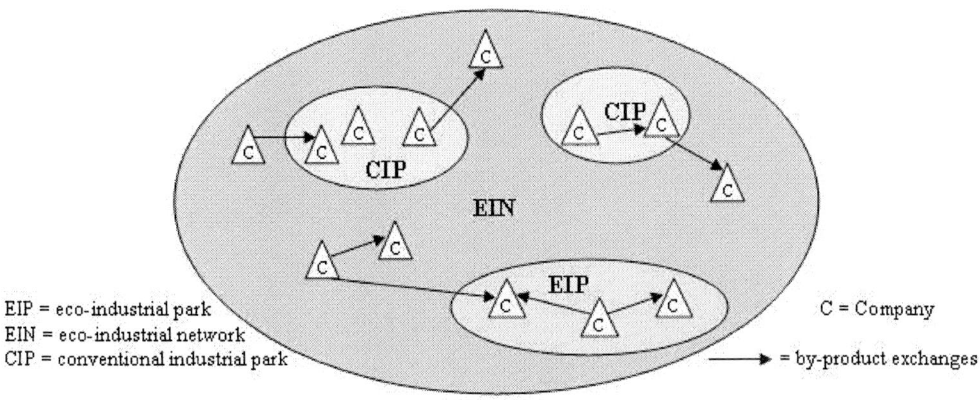

Figure 1. Eco-industrial network.

1.2. Principles of Industrial Ecology

Eco-Industrial Parks take up the principles developed by industrial ecology, which is the study of material and energy flows, with the aim to reduce significantly resource use and pollution. Indeed it proposes to apply to industrial systems and to their transformation-production cycles the same rules and principles governing the functioning of non-human biological systems that are the ecosystems, characterised by symbiotic relationships and by the absence of the idea of waste: each waste product is put back into the system, to produce energy or as a raw material, in order to start a process essential to maintain a global balance.

According to Allenby's scheme [4], traditional industry follows a linear process: the consumption of energy and materials to produce goods and services generates a significant amount of waste (Figure 2). Such a system could operate in a sustainable manner only if the resources fed and the space to dispose of the waste are unlimited.

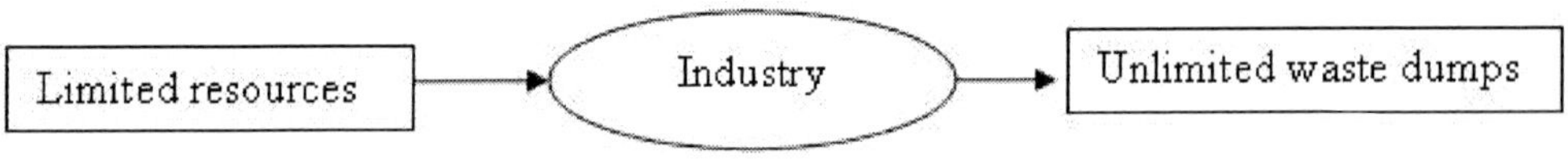

Figure 2. Traditional industry.

An ecological system, by contrast, is characterised by the right combination between dynamic balance and closed loop, Figure 3. In a dynamic balance, energy and waste are constantly recycled and reused by other organisms and processes within the system. In a perfect closed-loop system, only solar energy (or other renewable energy sources) should come from the outside, while all by-products should be constantly reused or recycled. Nevertheless, a total cycle closure cannot be achieved at the level of eco-districts, because at least the final products have to be released outside the system. This model brings about also the mutual dependence between the various parts of the system, along with problems and advantages connected to it.

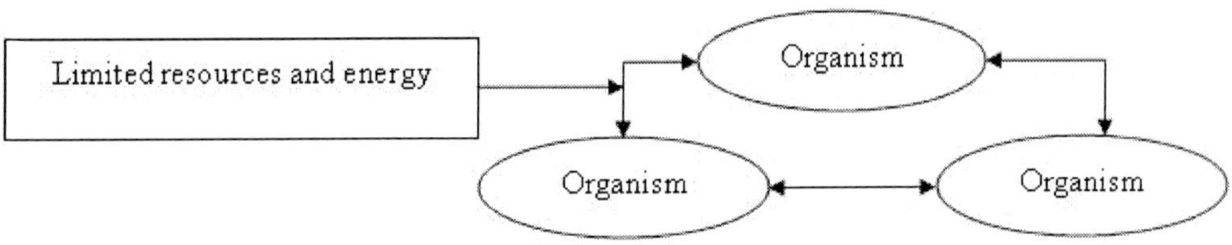

Figure 3. Ecological system.

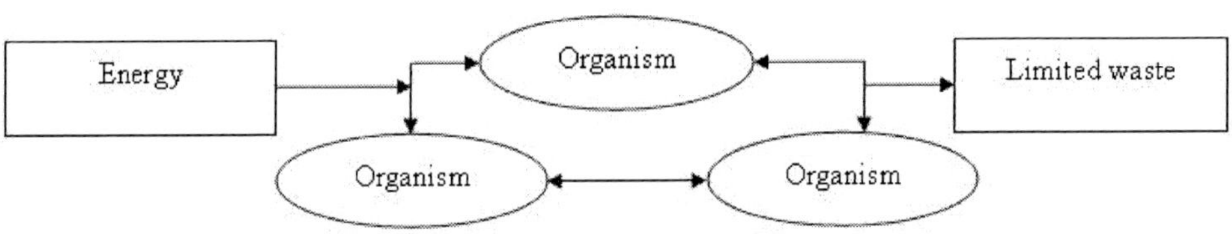

Figure 4. Eco-industrial parks.

Therefore the goal to be pursued is a limited use of inputs (virgin resources and materials), a limited amount of waste discharged from the system and, above all, a collaboration between the various components of the industrial ecosystem through the exchange of energy and materials, Figure 4. This model goes beyond the principle of "product stewardship", as it also takes into account the whole process, along with its waste, and it requires business cooperation, both vertical and horizontal.

There are therefore two possible approaches:

1) the creation of *symbiotic* production processes with regard to material flow (energy, heat, water, waste, emissions, people, goods...), whose aims of efficiency and effectiveness in the use of materials and energy are the basis of business cooperation;
2) the planning of services and infrastructures within the industrial parks applying the principles of environmental sustainability and closed-loop natural cycles.

Consequently, the main requirements for service sharing and for a successful industrial symbiosis are the following:

- complementary needs as far as material/energy are concerned;
- physical proximity in order to implement cascade systems of energy and water supply and to cut transport costs;
- presence of a homogeneous service demand in order to obtain economies of scale;
- importance of personal relationships between the various enterprises, also by means of associations.

Industrial parks are important tools of economic and social qualification of an area but, at the same time, they are potential sources of pressure on the environment and on surrounding communities.

1.3. Ecologically Equipped Industrial Areas

The notion of industrial park as an *eco-friendly place* (as well as an area of economic and urban development) has evolved in time and has led Italy to the construction of the so-called Ecologically Equipped Industrial Areas (in the year 1998). The Italian production system, organised into chains and districts, is characterised by a great flexibility and by the tendency to networking.

But as regards the environmental field, the Italian business sector, whose main feature is a high concentration of small and medium-sized enterprises, encounters a lot of difficulties in implementing preventive measures, due to a lack of culture in such issues.

Joint environmental management, dialogue with local authorities and business participation in the process are key elements in the enhancement of a new territorial governance framing and supporting a sustainable environmental policy on industrial areas. The Ecologically equipped industrial area may be regarded as a forum for environmental dialogue where all the stakeholders involved in its constitution, development and activities share their experiences, their resources and their goals and perform partnership actions aimed at complying with regulations but, in broader terms, at meeting the environmental needs and expectations of the resident industries and the local communities. Through a joint management of shared infrastructures and services, planned in concert with the resident group of enterprises, it is possible to create economies of scale allowing the environmental issues to be settled with a reduction in cost. The introduction of Ecologically Equipped Industrial Areas should not be viewed by stakeholders (enterprises, local authorities) as an external constraint hindering economic development but rather as a tool for revitalizing the area and increasing the enterprises production system's competitiveness. The aim of joint environmental management is to provide mutual benefits to industries, public authorities and local populations. The enterprise system is evolving and territories are being provided with policies and tools aimed at their strengthening. Industrial parks as well may take part in this change process, closely meeting businesses and citizens' needs and becoming one of the most effective strategies toward growing territorial competitiveness from an economic, social and environmental point of view.

The introduction of this new concept of industrial area, provided with technical and organisational tools aiming to minimize and manage jointly the pressures on the environment, responds to the need to replace the so-called *end of pipe* approach (combating pollution as a last stage of the process) with the precautionary principle preventing pollution. In particular, it is not meant just to supply industries with specific environmental equipment, as it has been so far, but to organize the industrial site so as to help, from an economic and technical point of view, each enterprise in the area achieve their environmental goals, be they prescriptive or voluntary.

1.4. Generation of Industrial Waste in Europe

Manufacturing industry waste comprises many different waste streams arising from a wide range of industrial processes. Some of the largest waste generating industrial sectors in Western and Central Europe include the production of basic metals, food, beverage and tobacco products, wood and wood products and paper and paper products [5].

It has been estimated that over 33 million tonnes of industrial waste were generated in Europe in 1998. Waste from the manufacturing sector continues to rise, despite national and international declarations to reduce waste from manufacturing industry, to introduce cleaner technologies and other waste minimisation initiatives and to work towards manufacturing practices that are sustainable in the long term.

The manufacturing industry has a central role to play in the prevention and reduction of waste as the products that they manufacture today become the wastes of tomorrow. Manufacturers can achieve innovative solutions:

- considering the impacts of their products throughout its life at the design stage of the product (LCA);
- using manufacturing processes that minimise material and energy usage;
- eliminating or reducing where possible the use of substances or materials hazardous to health or the environment;

- manufacturing products in such a way that they last longer and may be recycled or reused at the end-of-life stage.

EU and government policy across Europe is increasingly driven by the need to influence manufacturing practices in an effort to decrease the environmental impact of products during their manufacture, use and end-of-life.

The EU Strategy for Sustainable development emphasises the strategic target to break the link between economic growth, the use of resources, and the generation of waste.

Therefore natural resources and wastes is one of four key environmental priorities described in the 6th European Action Programme. The objective is to decrease the amount of waste generated and to achieve a relative decoupling between economic growth and generation of waste.

Europe aims to reduce the final amount of waste by 20% by the year 2010 and by 50% by the year 2050.

Figure 5 shows the amounts of manufacturing waste generated in nine Western European (WE) countries and seven Central and Eastern European (CEE) countries [6]. From 1996 to 2002 the generation of waste in the manufacturing sector increased by 15% in WE (from 123 Mtonnes to 141 Mtonnes) and by 42% in CEE (from 60 Mtonnes to 86 Mtonnes).

Due to lack of proper time series data from the large countries UK, France and Spain are missing in the indicator. The three countries are responsible for about half of the manufacturing waste generation in WE. The few data available from the countries indicate a slight decrease in waste generation.

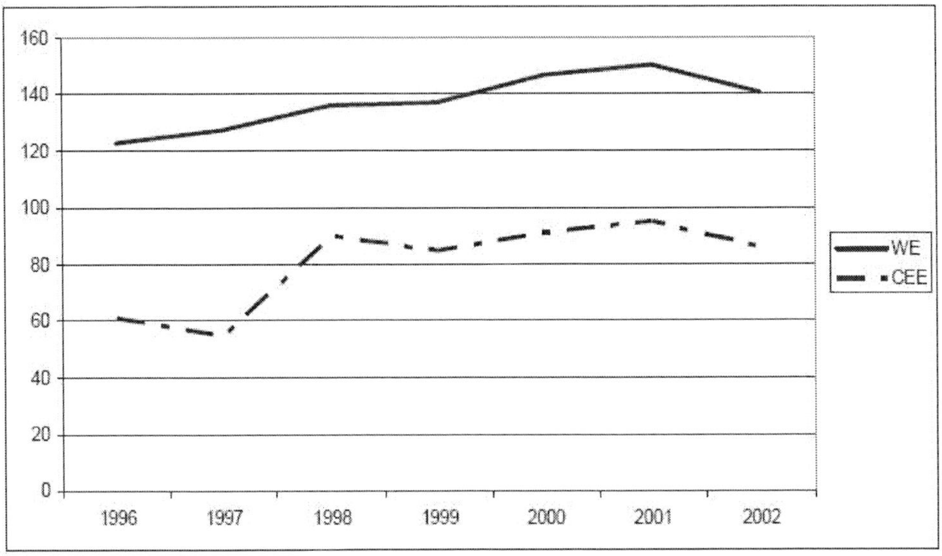

WE: Belgium, Germany, Denmark, Finland, Iceland, Italy, Netherlands, Norway and Portugal.
CEE: Czech Republic, Croatia, Malta, Poland, Romania, Slovakia and Slovenia.
Source: Wastebase/Eurostat NewCronos.

Figure 5. Manufacturing waste generation in Western Europe and Central Eastern Europe in Mtonnes.

Figure 6 depicts generation of manufacturing waste held against the gross value added in the manufacturing sector in the years 1996 to 2002 in the 13 European countries for which data are available. The gross value added by the manufacturing sector is steadily rising in the period (6%), but not as much as the generation of manufacturing waste (24%).

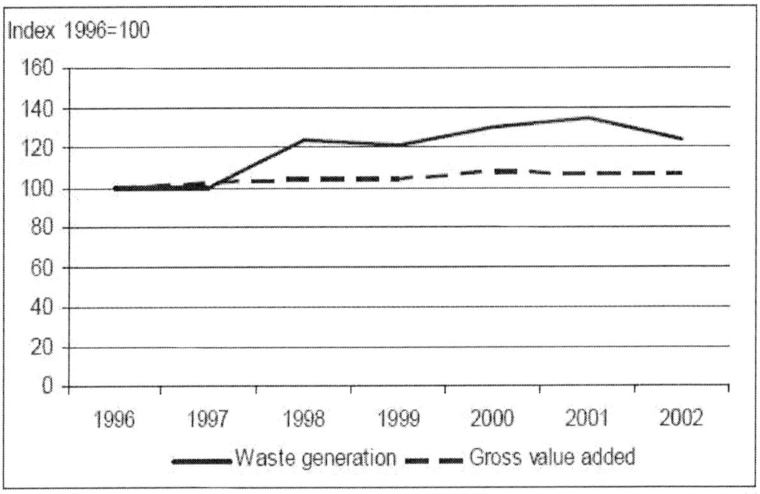

Countries included are Belgium, Croatia, Czech Republic, Denmark, Finland, Germany, Iceland, Italy, Netherlands, Norway, Poland, Portugal, Romania, Slovakia and Slovenia.
Values are indexed with 1996 = 100.
Sources: Waste data from Wastebase/Eurostat NewCronos., economic data from the World Bank.

Figure 6. Manufacturing waste generation in Europe compared to gross value in manufacturing sector.

This means that overall objective of decoupling waste generation in Europe from economic development has failed as regards manufacturing waste. The European manufacturing sector is very diverse and the individual countries have had very different trends in the period analysed. The growth at European level can mainly be explained by the massively rising waste amounts in the three countries Italy, Netherlands and Poland, while most other countries have stagnating or even falling waste amounts.

Figure 7 shows the national differences in manufacturing waste generation per capita as well as the temporal trends for 1996-2002 for a range of European countries. In nine countries a declining or steady trend can be observed, while seven countries have increasing amounts of manufacturing waste per capita. The manufacturing waste generation per capita is generally higher in the CEE countries (1000 kg per capita) than WE countries (ca. 700 kg per capita). Finland has a very high manufacturing waste generation per capita. However, Figure 7 does not reveal the reasons for the observed trends and distributions, and establishing causal relations to driving forces are beyond the scope of the indicator framework.

Theoretically, a given level of generation per capita could be explained by two factors:

1) the reliance of a country on the manufacturing sector (measured, e.g., as the share of the sector in the total economy);
2) the resource efficiency of the manufacturing sector and the structure of the manufacturing sector.

The above figure shows how important it is to also be aware of the trends inside the countries since this varies a lot and perhaps shows how the resource efficiency varies and changes within the countries.

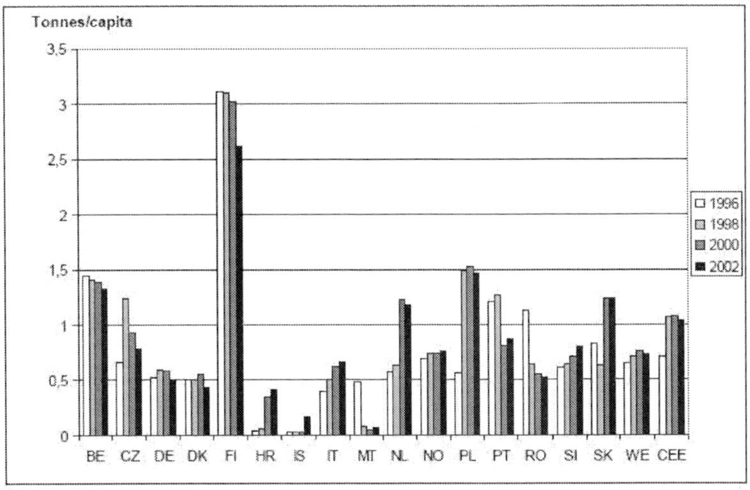

Sources: Waste data from Wastebase/Eurostat2004, population data from the World Bank.

Figure 7. Trends in generation of manufacturing waste per capita.

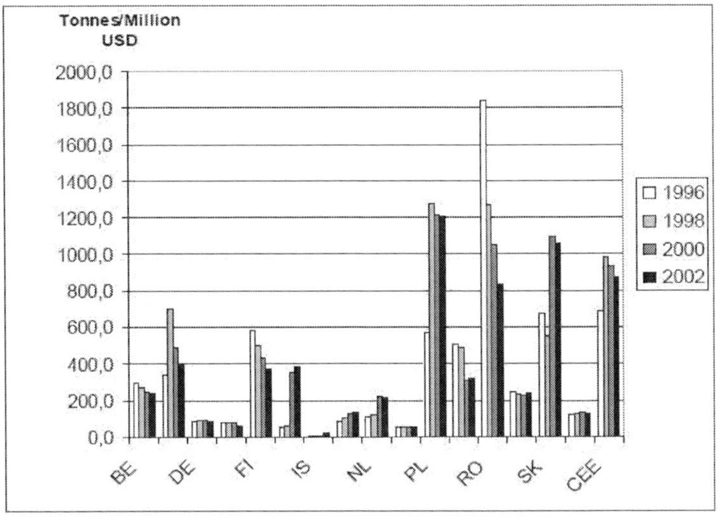

Figure 8. Trends in manufacturing waste generation per gross value added in the manufacturing sector.

Figure 8 shows the manufacturing waste generated per gross value added[1] in the manufacturing sector.

[1] In national accounts such as the United Nations System of National Accounts (UNSNA) or the NIPA's, gross value added is obtained by deducting intermediate consumption from gross output. Thus gross value added is equal to net output. (http://en.wikipedia.org/wiki/Value_added)

It can be seen that relatively small decoupling is seen in Belgium, Denmark and Germany while the waste intensity in the CEE countries (871 tonnes per million USD) are many times higher than the WE countries (129 tonnes per million USD). Romania, Poland and Slovakia have the highest waste intensity while Iceland, Norway and Denmark have the lowest. There is a falling tendency in five countries while the waste intensity is increasing in five other countries.

Manufacturing waste is typically heterogeneous and with differentiated range of potential impact on the environment. Although different wastes cause different environmental impact potential, there is also a quantitative aspect as more waste means potentially higher pressure on the environment.

The industrial structure, in terms of the prevailing sectors and branches, is an important factor for the amount and the kinds of waste generated, some sectors being the source of large quantities of hazardous waste with high environmental impacts. Waste also represents a loss of resources both in the form of materials and energy. Indeed, quantities of manufacturing waste can be seen as an indicator of the material efficiency of the manufacturing sector.

Excessive quantities of waste result directly from a large industrial production and inefficient production processes. Indirectly it reflects low durability of goods and unsustainable consumption patterns leading to high demand for manufacturing products and thereby potentially high waste quantities. Waste minimisation in manufacturing production processes is an increasingly important objective of manufacturing environmental strategies also in the framework of environmental certification. Different countries and industrial sectors are differently positioned in the process of optimising resource use, factory-level waste minimisation and the adoption of certification instruments.

2. The Main Final Thermal Treatments of Waste

The thermal treatment which takes out energy from waste includes also combustion and gasification processes. Combustion and gasification technologies of alternative fuels, like biomass, fuels from wastes and from other types of industrial wastes, are gaining importance. The energy conversion technologies are increasingly constrained by environmental regulations on regional, national and international level. So it is necessary to move towards advanced technologies like gasification systems in combined cycles.

Combustion and gasification processes of non-conventional solid fuels increasingly require the integration of different skills, due to the high variability of physicochemical parameters of the initial fuels. The nature of fuels influences choices in the processes and systems. The complicated fluid-dynamic behaviour of multiphase systems is another factor of influence [7]. So, when analysing the thermo treatment processes with energy recovery from solid urban wastes and industrial wastes it is necessary to look for their environmental and energetic performances.

The three technologies analyzed in this chapter are: a *waste to energy* plant of the latest generation, a *molecular dissociator* and a *gasifier* licensed in Italy. The three of them aim to dispose of every kind of waste, at the same time producing electricity and/or thermal energy but in different ways and with different results.

2.1. Waste to Energy Plant – System *a*

Although the traditional waste to energy plant uses a more obsolete technology, studying it helps to understand evolution and innovations in energy efficiency occurred in the last few years [8].The primary purpose of waste to energy plants, also called incinerators (waste burning plants) with energy recovery, is to burn waste (at 900°–1000°C) and make it biologically and chemically inert, by reducing its volume, with a great advantage that is the production of electricity and thermal energy. The heat produced by the combustion, generates steam which is used for energy production and for the heating [9].

Municipal waste incineration is a controlled combustion process which oxidizes the organic substances contained in waste, producing simple molecules in a gaseous state at standard temperature (gas): organic carbon is oxidized to carbon dioxide (CO_2), hydrogen to water (H_2O) and sulphur to sulphur dioxide (SO_2); generally the inorganic part of the waste is not subject to reactions and becomes a solid waste to be disposed of and/or to be recovered (ash or slag). Also unpleasant chemical transformations take place, such as the creation of nitrogen oxide (NO), hydrochloric acid and hydrofluoric acid, which toxic characteristics.

The cycle of wastes combustion is divided into the following phases:

- acceptance and storage;
- feeding;
- combustion;
- cooling of combustion smokes and warmth recovery;
- treatment of combustion rests: ashes, smokes, slags, waste waters.

Waste combustion phase is performed in furnaces and it involves three phases: drying, starting and combustion. In order to meet various needs, several kinds of furnaces have been introduced: moving grid, fluid bed and rotary drum. During combustion the smokes come out at a temperature of 1000°C, so it is necessary an intermediate phase of cooling in a combustion section and the extraction of heat from the smokes. But these smokes embed pollution elements. It is possible to treat these smokes. Polluting gas control equipment can be performed by mixing smokes with chemical elements (solids, liquids or gas) able to remove one or more polluting components. Often the absorption process on solids are used, like active carbon.

Another problem of wastes incineration is due to solid residual products and to their impact on the environment. The incinerators produce two kinds of solid wastes: slags discharged by furnaces, and fine particles (flying ash mixed to smoke) seized by smoke treatment. It is difficult to manage them because they embed large quantities of toxic compounds, like heavy metals and organ chlorinated compounds easily released in the environment through percolation.

2.2. Molecule Dissociation – System *b*

The molecular dissociation process includes the thermo chemical conversion of a carbon-composite into a burning gas [10].

Briefly, the process consists in a chemical splitting up of complex organic molecules into simpler ones through thermal gasification at low temperature (T = 400°–500°C) performed in controlled conditions of warmth and oxygen availability (max. 3–4% of O_2) inside a locked cell. Here, carbon chemical links are broken and simpler molecules take origin, thanks to warmth effect during starting phases of fuel use (i.e., methane gas). The process is very slow and the whole cycle lasts 24 hours; furthermore, the reduced turbulence and the flames absence lessen the powder presence in smokes. After breakdown, carbon matrix changes from a solid condition into a gas one, by producing a synthesized gas (syngas) with good combustible characteristics. The synthetic gas after depuration stage can be burnt in the presence of air (T≈1100°C) in a following stage, where there are thermodynamic cycles, like co-generator cycle. The systems can use several thermodynamic cycles, consequently they are more flexible in regulation between heat and electricity production. The remaining waste is composed of ash, inert and non toxic, and it can be disposed of in a dump or used for tiles construction or mixed with asphalt.

2.3. Gasifier Licensed in Italy – System *c*

A gasifier with a new technology was licensed in Italy. It is different from the traditional gasifier and the molecular dissociation thanks to the gasification reactor and to different conditions of management and control.

Gasification is a chemical decomposition produced by thermal energy. Without a surplus of air and oxygen, gasification produces thermo chemical decomposition of waste's organic matter and change this matter into *syngas* [11].

The system which produces the above-mentioned process involves a lot of sections, that can be listed according to their functions:

- *materials preparation* section: this involves the storage building and machinery where wastes are treated and stored (raw materials for the plant) in order to make their size suitable for feeding the reactor where the gasification process takes place. These treatments are usually meant to transform the waste into waste fuels (dry and crushed municipal solid waste).
- *gasification section*: this is the technological heart of the system. Thanks to the technology licensed in Italy, a specific reactor has been designed (with new characteristics), with vertical development, descending flow and best temperature of process at 1200°C. The waste is fed at the top of reactor and by falling down from top to bottom (as a fluid bed) is covered by flames come out from thermic lances which are tangentially in reactor. These lances produce high turbulences which create an immediate gasification. Thermic lances fed by pure methane and oxygen characterize this process which grant high ecological and energetic results. Another peculiarity of this reactor is that it can be fed continuously, wih all the wastes categories independently from their calorific value and chemical composition.
- *energy generation* section: it is the final step where the syngas is used. Nowadays the groups with endothermic/alternator engine are more used thanks to their high electric (about 38% compared to the fuel input) and global efficiency. Thermal recovery

takes place during the cooling cycle (engine side, smoke side and oil engine side) of the machine. This brings to total efficiency even higher than 80/90%. The temperature of the thermal energy recovery is about 80°C in case of water use, or 6–8 bar of pressure in case of steam production.

The solid wastes of gasification are transformed into glass, thanks to high temperatures, and are completely inert.

2.4. Comparison among Systems for Waste Disposal with Energy Recovery

It is necessary to point out which is the most efficient technology from an environmental and an energetic point of view. This comparison considers the following aspects:

- structure-process;
- performance;
- environmental impact.

From the structural-processing point of view, *system a* (*waste to energy plant* described on paragraph 2.1 and here called *system a*) reduces waste in ash, by using again combustion smokes to produce electric and/or thermal energy, whereas *system b* (*dissociator,* described in paragraph 2.2 and here called *system b*) and *system c (gasifier*, described in paragraph 2.3 and here called *system c*) transform (in different ways) the raw material into a synthetic gas used then as a fuel.

The main body of *system a* is the furnace of combustion, in *system b* is the cell and in *system c* is mono-tube reactor with vertical development, the last two produce syngas. Another important aspect is the very high chimney for *system a* compared to the chimney a few meters high for *system b.*

Despite the fact that the final product is the same, there are a lot of differences between *b and c systems:* the main body of the system, the working temperature, the procedures and the scheme for syngas production.

System b works at about 400°C, in presence of oxygen at 4–5%, so the raw material transformation is exothermic; so it only triggers the reaction with a moderate electric resistance (200-300kW per 100 t/g of waste).

System c has a different mechanism. Its reactor has a variable thermic profile (optimum temperature at 1200°C). The waste falls down in the reactor from top to bottom, and it is burnt by flames generated by thermic lances which are fed with methane and oxygen. This lets the process be controlled, independently from changes of waste calorific value, but with unavoidable economic consequences (262 m^3/h of methane and 560 m^3/hof oxygen per 3 t/h of input waste are necessary).

System b has a lower self-consumption (less than 2%) and it can treat any kind of waste without a pre-treatment. Furthermore, both the gasification technologies dimensions fill a minimum space, and have a modular process; *system a* does not present these advantages. The energetic results depend on the choice of the system for the production of electric and/or thermic energy. Analysis results, listed in Figure 9, show how syngas brings higher efficiencies. Moreover, *system b* produces very low percentages of inert ashes, about the 3–4%, compared to 30% of toxic wastes produced by *system a.*

For systems *b* and *c* it's important to emphasize the conversion energetic efficiency, that is 0.98 for the first and 0.68 for the latter, and also the percentage composition of syngas, (see Figure 10).

System c offers a cleaner gas compared to *system b,* with high concentrations of hydrogen and carbon monoxide, and low percentage of nitrogen and negligible sulphur compounds.

ENERGY EFFICIENCY		
TECHNOLOGIES	**SYSTEM TYPE**	**EFFICIENCY**
Molecular Dissociator	Steam Turbine System	28%
Molecular Dissociator	Combined cycle system	33%
Waste to energy plant	Steam Turbine System	26%

Figure 9. Energy efficiencies (expressed in percent) of system a and b.

SYNGAS COMPOSITION ITALY LICENSED GASSIFICATOR	
Elements	**After Depuration**
H_2	42,18%
CO_2	10,95%
CO	31,38%
N_2	10,60%
CH_4	4,25%
H_2S	0,68%

SYNGAS COMPOSITION	
Elements	**After Depuration**
H_2	14% - 18%
CO_2	7% - 10%
CO	23% - 27%
N_2	42% - 55%
CH_4	1% - 3%
Dust	< 50 mg/m^3
Sulphur comp.	< 1150 mg/m^3
TAR Remaining	< 5 mg/m^3
Nitrogen comp.	< 55 mg/m^3

Figure 10. Percentage composition of syngas.

Despite the sophisticated instruments for the reduction of smokes, from an environmental point of view, the smokes of *system a* present organic material deriving from uncompleted combustion or thermic synthesis reaction. *Systems b and c* generate syngas in good conditions, so they do not give rise to hazardous dioxins and furans. Experimental data demonstrate that emissions to atmosphere generated by the combustion of the synthetised gas of *system b and c* are lower than emissions produced by *system a*, especially when it comes to sulphur, nitrogen and chlorine compounds.

3. Environmental and Economic Assessment Methods

In the industrial areas analysis, the relationship between industry and its local environment misses a main dimension of sustainable development: the environmental protection, that is the connection between products and pollution phenomena linked to them.

The main purpose of a sustainable development consists in granting a high quality level of life to all people, by taking less natural resources and using them as efficiently as possible. In other words the energetic resources management needs measures and means related to industrial systems eco-efficiency.

The environmental and economic assessment methods let the industries analyse the environmental matters considering not only the production but also the resources that is possible to re-use and save (energy, materials, water, natural resources).

The decision-taking processes in the environmental field depend on the purposes to be attained, as those purposes determine the demands which to answer to, the actions to undertake and the analytical and procedural means to use [12].

A large number of instruments have been created as methodological aid for strategies and techniques of environmental management. The choice of the most appropriate instrument depends first of all on the object to be studied, on the purpose of decision process, the space and time context of the project, the available information about environment impacts, and on the doubts about costs and benefits. Anyway, a unique method to establish the most appropriate instrument does not exist, therefore often multiple instruments with the same technical aids and the same information are used in an integrated way. Life Cycle Assessment and Cost Benefit Analysis (CBA) are the major instruments, which emphasise environmental and socio-economic aspects in order to make the best choice.

3.1. Life Cycle Assessment

Life cycle assessment (LCA) is the broadest indicator and an internationally standardized method (ISO 14040 and ISO 14044). It not only evaluates the impact on climate change, but also other impact categories such as acidification potential, eutrophication potential, ozone depletion potential, and ground level ozone creation. For each of these impact categories, the product or system is evaluated over its complete life span, from the extraction of raw material and manufacturing, to the use of the product by final consumers and end-of-life processes like recycling, energy recovery, and ultimate waste disposal. The ISO standards provide robust and practice-proven requirements for performing transparent LCA calculations. Moreover, one can make use of extensive databases containing life cycle profiles of many goods and services, as well as many of the underlying materials, energy resources, transport systems, etc. Nevertheless, LCA calculations remain very complex [13].

In order to achieve the established purposes in a transparent and comprehensible way, it is necessary to take into consideration and describe the following elements:

- The function unit: measurement of input and output flow of the product system;
- The product system: an ensemble of processes linked by flows of matter, energy, wastes;

- The system's borders: selection of productive processes and environments;
- High quality and reliability of data (data bank, temporal, geographic and technological coverage).

The inventory analyse involves the construction of reality analogical model. First of all the flow-chart including the product system's processes is necessary [14]. Consequently, the second step is data collection, that is direct surveys, estimations, medium values or values from literature and data bank. The inventory aim is to get objective data, which, later on, would be processed in order to take useful assessments for an improvement of processes analyse (see Figure 11).

The environmental and energetic allocation could be carried out as:

- physical quantity, such as mass, volume and energy;
- economic values of the products.

The results are estimated and associated under their effects on the environment, such as greenhouse gas effect or stratospheric ozone deletion. Environmental effects can be on global, regional, local scale: for example GHG effect has global effects, while a noise has local effects.

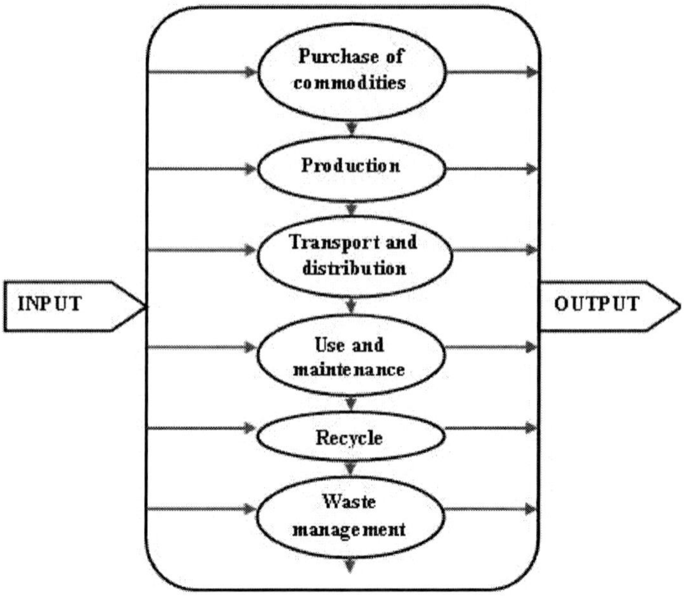

Figure 11. Inventory analysis.

Nowadays in life cycle analyses the main impact categories are:

- greenhouse effect;
- stratospheric ozone deletion;
- acidification;
- eutrophication;

- photochemical oxidant formation;
- human and environment toxicity;

The environmental impacts quantification is performed in the phase of distinction thanks to scientific models and correspondence's factors, on an international level (i.e., greenhouse effect follows global warming potential, that is, the equivalent CO_2 quantity). To synthesize the analysis and compare it with more productive systems, it is necessary to make the data normal (i.e., to estimate the pro-capite environmental charge) and to carry out a weighing in order to get a single index of impact. In the last phase of LCA, that is the interpretation, the results of inventory analysis are correlated with those ones of impacts analysis, to propose better measures for a reduction of environmental charges.

Life cycle assessment is used for more aims: processes' improvement, products innovation under new standards of production, assessment of carbon footprint, progress of environmental policy strategies.

On a planning and organizing level it is very hard to reach a production improvement, which consists in choosing the solutions to be applied to productive system and suitable to maximize the global environmental energetic efficiency.

LCA application does not always guarantee a reduction of energy consumption or emissions, but it allows people to evaluate a service or a product in a total way, in order to avoid a wrong interpretation, i.e. an improving intervention that moves a problem from the analysed area to another one.

3.2. Cost Benefit Analysis CBA

CBA is one of the most popular method for investment assessment, especially when they refer to support measures on territory; it chooses the best proposal, or, in case the proposal is only one, it verifies the project costs, that should be lower than the benefits, in order to improve the general social and economic welfare of the project context [15].

CBA is divided into three phases:

1. *identification of all positive and negative effects of the project.* The negative effects involve: water consumption, wastes production, etc. The positive effects could be: employment, water availability for civil agricultural or industrial use, the creation of a green area or a system for waste water conditioning.
2. *quantification of effects from the previous phase.* The costs and the benefits of a project are expressed in monetary terms, while the other effects are expressed in their own measure units (jobs number, decibel of produced noise, etc.).
3. *interpretation in monetary terms of all the effects measured in quantification phase.* There are direct financial costs and benefits (expenses and revenues of project) and indirect economical ones (damages and benefits caused by project to other activities).

CBA must pay attention to the beneficiary of project benefits, and to the supporters of direct and indirect costs. When the best project is chosen an ongoing process is established, and it involves more phases, including *identification and assessment* which are the most important. Identification means to consider demographic, social, cultural and economic aspect of social groups.

It is important to include stakeholders in decisional process as they could be stimulated through offsetting measures.

To let the changes in life and habits be accepted, it is important to give population a right information about benefits project, and to establish an active cooperation between population and public decision-maker (particularly regarding rural population); furthermore, identification must show eventual project damages on environment, and suggest some ways to reduce them (external effects).

The more the identification reaches a high and complete level of information, the less the decision-makers have the possibility to discuss faults. The identification phase requires a lot of time in order to avoid errors, as they are expensive and resources are very few.

After realizing the project, benefits could be:

a) *transitory* – they can be carried out only during project realization;
b) *lasting* – linked to the entire life of the project. These lasting benefits are direct, when they give advantages to receivers, and indirect, when they are in favour of the community;
c) *concrete* – they can be quantified;
d) *intangible* – of intangible quantification.

And costs can be divided into:

a) *monetary* – supported directly by public administration and related to complete realization of project, its maintenance and management;
b) *social* – supported by community. They are primary if they refer to projects' costs that the community supports by renouncing other alternative projects; or secondary if they refer to external effects on people or things, but they have no influence on the market.

4. Case Study

The examined case study is an industry leader in the world for production of polyester nonwovens with high toughness by continuing thread and flake. The firm produces an industrial fabric that is a supporter for bituminous membranes destined to waterproofing of roads and roofs, as PET bottle after use. This firm is located in the south of Italy within a technological park that is managed by a company that provides services for the firms in the industrial area, such as electric and thermal energy, technical gas (i.e., steam overheated) and industrial water, with advantageous conditions for the industries, and also sustains the territory industrialization by giving services and utilities similar to the most industrialized areas of the country. Furthermore, it treats the waste of the industrial activities of the industrial area, and also waste coming from activities of regional and extra-regional territories with specific authorization codes (EWC European Waste Catalogue).

The firm in this case study is a partner of the manager company and purchases from it steam, water and electric energy. Before the final discharge, it sends the wastes to the treatment plant [16][17].

The industrial process of nonwovens production is divided into the following phases:

- bottle washing and flake production: from plastic bottles PET material is selected and sent to a washing centre and then to the mill to be reduced into little pieces; at the end bottles are centrifugated to set apart the water, and sent to a washing section. The obtained product is called flake.
- polyester drying-up and extrusion by forming threads set in a random way on a sucked carpet: flakes are heated until they become crystals, then they are dried and extrapolated. After extrusion, the melted polymer gives rise to a film;
- mechanical binding process (needlepunch technology);
- glass thread inclusion between the two films and further mechanical binding;
- thermo-fixture and longitudinal cut, and consequent selvedge production;
- thermal consolidation and product incorporation in resins.

To summarize, the analysis of life cycle defines the functional unit as 1 m^2 of fixed and reinforced nonwoven, and considers one year's time for the system and the process.

In the following inventory analysis, data are collected, measure units are converted, and data refer to their functional unit, as shown in Figure 12.

The diagram in Figure 12 summarizes resources, energy consumption, all emissions (to air, water and ground) and production of waste.

Thanks to software, it has been possible to assess the impacts on the environment (i.e., CO_2eq, CH_4, SO_X e PM10, COD, BOD and solids).

Industrial waste referred to in the considered firm are divided into two categories:

- waste products: these are not dumped, but used again in the production cycle or sold for other purposes (polymer badge, nonmelted selvedges, melted selvedges, felts, roll line starts);
- wastes of various origin (residues of the washing system, iron waste, wood waste, paper, some oils, etc.)

For an economic-environmental assessment, it is necessary to incorporate the system analysis with a cost/benefit analysis by emphasising emissions and energy savings.

If the analysis is extended to the whole area where the firm is located, the processes of all the productive activities have to be analysed in order to find the the best area for fuel co-production.

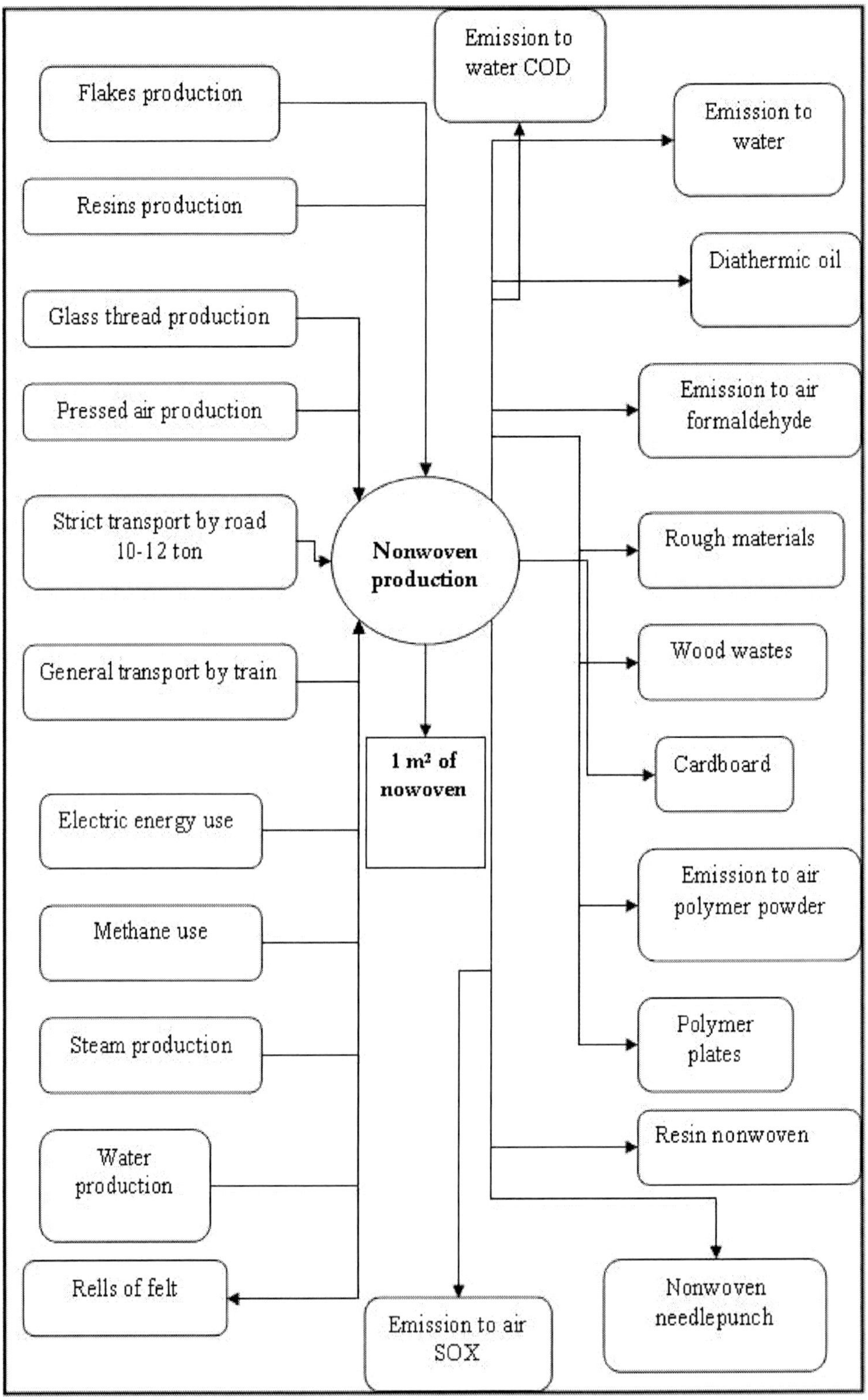

Figure 12. Flow diagram of nonwoven production.

To carry out an energy system from wastes in an industrial park, it is necessary:

- to verify the area and examine the physicochemical characteristics of the wastes;
- to find out the critical elements of the system;
- to study the draw-plate for waste treatment after production and their technologies;
- to make a hypothesis about the logistics of the waste collection;
- to give the system dimensions;
- to assess possible environmental effects in the neighbouring area (air quality; impacts on territory, on noise, on waters, on ground, etc.)
- to emphasize relief and reward measures of impacts, both environmental and social;
- to study the economic/financial feasibility and analyse the economic new effects in the industrial area.

The main benefits of this system are to close the waste cycle at local level, and to use energetic and economic benefits from waste use for energy purposes. Furthermore the disposal of wastes at a local level means a reduction of costs, as otherwise transport costs, eco-tax and local duties must be introduced.

CONCLUSION

Because of the continuing increase in population, the resources of our natural environment are limited: most of the materials we use to create products come from natural sources, and several production systems stil require virgin products rather than recycled ones..

To overcome the crisis of waste management, it is necessary to introduce a *Zero Waste* strategy. This consideration involves rejecting the whole notion of waste, and moves the problem towards the starting point of an industrial eco-system [18].

The traditional industrial system is based on a unique direction: *take-transform-dump* Materials are taken from the earth's crust, moved to firm, used to produce finished goods and waste, destined to the disposal or waste to energy plant. Extraction, production, transport, and removal of resources are causes of environmental destruction and global heating.

It is necessary to revise the unidirectional industrial system and form a closed circular one by recycling the resources dumped from community to industries. The economic and social system must know that the circular course of nature is the most efficient, less expensive and more profitable, and avoids the environment's deterioration.

In the end, it is important to integrate the responsibilities of community, industry and government in order to make a zero wastes strategy feasible.

REFERENCES

[1] Corsari, L. and Stracchini, V. (2006). Linee guida per la realizzazione di Aree Produttive Ecologicamente Attrezzate della Provincia di Bologna. Bologna: Delibera della Giunta Provinciale N. 407 del 21 novembre 2006, in Italian.

[2] Lowe, E. (1997). Creating by-product resource exchanges: strategies for eco-industrial parks. *J.Cleaner Prod.*, Vol. 5 n. 1-2.

[3] Lowe, E. (2001). *Eco-industrial Park Handbook for Asian Developing Countries.* A Report to Asian Development Bank, Environment Department. Oakland,CA: Indigo Development.

[4] Allenby, B. (1992). Industrial Ecology: The Materials Scientist in an Environmentally Constrained World, *Materials Research Bulletin*, 17 (3), pp 46-51.

[5] EEA European Environment Agency (2006). *Generation of manufacturing waste.* Indicator fact sheet.

[6] European Topic Centre on Resource and Waste Management/EIONET. Industrial waste. 11/04/2008. Available from: http://waste.eionet.europa.eu/waste.

[7] Chitone, R. and Russo, G. (2001). Trattamento termico di rifiuti solidi industriali e civili con recupero di energia. *Ricerca e Futuro*, Number 20, in Italian.

[8] Cammarata, G. (2006). Impianti di Termovalorizzazione, Dipartimento di Ingegneria Industriale e Meccanica-Sezione di Energetica Industriale ed Ambientale-Univerisità degli Studi di Catania, Materiale didattico a sostegno del corso di Complementi di Impianti Termotecnica, in Italian.

[9] Arena, U. and Mastellone, L. (1998). Centrali di termovalorizzazione di rifiuti solidi urbani. *La Termotecnica*, 48 (1), in Italian.

[10] BIODEC and EPPM AG (2008). Gasification System- Relazione sulla dissociazione molecolare. Private communication.

[11] Ecologia Informatica (2008). The Waste Remedy. Private communication. Available from http://www.ecologiainformatica.com

[12] Breedveld, L. (2006). Analisi del Ciclo di Vita (LCA): *Metodologie di Valutazione Ambientale*. Milano: Franco Angeli, in Italian.

[13] SETAC Society of Environment Toxicology and Chemistry (1990). *A technical Framework for Life-Cycle Assessment.* Workshop report from Pensacola FL USA; August 18-23.

[14] ISO 14040 (2006). Valutazione del ciclo di vita, principi e quadro di riferimento; CEN, EN ISO 14040:2006.

[15] Brisighella, L. and Bonanno, C. (2006). Tecniche di valutazione: Metodologie di Valutazione Ambientale [DVD]. Milano: Franco Angeli, in Italian.

[16] The Freudenberg Politex Group (2007). *Gestione Rifiuti*. Private communication.

[17] The Freudenberg Politex Srl (2006). Documento di Valutazione del rischio. Private communication.

[18] Incerti, M. (2008). *Rifiuti zero? Serve realismo.* Il resto del carlino, 26/01/2008, in Italian.

In: Industrial Pollution Including Oil
Editor: Harry Newbury and William De Lorne ISBN 978-1-60456-917-9

Chapter 7

CARBON FOOTPRINT ASSESSMENT: A TOOL FOR PROMOTING LONG-TERM POLLUTION EMISSION REDUCTION AND SUSTAINABLE DEVELOPMENT?

Silvana Kühtz*
Università della Basilicata, Facoltà di Ingegneria DIFA Lazzazzera, 75100 Matera, Italy

I think that when sustainability begins to drive business practices instead of greed, we'll evolve as a species.

— Tom Kay

ABSTRACT

With the growing interest in global warming, climate change, and the influence that human activities have on our environment, there are increasing opportunities for firms to assess their carbon footprint and do something to transform behaviours and invest in the environment with concrete actions.

Moreover, an increasingly common market-based mechanism is the carbon market, which links marketplace values with the sequestration of carbon. In this chapter, we highlight some of the carbon footprint assessment uses and the market-based mechanisms (especially voluntary ones) that are effective both for the firms and for nature, in that they allow benefits for both.

One research program conducting designs is a voluntary approach for firms and individuals to calculate and reduce their carbon footprints. We have analysed some real cases that have chosen to adhere to an Italian carbon footprint assessment program that entails forestry management and sustainable development behaviour programs.

Keywords: carbon footprint, carbon market, offset, Italy, sustainable development, SD education

---

* silvana.kuhtz@unibas.it
fax: + 39 06 62204861

1. INTRODUCTION

In the past few years, successful companies have expanded their field of vision and their responsibilities to look at the processes and operations of the companies that they buy from and companies that they sell to. This supply chain analysis has allowed them to make better, more informed decisions to run their operations and limit their impacts. Many benefits have been gained: increased efficiency, reduced waste, and lower capital investments are just a few (CarbonTrust 2006b; Albino and Kühtz 2004; Beamon 1999).

The analysis of materials and energy flows within supply chains can be conducted with input-output accounting models based on production processes (Albino et al. 2003b; Albino et al. 2002; Lin and Polenske 1998). A concept of environmental impact was then drawn, based on the quantitative measures of all of the resources used (materials, including water, and energy) and all the intermediate outputs and wastes produced (including CO_2 emissions) to produce the final products destined for the market. Then, for each unit of product sold to the market, the local or global environmental footprints were defined as the quantity of resources used and of wastes produced inside the supply chain or wherever those resources were used and wastes produced, respectively.

The pressure of population growth and of economic intervention on the natural environment, whether in terms of consumption of non-renewable resources or of reduction in the resilience of natural environment, have in the last twenty-five years led to an explosion of environmental problems that have been building up for a long time.

The latest scientific evidence provided in the assessment report from the Intergovernmental Panel on Climate Change (2007) makes clear that the time for action is now. More than 2,500 of the world's leading scientists from over 130 countries concluded that climate change is happening and that human activities are now very likely to be the cause. Twenty-first century organisations will then need to transform themselves in order to adapt to the challenges, risks and opportunities posed by climate change (Whitney et al. 2007).

In Italy, to foster sustainable development and implement environmental voluntary standards in small and medium enterprises, some regions (e.g., Piemonte, Toscana and others) have launched financing campaigns to fully or partially cover the costs of environmental certifications (i.e., EMAS *Environmental Management Auditing Scheme*, ISO 14001:2004 *International Organization for Standardization*, Ecolabel…). Other national Italian schemes include tax reductions for the enterprises or individuals who embark on projects that entail the substitution of renewable for exhaustible resources or similar interventions in energy savings.

The priority, both for economical and ecological reasons, are then to reduce and eliminate carbon dioxide and other greenhouse gas emissions wherever possible. As firms and individuals, there is much we can do immediately to transform some of our wasteful habits and use now the best available technologies. But the issue is complex, and not amenable to any single or short-term solution. It will take new processes, new inventions, and new attitudes to tackle the issue of protecting our planet (Dembo and Davidson 2007). However, there is something else we can do while we invent new generations of plants and fully clean technologies: we can, i.e., (i) calculate our carbon footprint; (ii) plan the immediately feasible emission reductions; (iii) engage our employees in reduction plans, and life-long learning plans; and iv) offset the remaining emissions in the compliance or/and voluntary carbon market (a way to put a price on carbon and to give real value to natural resources).

2. Carbon Footprint

A new idea will not necessarily be a new idea, an absolute newcomer in the world of thinking: it could more simply be a new way of connecting pieces of knowledge that are already well known. This is the case of carbon footprint.

The term "ecological footprint" was coined by William E. Rees, a planner at the University of British Columbia — who had previously used the term "appropriated carrying capacity" — after a computer delivery man told him that the new machine, which took up less space than his old model, had a "smaller footprint" (Wackernagel and Rees 1996).

Both the media and various scientific references are increasingly using the term "carbon footprint" in articles about the need to mitigate climate change by reducing our carbon dioxide emissions. Footprints are spatial indicators, measured in hectares or square metres. Instead, the property that is often referred to as a *carbon footprint* is actually a "carbon weight" of kilograms or tonnes per person or activity (Hammond 2007).

The most important greenhouse gas arising from human activity is carbon dioxide (CO_2). Virtually all human activities cause CO_2 emissions that lead to climate change (Kyoto Protocol, 1997). Understanding and addressing the full range of our impact is crucial for the effects of climate change to be minimised.

Carbon footprints are sometimes calculated to include CO_2 only and are expressed in tCO_2 (tonnes of CO_2) (Wiedmann and Minx 2007), but typically are calculated to include all greenhouse gases and are expressed in tonnes of CO_2 equivalent [1](tCO_2e).

A recent report from the Carbon Trust (Carbon Trust 2007) defines *carbon footprint* as "the total set of greenhouse gas emissions caused directly and indirectly by an individual, organization, event or product." Similarly, the *Dictionary of Sustainable Management* defines it as "the total amount of greenhouse gases emitted directly and indirectly to support human activities, usually expressed in equivalent tons of either carbon or carbon dioxide" (Presidio School of Management).

We prefer these definitions, as they allow us to immediately relate the carbon[2] footprint (of an individual, a company, an organization) with the carbon market mechanisms that buy and sell tonnes of CO_2 equivalent. Therefore in the course of this chapter we consider the *carbon footprint* a measure of the amount (expressed in units of CO_2 equivalents) of greenhouse gases that enter the atmosphere as a result of a given activity or product. We call it a footprint because it's like the mark you leave on the earth as you go about your daily activities, whether you are a firm, an organization or an individual. A carbon footprint includes activities that result in direct and indirect emissions, and therefore it is closely related to where the boundaries for particular activities are drawn. Direct emissions, or those under full control of an organisation, are always included within the company's carbon footprint (including emissions associated with electricity consumption). It is really the

---

[1] The greenhouse gases are converted to a carbon dioxide equivalent, that includes the global warming effect of the other greenhouse gases compared with carbon dioxide. Under Article 5 of the Kyoto Protocol, the global warming potentials (GWPs) given in IPCC (1996a) for a time horizon of 100 years is used to convert the gases in the Protocol to the common unit of CO2 equivalents. When the emissions of the various gases are multiplied by their respective GWPs, the emissions are converted to a common unit, often called CO2-equivalents. The common definition of the CO2-equivalent is: CO2-eq(H) = GWPi (H)·Ei where GWP(H)i is the global warming potential of gas i; Ei represents the emission of gas i measured by mass; and CO2-eq(H) is the CO2-equivalent amount of gas i using GWPs for a time horizon H.

[2] "Greenhouse gas" and "carbon" are in fact often used interchangeably in the terminology.

inclusion or exclusion of indirect emissions that differs when calculating an organisation's, product's or activity's carbon footprint. Indirect emissions, or those emissions not controlled or fully controlled by an organisation, can include emissions from travel (e.g., business flights), supply chain emissions, employees commuting to work and so on.

For a single product it is possible to calculate its carbon footprint as well, it is defined the *carbon dioxide equivalent* emitted across the supply chain for a single unit of that product. For example, the carbon footprint of a can of soft drink is the total net amount of carbon dioxide equivalent emitted to produce, use and dispose of a single can (Carbon Trust 2006 b).

Establishing the carbon footprint of an organisation can be the first step in a programme to reduce the emissions it causes (Carbon Trust 2007) and to plan offset programs for those emissions that cannot be immediately eliminated. This reminds us the adagio *first measure then manage*, i.e., "If you can't measure your carbon impact you can't manage emission reduction plans."

We will highlight in the course of this chapter that it is becoming evident that knowing (i.e., measuring) your carbon footprint is a first necessary step towards sustainable development. Not sufficient, of course, as other fundamental steps are: planning reductions and acting accordingly, as well as educating to new behaviours and offsetting, to cite just a few.

So measuring the right things in the right way will help guide our actions. With that said, effective action does not always need measurement: doing everything we can to make sure that our future planet supports a diversity of life and prosperous societies should not be limited by possible difficulties in measuring/calculating key indicators.

2.1. Carbon Footprint Calculation: Life Cycle Assessment (LCA)

To calculate carbon footprint, sound environmental accounting methodologies are available (ISO 14040:2006, 14044:2006 and 14064:2006). With input-output approaches based on processes, for a given final product output, materials, energy and waste (pollution) flows are computed thus providing a measure of resources consumption and of the environmental impact of production processes (Albino and Kühtz 2003).

A typical tool that can be used is the life cycle assessment (LCA) applied to the whole supply chain. It gives a measure of the total emissions associated with the provision of individual products and services. Typically, this includes raw material extraction, manufacturing, distribution, retailing, consumption, disposal and recycling, that is, includes direct and indirect emissions.

Analysing carbon emissions in this way provides also a powerful insight into how also the daily decisions by individual consumers drive economy-wide carbon emissions. Recently many studies have underlined that it is possible to reduce the carbon emissions in the products, by assessing and optimising emissions across full product supply chains. It is a powerful tool for both government and business in the drive to a low carbon economy (Carbon Trust, 2006a).

Life Cycle Assessment (LCA) is the broadest indicator and an internationally standardized method (ISO 14040:2006 and ISO 14044:2006). It not only evaluates the impact on climate change, but also other impact categories such as acidification potential, eutrophication potential, ozone depletion potential, and ground level ozone creation. For each of these impact categories, the product or system is evaluated over its complete life span, from the extraction of raw material and manufacturing, to the use of the product by final consumers and end-of-life processes like recycling, energy recovery, and ultimate waste disposal (SETAC 1990). The ISO standards provide robust and practice-proven requirements for performing transparent LCA calculations. Moreover, one can make use of extensive databases containing life cycle profiles of many goods and services, as well as many of the underlying materials, energy resources, transport systems, etc. Nevertheless, LCA calculations remain very complex (Baldo et al. 2005).

Some authors have found some definitional conflicts in the literature for *carbon footprint* (Johnson 2008). They should be seen not as an aberration, but as an inherent aspect of carbon foot-printing, life-cycle assessment, material flow analysis and other such methods of evaluating products. Johnson underlines (2008) that it was precisely due to definitional conflict that LCA was invented in the first place. Known in the early 1970s as Resource and Environmental Profile Analysis (REPA), it was first used to refute claims that throwing away soft-drink bottles always causes more negative environmental impact than recycling them (DEFRA 2005).The argument was very much based on definitions, i.e., the REPA defined a broader life cycle than was usual at the time precisely in order to make its case.

So, although we will not go through the issue of standardization in the course of this chapter, it is important to underline that for a footprint to be comparable, definitions must be sensible and transparent (i.e., reproducible by others). The rigour and appropriateness of any definition will depend on: (1) significance of the definition to the expected results; (2) the transparency of the method applied and (3) the availability of data. Regardless of the definition or standard used, the key is to state clearly what has and has not been included within the carbon footprint. This helps to explain what aspects of an organisation, product or activity have been considered in a carbon neutral claim, and which have been left out of scope and why.

2.2. Carbon Footprint and the Kyoto Protocol

The Protocol is one of the most complex treaties ever negotiated, we will not go through all of the complex mechanisms and implications: there are entire books and reports dedicated to the Protocol itself (e.g., Brack et al. 1999). Anyway, we will highlight some of its main issues.

In its core lie legally binding targets for the world's wealthier countries to reduce greenhouse gas emissions over 2008–2012. Countries with Kyoto targets can achieve these targets through domestic efforts. They can also reduce greenhouse gases (GHG) in other countries at lower cost than at home by making use of the three flexible mechanisms set out in the Protocol: joint implementation (JI), the Clean Development Mechanism (CDM) and international emissions trading (ET). Because the geographic location of GHG emissions is environmentally irrelevant, cost-effectiveness considerations speak for allowing countries to take credit for overseas actions that curb GHG emissions at source or enhance the removal of GHGs by sinks. CDM and JI have strict rules as to what can qualify as an acceptable project, and clear methodologies for ensuring that the emissions reductions are real, verified, additional (they wouldn't have happened anyway) and avoid leakage (they don't cause more emissions some where else). The carbon markets established by these schemes interact in complex ways. Cost-minimization is the core motivation for trades among and across schemes, particularly at the entity level, but legal and institutional provisions, and broader political and strategic considerations, also play a significant role in determining whether, where, when and how trades take place (Yamin and Depledge 2004).

The Kyoto Protocol defines legally binding targets and timetables for cutting the greenhouse-gas emissions of industrialized countries that ratified the Kyoto Protocol. Accordingly, from an economic or market perspective, one has to distinguish between a *mandatory market* and a *voluntary market.*

2.3. Carbon Market Theory

The term carbon market refers to the buying and selling of emissions permits that have been either distributed by a regulatory body (the mandatory market) or generated by GHG emission reduction projects (voluntary market). Six GHGs are generally included: CO_2, methane, nitrous oxide, sulphur hexafluoride, hydro fluorocarbons, perfluorocarbons. GHG emission reductions are traded in carbon credits, which represent the reduction of GHGs equal to one metric tonne of CO_2 equivalent.

Many think markets for emissions reductions are among the most innovative and cost-effective means society has of creating a market pull for new clean technologies while, at the same time, putting a price on pollution and thereby providing incentives for people to emit less (Bayon et al. 2007). The theory is that carbon markets are able to channel resources toward the most cost effective means of reducing GHG emissions. At the same time they punish monetarily those who emit more than an established quota, and reward those who emit less.

By turning units of pollution into units of property, the system makes it possible to exchange pollution all over the world. In a world where pollution has no price, the default decision will always be to pollute. In a world where pollution costs between 10 and 30 € a tonne, the decision is no longer quite so easy.

Once markets take shape, emitters have a variety of options available to them. If they believe they can reduce emissions (or to say it in different words, reduce their carbon footprint) cheaply by changing production processes or experimenting with new technologies, they have an incentive to do so. If they believe they can change their production process but that this will take time, emitters can purchase credits up front in the hopes that, down the line, they will be able to make them back through emissions reduction technologies. If, on the other end, emitters believe they will emit more in the long run, they can buy credits now for use later. Basically the system enables the trading of emissions across temporal as well as geographic boundaries.

Some people dispute the benefits of the carbon market and of the flexible mechanisms, and others feel that markets allow companies to 'greenwash' previously tarnished environmental reputations without changing their behaviour in important ways. Let's be clear: of course the market and voluntary offsetting schemes do not solve the problem of global warming and increased industrial pollution on their own. But it is also true that we are not going to solve the problem without them either.

The mandatory market involves *project based transactions* (emission credits are the result of a specific carbon offset project) and *allowance based transactions* the trading of issued allowances created and allocated by regulations under a cap-and-trade regime. The authority caps the quantity of emissions that participants are permitted to emit and issues a number of tradable allowance units equal to the cap. The Kyoto Protocol created the three flexibility mechanism cited above (ET, CDM and JI) which are the basis for the regulated international compliance market.

The carbon market grew in value to an estimated US$30 billion in 2006 (€23 billion), three times greater than the previous year. The market was dominated by the sale and re-sale of European Union Allowances (EUAs) at a value of nearly $25 billion under the EU ETS (€19 billion). Project-based activities primarily through the Clean Development Mechanism (CDM) and Joint Implementation (JI) grew sharply to a value of about US$5 billion in 2006 (€3.8 billion) (Capoor and Ambrosi 2007).

2.4. The Voluntary Market

Unlike the regulated market, the voluntary market does not rely on legally mandated reductions to generate demand[3]. As a result, the market suffers from fragmentation and a lack of widely available impartial information. This can be attributed to the fact that it is partially composed of deals that are negotiated on a case-by-case basis, and that many of these deals neither require the carbon credits to undergo a uniform certification or verification process nor register them with any central body (Bayon et al. 2007). On the other hand there is also an added value as these projects avoid bottlenecks (so common in, e.g., CDM projects), and transaction costs, and are characterised by flexibility and innovation.

The voluntary carbon markets have seen dramatic growth since 2003-2004. There is something extremely appealing about a business model that caters to the better instincts of mankind to make the world a better place. Inevitably, a wide-open space such as this attracts some bad apples that create problems for other players in the segment. This is why it is critical that responsible market players develop a simple but credible standard for voluntary emission reductions (Capoor and Ambrosi 2007).

This is important since a significant number of companies are now offering project-based emission reductions (carbon offsetting schemes) to individuals that may wish to offset GHG emissions linked to their way of life (residential energy use, commuting, travel); to customer-facing companies that wish to offset GHG emissions from their operations or from specific products or events (sports, concerts, conferences but also travels, mortgages, utility bills, shipping and other goods and services their customer may wish to render carbon neutral); to high emitting companies that may wish to voluntarily offset the GHG emissions from some portion of their activities that they cannot immediately reduce through their operations.

Credibility is important to both the responsible individuals and corporate houses for whom reputation is an important motivator to reduce their climate footprints. A credible voluntary standard will do more to attract value to this segment than any other action. Credibility

[3] There are some exceptions of legally binding voluntary market. Players in this market form part of a scheme in which they voluntarily set themselves legally binding GHG emissions reduction targets. This is the case in the Chicago Climate Exchange (CCX). The CCX is a voluntary system (based in Canada, the US and Mexico) which sets self-imposed, legally binding reduction targets on its members. The system is a hybrid between a cap-and-trade and project based scheme. Participant members have an emissions cap based on historic (1999-2002) emission levels. To comply, members can either reduce their emissions below their cap or purchase offset credits generated from projects in the US, Brazil, Canada or Mexico that focus primarily on landfill and agricultural methane destruction, or carbon sequestration in soils or forest-based biomass.

is also critical to the extent that suspicion may spread to the (regulated) CDM and JI market. In order to be credible, a voluntary standard does not need to be exactly the same as the complex standards created by the CDM Executive Board. It could, for example, have a much more simple and intuitive standard. It may, for instance, require third party verification; perhaps this could be community-based and enforced. One can hope that the market would sort this out and allow the best standards to flourish. Alternatively, it might be worth considering a process such as the Gold Standard (endorsed by some 40 NGOs), the Voluntary Carbon Standard (developed by the Climate Group, the International Emissions Trading Association (IETA) and the World Economic Forum Global Greenhouse Register) or other, to synthesize an acceptable standard from the ones already proposed (Capoor and Ambrosi 2007).

Some of the more optimistic estimates for the size of the voluntary market by 2010 are as high as 400 MtCO2e (or almost as high as the CDM market is today). U.S. analyst Trexler estimated that US demand alone for offsets under the voluntary market could almost double annually from today to 250 MtCO2e by 2011. While such numbers may be hard to imagine today when the voluntary retail segment accounted for only about 20 MtCO2e in 2006, such a future is certainly possible (Trexler 2007; ICF 2006).

2.4.1. Voluntary Market and Responsibility

As Amartya Sen underlines in his book *Development as Freedom*: "As people who live – in a broad sense-together, we cannot escape the thought that the terrible occurrences that we see around are quintessentially our problems. They are our responsibility-whether or not they are also anyone else's." And specifically in the context of environmental protection "the environmental challenge is part of a more general problem related to resource allocation (…). We have to examine the part that can be played by the development of social values and of a sense of responsibility that may reduce the need for forceful state action."

The growth of the voluntary markets is a welcome indicator of the appetite that ordinary individuals and companies across the world have to take personal responsibility for the problems of our planet. Their rapid growth is also a reflection on how much more their governments could be doing to reduce emissions. It also reflects that other governments need to participate in the effort to spur innovation towards a low-carbon future by taking steps to reduce their carbon footprint.

Many non-profit and academic organizations are supportive of the voluntary carbon market because it provides individuals (and also corporations and large organizations) with a means of participating responsibly in the fight against pollution and the spoiling of our planet in a way that the compliance market often do not.

Especially forest projects flourish in the voluntary markets. A major criticism concerns the imprecise nature of GHG sequestration quantification methodologies for forestry projects. However, we have to note the community co-benefits that forestry projects foster. Project types in the voluntary market range from avoided deforestation, afforestation/reforestation, industrial gas sequestration, increased energy efficiency, fuel switching, methane capture from coal plants and livestock, and even renewable energy.

2.5. Carbon Footprint, Carbon Market and Education

Carbon offsets (the counterbalancing of emissions via actions that reduce emissions elsewhere from where they are produced) continue to attract controversy.

We agree that what we need to do is to cut our emissions (i.e., reduce our carbon footprint with clear actions, e.g., reduce our consumes here and immediately), but to do this the public have to be informed and properly educated. Yet, in many countries administrations tend to value *hard* technical fixes for problems over *soft* social and cultural approaches that rely on transforming behaviour. Energy policies are often driven by the idea that technology and the market will provide for the future, rather than any kind of change in consumption and lifestyle.

We see both the carbon footprint indicator and the market mechanisms as *important opportunities* for educating the public about the planet's main issues and their potential active role in addressing the problem. Integrated approaches that promote new behaviours are needed. In order to do this, the role of research along with formal and widespread informal education for sustainable development and cleaner production needs to be central in firms and administrations strategies (Kühtz 2007). The carbon footprint and the market mechanisms (we consider in particular the voluntary market examples) have then the potential to become an active driver of change today and a real opportunity of widespread and interdisciplinary education for sustainable development in action.

Moreover, we have to underline and repeat that even with the best efforts of organizations and individuals, reducing the carbon footprint does not mean eliminating all the pollution they (we) produce, therefore after reducing the alternatives are to ignore the remaining emissions or to offset them in a way or another (Dembo and Davidson 2007) this is why we see a clear link amongst carbon footprint and market mechanisms.

Further, even those who are not convinced by the wealth of scientific evidence that our planet is at stake can realize that the need to take action now is a clear business imperative. World capital markets and governments have decided that businesses need to manage their carbon footprint as effectively as they do their finances.

Moreover, these are some undisputable facts that regard our planet: the earth's natural resources are limited, meaning not only that energy costs will continue to rise, but also that efficient energy production, responsible use, and innovation to find alternative fuel sources are necessary; in the "war for talent" an organisation's environmental credentials are no longer just a "nice-to-have"; citizens and customers are increasingly demanding "green" products and services.

3. What Carbon Footprint Is Used for So Far

"Just as new dieters keep a food diary, an excellent way to start reducing your emissions is by using a carbon calculator" in the *Rough Guide to Climate Change* (Henson, 2007) there is a whole chapter entitled *Reducing your carbon footprint and lobbying for action*. It suggests there to use the carbon calculators available on-line to calculate your personal carbon footprint, and then it gives a number of low-carbon hints to transform usual daily behaviours, and to engage in offsetting schemes. The book we are referring to is climate neutral itself.

If one browses in the web, finds out that many companies and organizations in the world are posing themselves these questions: How can an organisation start the process of green transformation and achieve the resulting benefits of efficiency, green reputation and cost savings? What is the method that a project team could follow? Which carbon offsetting scheme to follow to become carbon neutral? (e.g. Lanyon-Hogg 2007; Whitney et al. 2007).

And again browsing in the web one finds out that the first answer of companies/organizations and of their consultants is: *to be able to become carbon neutral, an organisation first needs to know its carbon footprint: the emissions for which it is responsible* (Carbon Trust 2006c).

Basically the carbon footprint is becoming the first indicator of pressure on the environment that the company should calculate when greening issues arise (Ecosystem marketplace 2008).

We can say that the carbon footprint properly calculated allows for a number of advantages: allows companies to display legitimate environmental credentials and then plan long term emissions reduction plans; allows consumers to choose lower carbon emission products and services (the potential and implications to a company's sales and profits of these choices will promote investment in cross supply chain efficiencies and investments in i.e. energy reductions), and allows companies and consumers to understand the CO2e implications of choices they make and change lifestyle and or consumption behaviours. Awareness can promote sustainable development. There are many firms that are making declarations of carbon reductions and are making of these statements a very important corporate issue. However, as more high profile companies adopt carbon-offsetting policies, it is possible that offsetting could, in future, become a popular trend or be integrated as a key component of corporate social responsibility. Carbon offsets are also being used as a marketing tool for brand enforcement and loyalty. Some firms choose to pay for the offsets themselves, while others pass on the costs to the customer.

A number of producers and retailers especially in countries that are taking the lead of these issues, like UK, are currently exploring and piloting carbon labelling (White 2008). When UK consumers were asked if they would value information regarding a product's carbon footprint when making a buying decision, a majority of 56% stated that they would value this information (Wheatland 2007). Equipped with reliable information regarding the carbon footprint of a product or service, the study found that close to half of consumers (44%) would change their buying behaviour in some way, and 43% would even pay more to reduce their carbon footprint. Looking to the future, it is manufacturers and producers who consumers believe should take the lead in minimising the carbon footprint of the products and services they consume.

Various governments, eager to demonstrate their responsibility, have been developing plans to purchase carbon offsets, particularly for air travel. In the UK, the *Department for Environment, Food and Rural Affairs (DEFRA)*, the *Foreign Commonwealth Office (FCO)*, and the *Department for International Development (DFID)* announced plan to offset their air travel emissions. In addition to government departments, various cities, such as Rotterdam and Portland have made pledges to reduce CO2 emissions or become CO2 neutral.

A recent study conducted by SEI for WWF living planet entitled significantly *The right climate for Change – using the carbon footprint to reduce CO_2 emissions* (WWF/SEI 2007), is a guide for local UK authorities to use Carbon Footprint as an indicator to investigate how consumption behaviours can change carbon footprint locally. Carbon footprint as a top level indicator is linked to local monitoring and to a number of local agendas including access to services, planning decisions, health, fuel poverty, housing conditions and waste management. Awareness can promote CO_2 emissions reductions.

In Italy there is not yet such a thing like a national monitoring scheme of carbon footprints, and companies that are starting to *go green* are still a minority. But there is no doubt that Italy as all other EU countries will have a huge influence on the international bid to be leaders in combating polluting emissions. The EU carries a huge responsibility for CO2 emissions in the rest of the world due to its carbon-intensive consumption and trade patterns.

4. Research and Case Study

In the past recent years we have developed a supply chain methodology to help companies monitor their materials and energy flows and identify emissions reduction savings opportunities across the supply chain (Albino and Kühtz 2004). We have recently expanded the methodology to build the full carbon footprint of products drawing on standard life-cycle analysis techniques (LCA) and combining the carbon footprint with forestry offsetting projects located in the South of Italy.

The idea that was initially only a research project, developed then in a voluntary offsetting project named *FilieraCO2-fai la differenza* (internationally registered trademark, that literally means: *SupplyChainCO2-make the difference,* Figure 1).

Figure 1. The logo of the project FilieraCO2.

It involves different actors: an Italian organization that has acquired the trademark; a consulting company that provides services to the Agricultural and Forestry Industry (Italy based); an independent third party verification authorised by the Italian Ministry of Agricultural Policies-Decree Mipa; and our University Department that develops the methodology, the calculations and the validation of the whole CO2 supply chain.

There are various steps that lead to our certified offsettings: initial analysis and engagement (both for the firm and the land owner where the trees are planted); construction of the carbon footprint; opportunity identification and prioritisation; presentation of results and implementation planning; design of the offsetting programme and of forestation planting and trees management; sustainable development education schemes; registration and certification.

The reasons for which organizations might want to calculate their carbon footprint and offset their emissions are multiple: As part of an environmental strategy; As part of a carbon neutral strategy driven by CSR reasons or brand positioning; As a way to anticipate future legislation or to gain experience in the carbon market.

Whatever the case, defining the reasons will help to identify the type of offsets that organisations should acquire if they decide to offset at all. For example, an organisation driven by CSR reasons, is likely to acquire offsets from small scale projects with a strong sustainable development component, like ours.

4.1. FilieraCO2 and Forestry

The true meaning of life is to plant trees, under whose shade you do not expect to sit.
- Nelson Henderson

Offsetting is a short-term fix for a deep and long-term problem. Nevertheless, by pricing carbon offsetting brings market forces to bear on our most urgent challenge. It is not a perfect solution, or a total solution. It is an interim measure while we find and implement other solutions, it buys time while homes are insulated, heat-and-power clean plants are constructed and new green technologies are invented and implemented (Dembo and Davidson 2007). FilieraCO2 responds to high standards in that it is verified by an accredited third party according to a standard; it ensures the reductions achieved are maintained over time; it avoids offsets being used or counted more than once, as it operates a registry where credits get accounted, registered to buyers and retired.

FilieraCO2 offsetting scheme although is planning other kind of offsetting projects, is concentrating mainly on trees (especially in downgraded Italian territories) for many reasons listed hereafter. They in some cases also represent an answer to concerns raised by critics.

1. According to IPCC, land use change, primarily deforestation, has caused a third of anthropogenic gas emissions since 1850, and the World Resources Institute calculates that deforestation and forest harvest currently contributes to just over 20% of annual green house gas emissions. A recent research carried out at Wageningen University (van Minnen et al. 2008) states that: "Carbon sequestration in plantations can play an important role in mitigating the build-up of atmospheric CO_2. The actual magnitude depends on natural and management factors, social barriers, and the time frame considered. In addition, there are a number of ancillary benefits for local communities and the environment. Carbon plantations are, however, particularly effective in the long term. Furthermore, plantations do not offer the ultimate solution towards stabilizing CO_2 concentrations but should be part of a broader package of options with clear energy emission reduction measures."

2. Offsets allow us to reduce emissions above and beyond the levels set by the political agreements. Voluntary offsets complement compliance market reductions; they go above and beyond existing government policy and targets, helping to drive action while the politics catches up.
3. Tree planting and forest conservation is the only viable mechanism that we have for actually taking carbon out of the atmosphere[4]. Moreover, forests are more than just carbon farms. Provided tree planting and forest conservation schemes are carried out with sensitivity and understanding, they bring many other benefits. They provide biodiversity, they prevent soil erosion, they help purify water sources, they provide wildlife habitats and shelter and recreation. If managed properly they provide building materials (that store carbon instead of releasing it) and much more.
4. Modern advances in computing, both on the firm side (polluter) to calculate its emissions using LCA software and updated database, and on the trees (carbon sink) side, with CO2 modelling (as the one we have developed, *CO2 seize*) allow accurate calculation of carbon footprint, and of carbon content of trees, plus how this carbon will be stored and released over time.
5. Also, exposure to some risk (as fire) is not, according to us, an argument not to plant trees. It's an argument to manage risk using modern, tried and tested tools.
6. The areas where FilieraCO2 is primarily planting trees and managing existing forests are the Regions in the South of Italy, as they are often areas where natural resources and natural beauty can still be protected and fostered, but too often are instead neglected. Smallholder farmers that enter FilieraCO2 registry are rewarded economically when their 'credits' are sold via FilieraCO2, this way they are encouraged to give value to their forests and are to improve the practice of sustainable forestry management. The fact that Italian industries (generally located in the north of Italy) invest in this program to reduce their carbon footprint can be a way of nurturing and strengthening partnerships between the developed industrial North and the agricultural beautiful South, and encourage a real sustainable development.
7. Also, it can represent a way to foster sustainable tourism in the South of Italy and sustainable development education programs can be implemented. Combining public programs and incentives with private investments can help increase the impact and resulting benefits. Efforts to value ecosystem services can help raise public awareness, support private lands stewardship, encourage ecosystem restoration and protection, and begin to emphasize the need to reduce consumption and manage human impacts. Offsetting via planting/managing trees can educate people and businesses about their impact on the climate and can reinforce broader behaviour to reduce emissions. In combination with other policies they are an essential component of carbon neutral goals.
8. It is well known that urban air quality can impact human health. Allergies, asthma, bronchitis, and related conditions have all been linked to airborne pollutants. The good news is that there is strong evidence of the mitigating impacts that land use, open space, and vegetation management can have on local and regional air quality. Urban trees provide distinct benefits related to air quality: Microclimate effects,

[4] In fact, the technology for capturing CO_2 and pumping it into rock formations is ready, but it is costly and there are many questions as to whether it is safe, etc.

> including temperature reductions; Removal of air pollutants; Reduced energy consumption in heating and cooling buildings. Various studies have quantified the role of urban trees in pollution removal. Pollution removal by urban trees in the United States has been estimated at 711,000 metric tons annually, a service valued at $3.8 billion (Fernholz 2008; Novak et al. 2006).

Taking inspiration and modifying a bit the sentence posed at the beginning of this paragraph (that is in turn taken from an ancient Greek saying) we could say that an appropriate definition of *sustainable development* is:

Sustainable development is in place when people plant trees under whose shade they do not expect to sit.

4.2. Case Studies and Education

When taking socio-economic, strategic and political decisions, managers, entrepreneurs and the public in general now find themselves having to consider them from the point of view of sustainable development as well. However, sustainable behaviour is not automatic, it is innovative and as such has to be developed and strengthened by research, transferred and adopted by means of education. It is increasingly clear that while the achievement of sustainable development may require an increase in specialised and scientific knowledge it depends to a larger extent on increasing the awareness of individuals and communities. Winning this battle requires an approach to education and communication that promotes the right values, particularly the values of honesty and fairness, and an awareness that we all have a part to play in the fate of the planet (UNESCO 2000).

Italy is not as advanced as for example UK. Many barriers and objections still need to be overcome so that daily behaviours can be transformed in the view of the realization of a sustainable development (Kühtz 2007).

Involving in the first place only the University, via a research project we have evaluated the carbon footprint of a company that produces polyester nonwovens for roofing membranes, Freudenberg Politex (unit located in Pisticci, MT, Italy). Our supply chain analysis has made clear to the firm (that has the aspiration to become a zero emissions company) that measuring emissions is a critical first step toward offsetting emissions, whether some or all, and toward new critical choices. The next step of our research project is to identify the most attractive opportunities for energy and materials reductions, for sustainable development education and how those efforts interplay and affect other corporate objectives.

Our team has developed a Green Edu FilieraCO2 programme which is the first step in the strategy for change. If it is people who are largely responsible for the way that energy is used – or wasted – in the working environment, it follows that one of the biggest challenges is to change the way that people work.

Although FilieraCO2-fai la differenza is newly born some interesting cases have already been certified and many others are in due course. We will cite hereafter only some of the certified cases, those with a big impact in terms of education and diffusion of a new mentality.

Two elementary schools and a primary school (located in Bari, South of Italy) have embarked with FilieraCO2 in a long term education program that has involved the teachers on climate change issues. The pupils (around 1000) and parents have been involved as well via the teachers and via ad hoc laboratories; by the end of the school year the school intends to offset its emissions and reduce where possible the energy consumptions for next year.

The Orchestra del Teatro Olimpico (Vicenza, Italy) has chosen to offset its emissions due to its 2008 concerts, by caring for a forest area located in the very south of Italy.

A documentary movie about the effects of global warming from the nature point of view, entitled *Going North* and launched in 2008, offsets its CO2 emissions through forestation projects located in the very South of Italy.

Acquiring credible good quality offsets is crucial when defining an offsetting strategy, this is why we endorse this project as scientific academic team and are devoting research resources to it. As the voluntary market remains largely unregulated, buyers should carry out their own due diligence to assess the integrity and credibility of the offsets they are buying.

Demonstrating a commitment to reducing carbon emissions and purchasing offsets is a way for firms to boost their "green" image as an environmentally responsible company. This is a relatively new attitude for Italian companies and products.

Conclusion

Both the media and various scientific references are increasingly using the term "carbon footprint" in articles about the need to mitigate climate change by reducing our carbon dioxide emissions. We have gone through some definitions and have chosen the one that accounts for greenhouse gases.

Anyway, regardless of the definition or standard used, the key is to state clearly what has and has not been included within the carbon footprint. This helps to explain what aspects of an organisation, product or activity have been considered in a carbon neutral claim, and which have been left out of scope and why. Basically the carbon footprint is becoming the first indicator of pressure on the environment that the company calculates when greening issues arise. Awareness can promote sustainable development. We have analysed some real cases that have chosen to adhere to an Italian carbon footprint assessment program that entails forestry management and sustainable development behaviours programs and that we are promoting as a university. *Filiera CO2-fai la differenza* is a new Italian offsetting project that provides many significant benefits that allow: CO_2 emissions to be offset; the promotion of social sustainable behaviour; a contribution to preventing hydrogeological risks so common in Italy; wide areas to be protected, particularly in southern Italy, where land is often impoverished and abused but presents huge potential; eco-compatible forestation and farming to be promoted; and local biodiversity, wooded areas and eco-compatibility to be protected (FilieraCO2.it).

To plant trees is to give body and life to one's dreams of a better world.

—Russell Page

ACKNOWLEDGMENTS

The author wishes to thank the PhD student, Francesca Intini, for her friendly support in the writing of this chapter; Riccardo Sollini, President and CEO at Freudenberg Politex Italy and Massimo Migliavacca R&D director at Freudenberg Politex Italy, for trusting that research and academic work are fundamental for the development of a sustainable entrepreneurial future.

REFERENCES

Albino, V., Izzo, C., and Kühtz, S. (2002). Input-Output Models for the Analysis of a Local/Global Supply Chain. *International Journal of Production Economics,* volume 78 n. 2, 119-131.

Albino, V., Dietzenbacher, E., and Kühtz, S. (2003a). Analyzing Material and Energy Flows in an Industrial District using an Enterprise Input-Output Model. *Economic Systems Research*, vol.15, n.4, 457-480.

Albino, V., and Kühtz, S. (2003b). Assessment of environmental impacts of production processes in the industrial districts using input-output modeling techniques. *Jl of Environmental Informatics*, volume 1 n.1, 7-20.

Albino, V., and Kühtz, S. (2004). Enterprise input-output model for local sustainable development – the case of a tiles manufacturer in Italy. *Resources, Conservation and Recycling,* vol. 41, n.3, 165-176.

Baldo, G.L., Massimo, M. and Rossi, S. (2005). *Analisi del ciclo di vita (LCA) - Materiali,* prodotti, processi produttivi. Milano, Italia: Ed Ambiente.

Beamon, B.M. (1999). Designing the green supply chain. *Logistics Information Management*, volume 12, 332-342.

Brack, D., Grubb, M., and Vrolijk, C. (1999*). The Kyoto Protocol A Guide and Assessment.* Royal Institute Of International Affairs Chatham House, London UK, 342 pp.

Capoor, K., and Ambrosi, P. (2007). *State and trends of the carbon market 2007.* The World Bank, Washington DC USA.

Carbon Trust (2006a). The carbon emissions generated in all that we consume. www.carbontrust.co.uk.

Carbon Trust (2006b). Carbon footprints in the supply chain: The next step for business. *Report Number CTC616*, November 2006, The Carbon Trust, London, UK. www.carbontrust.co.uk.

Carbon Trust (2006c). The Carbon Trust three stage approach to developing a robust offsetting strategy. *Report CTC621,* November 2006, The Carbon Trust, London, UK. www.carbontrust.co.uk.

Carbon Trust (2007). Carbon foot-printing. Report Number CTV033, August 2007, The Carbon Trust, London, UK. www.carbontrust.co.uk.

DEFRA (2005). Potential for a 'Product Body' and for Product EIA in the UK. UK Department of Environment Food and Rural Affairs. Contract Reference EBCD3, March.

Dembo, R., and Davidson, E. (2007). Everything You Wanted to Know About Offsetting But Were Afraid to Ask, a zerofootprint publication.

Ecosystem Market place and Business for Social Responsibility (2008). Offsetting Emissions: a business brief on the voluntary carbon market. Available at: http://www.ecosystemmarketplace.com/media/pdf/BSR-EMVoluntaryCarbonOffsets.pdf.

Fernholz, K., Bowyer, J., Lindburg, A., and Bratkovich, S. (2008). Ecosystem markets – New mechanisms to support forestry. Report of Dovetail Partners inc.

FilieraCO2, www.filieraco2.it.

Hammond, G. (2007). Time to give due weight to the 'carbon footprint' issue. Nature 445 (7125): 256. http://dx.doi.org/10.1038/445256b.

ICF (2006). Voluntary Carbon Offsets Market: Outlook 2007. ICF International: London, UK.

IPCC (1996a). Climate Change 1995: The Science of Climate Change. Cambridge University Press, Cambridge.

IPCC (2007). Climate Change 2007: Synthesis Report. Contribution of Working Groups I, II and III to the Fourth Assessment Report of the Intergovernmental Panel on Climate Change [Core Writing Team, Pachauri, R.K and Reisinger, A. (eds.)]. IPCC, Geneva, Switzerland, 104 pp.

Johnson, E. (2008). Disagreement over carbon footprints: A comparison of electric and LPG forklifts, *Energy Policy* vol. 36, n. 4, April, 2008, 1569-1573.

Kühtz, S. (2007). Adoption of sustainable development schemes and behaviors in Italy: barriers and solutions - what can educators do? *International Journal of Sustainability for Higher Education,* volume 8 n. 2, 155-169.

Kyoto Protocol (1997). Third Conference of the Parties to the UNFCCC (COP 3) in Kyoto, Japan, on 11 December.

Lanyon-Hogg, R. (2007). Cutting the carbon footprint of IT - How to deliver measurable savings – an IBM study with the support of Defra. *IBM Global Business Services*, available at www.ibm.com.

Lin, X., and Polenske, K.R. (1998). Input-Output Modelling of Production Processes for Business Management. *Structural Change and Economic Dynamics*, volume 9, 205-226.

Nowak, D. J., D. E. Crane, and Stevens, J. C. (2006). Air pollution removal by urban trees and shrubs in the United States. USDA Forest Service, Northeastem Research Station. *Urban Forestry and Urban Greening*, volume 4, 115-123.

Presidio School of Management, http://www.sustainabilitydictionary.com

Sen, A. (2000). Development as freedom. NY USA: Anchor Books.

SETAC Society of Environment Toxicology and Chemistry (1990). *A technical Framework for Life-Cycle Assessment*, Workshop report from Pensacola FL, USA August 18-23.

Taiyab, N. (2006). *Exploring the market for voluntary carbon offsets*. International Institute for Environment and Development, London UK.

Trexler, M. (2007). "US Demand?", presentation at the Point Carbon Carbon Market Insights 2007, Copenhagen, 13-15 March.

UNESCO (2000). World education report.

van Minnen, J.G., Strengers, B.J., Eickhout, B., Swart, R.J., and Leemans, R. (2008). Quantifying the effectiveness of climate change mitigation through forest plantations and carbon sequestration with an integrated land-use model. Carbon Balance and Management 3:3 doi:10.1186/1750-0680-3-3, available at: http://www.cbmjournal.com/content/3/1/3

Wackernagel, M., and Rees, W.E. (1996). *Our Ecological Footprint - Reducing Human Impact on the Earth.* New Society Publishers Gabriola Island, B.C., Canada.

Wheatland, J. (2007). The L.E.K. Consulting Carbon Footprint Report 2007. Carbon Footprints and the Evolution of Brand-Consumer Relationships, www.lek.com.

White, R. (2007). *Carbon labelling: evidence, issues and questions*, UKERC, www.ukerc.ac.uk.

Whitney, G., Stephenson, M., and Jackson, A. (2007). Gorilla, Peacock, Ostrich or Mule. How is your organisation reacting to one of the world's greatest challenges? IBM Global Business Services, available at: http://www-304.ibm.com/jct03004c/easyaccess/fileserve?contentid=112746.

Wiedmann, T., and Minx, J. (2007). A Definition of 'Carbon Footprint', ISA UK Research Report 07-01 available at www.isa-research.co.uk.

WWF-UK/SEI (2007). *The right climate for Change – using the carbon footprint to reduce CO_2 emissions – a guide for local authorities.* Godalming, Surrey UK: WWF-UK, Panda house.

Yamin, F. and Depledge, J. (2004). *The International Climate Change Regime: A Guide to Rules, Institutions and Procedures.* Cambridge UK: Cambridge University Press.

In: Industrial Pollution Including Oil
Editor: Harry Newbury and William De Lorne
ISBN 978-1-60456-917-9

Chapter 8

INDUSTRIAL POLLUTION AND CONGENITAL MALFORMATIONS

Jean D. Brender[1], F. Benjamin Zhan[2], Peter H. Langlois[3] and Lucina Suarez[4]

[1]Department of Epidemiology and Biostatistics
Texas AandM Health Science Center, School of Rural Public Health
College Station, Texas (USA)
[2]Texas Center for Geographic Information Science
Department of Geography, Texas State University
San Marcos, Texas (USA) and
School of Resource and Environmental Science
Wuhan University, China
[3]Birth Defects Epidemiology and Surveillance Branch
Texas Department of State Health Services, Austin, Texas (USA)
[4]Epidemiology and Surveillance Unit
Texas Department of State Health Services
Austin, Texas (USA)

ABSTRACT

Congenital malformations remain the leading cause of infant mortality, and etiologies for the majority of these conditions are unknown. Several studies on the relation between maternal residence near hazardous waste sites and industrial facilities and risk for birth defects in offspring, show positive associations between these sites and some malformations such as chromosomal anomalies and neural tube defects. This chapter provides a systematic review of published studies on the relation between maternal residential proximity to hazardous waste sites and industrial facilities and risk of birth defects in offspring. Without exception, exposure assessment was a challenge in these studies. Frequently, investigators measured the distance between residence and industrial facilities and waste sites. Residential proximity to such sites does not necessarily equate with chemical exposures at levels that might be considered teratogenic. Newer approaches such as interpolation and dispersion models show promise for refining estimates of chemical exposures in residential areas. Any relation between living in the vicinity of chemical emissions and birth defects likely is confounded by other factors. Maternal race/ethnicity, education, and occupation have been found to be associated with living near such facilities and sites, factors that are related to risk of some birth defects. Variability of risk effect estimates by geographical region and maternal characteristics (e.g., maternal age) may also complicate interpretations of associations between residential proximity to industrial emissions and birth defects .Because of ongoing public concern that industrial pollution might be linked to birth defects, better studies are needed to examine these associations. Such studies should be population-based to assure adequate statistical power, variability of exposure, and generalizability to other populations. More precise methods of exposure assessment coupled with the careful collection and consideration of potential confounding factors will help elucidate whether ambient exposures to air toxics and other types of industrial pollution pose teratogenic risks.

INTRODUCTION

This chapter will focus on the relation between maternal exposure to industrial pollution and birth defects in offspring. We will review the studies published and their limitations, describe one approach for linking environmental and birth defect data, discuss other approaches for exposure assessment, offer some conclusions based on the extant data, and provide recommendations for future research.

Birth defects (defined here as major structural defects or chromosomal anomalies) occur in approximately three percent of all live births, and are a leading cause of infant mortality in the United States [Honein et al., 1999; Petrini et al., 2002]. Birth defects and genetic disorders result in substantial morbidity and account for 12% of all pediatric hospitalizations [Yoon et al., 1997].

Despite significant morbidity and mortality associated with birth defects, the causes of two-thirds of these defects remain unknown [Nelson and Holmes, 1989; Schardein, 2000]. It was once thought that the placenta was a barrier between the developing embryo and the external environment. We now know that is false, since birth defects may be caused by a variety of maternal exposures such as medications [King et al., 1996; Schardein, 2000], cigarette smoking [Wyszynski and Beaty, 1996], and air pollution [Ritz et al., 2002; Gilboa et al., 2005]. A poll taken in 1999 found that 87% of Americans believed that environmental factors such as pollution cause increased rates of diseases and health problems [Pew Charitable Trusts, 1999], and many believe these factors also caused birth defects.

Latency is the time between exposure and health outcome. The latency for many health outcomes is years or decades (e.g. for cancer). Since many people change residences in that time period, the resulting exposure misclassification makes it more difficult to detect any association between exposure to environmental sources and health outcome. However, latency for most birth defects is less than nine months, greatly reducing misclassification of exposures.

The study of birth defects (teratology) reveals other interesting principles [Schardein, 2000]. The embryo usually has a greater susceptibility to chemicals than the adult, and a teratogen (something causing a birth defect) need not be deleteriously toxic to the mother. For an agent to cause a birth defect, it must be present during the time of organ formation or development (organogenesis) or when genetic activity required for later organogenesis occurs (e.g., chromosomal dysfunction during the formation of sex cells). While that may be quite short (e.g. 1 week for neural tube defects), timing is less relevant for continuous or chronic exposures. Finally, species vary considerably in their sensitivity to teratogenic influence of a given chemical, so laboratory toxicology studies must be complemented by epidemiologic studies in human populations.

Birth defects present some challenges from an epidemiologic perspective. While birth defects of any type comprise 3% of live births, individual birth defects are relatively rare. The fetus is exposed to an array of unknown genetic and environmental factors, and the biologic mechanisms that cause most birth defects are unknown. The defects identified at birth represent only the birth prevalence, not the true incidence of the condition since many conceptions resulting in birth defects are lost before term pregnancy and are unrecognized [Yoon et al., 2001]. Because they are rare, the usual epidemiologic approach is a case-control study, which may have attendant problems such as selection bias and especially recall bias. Also, it may take a very long time to accrue sufficient cases to study exposures around specific sites such as factories or hazardous waste sites. Because many birth defects exhibit a range of severity, minor cases may go undetected or may be detected only by those using new or expensive technology (e.g. detecting minor heart defects using Doppler echocardiography) that in turn may lead to artifactual clusters and trends [Langlois and Scheuerle, 2007].

Nevertheless, because (a) birth defects have such great impact on infant mortality, (b) the etiologies of the majority remain unknown, (c) they exhibit a relatively short latency between exposure and health outcome, and (d) they may result from very low exposure levels, birth defects are ideal for studying the association between industrial pollution and adverse health outcomes. The Agency for Toxic Substances and Disease Registry (ATSDR) has identified birth defects and reproductive disorders as one of its seven priority health conditions [Sever, 1997].

Hazardous Waste Sites and Birth Defects

Dependent on site conditions, living near hazardous waste sites can pose potential health hazards from exposure to site contaminants in air, dust, soil, and water. Since the early 1990s, over 20 studies have been published on the relation between potential exposure to site contaminants at hazardous wastes sites and birth defects in offspring. Table 1 lists these studies by year, country, type of study design, malformations studied, methods of exposure assessment, and findings.

Most studies have been conducted either in the United States or Europe and have used population-based birth defect registries to identify birth defects in the relevant geographic locations and time periods. With the exception of two studies in Table 1 [Eizaguirre-Garcia et al., 2000; Dodds and Seviour, 2001] that focused on populations living around a single waste site, these studies examined multiple waste sites. Examination of populations around numerous sites increased statistical precision, but decreased the likelihood of identifying specific site contaminants and/or site conditions that might increase risk of birth defects due to the heterogeneity of site conditions and contaminants.

Table 1. Studies of maternal residence near landfills and hazardous waste sites and congenital malformations

Reference, Year, Country	Study design, Regional description	Congenital malformations included	Exposure description	Findings
Geschwind et al., 1992 USA	Population-based, case-control study; New York State live births, 1983 – 1984	All birth defects combined; malformations grouped into seven general categories	Exposure risk index that incorporated distance from and the hazard ranking score for each hazardous waste site within 1-mile radius of birth residence	Maternal proximity to waste sites slightly associated with congenital malformations in offspring (OR 1.1, 95% CI 1.1, 1.2) and high exposure risk was more strongly associated with these defects (OR 1.6, 95% CI 1.3, 2.0), especially for defects of the musculoskeletal and integument systems; proximity to sites with pesticides associated with defects of musculoskeletal system (OR 1.2, 95% CI 1.1, 1.4); proximity to sites with plastics associated with chromosomal anomalies (OR 1.5, 1.0, 2.1). Modest effect for central nervous system defects in offspring with maternal residence within 1 mile (OR1.3, 95% CI 1.1-1.6) as well as high exposure risk category (OR 1.5, 95% CI 0.7-3.2) adjusted for age, race, education and other variables.
Shaw et al., 1992 USA	Population-based, case-control study Five-county San Francisco Bay Area live births and fetal deaths 1983 - 1985	All congenital malformations (grouped in 10 malformation groups) except those considered to be inherited or could be attributed to another exposure	Mother's residence at the time of delivery in a census tract with one or more sites with documented environmental contamination	Few associations noted between malformations studied and maternal residence within a census tract with site(s) containing environmental contaminants; elevated risk (OR 1.5, 95% CI 1.1, 2.0) for heart/circulatory defects in offspring of mothers who resided in census tracts with sites with evidence of potential human exposure
Croen et al., 1997 USA	Population-based, case-control study California live births, fetal deaths, and terminations, 1989 - 1991	Neural tube, conotruncal heart, and oral cleft defects	764 hazardous waste sites classified with respect to human exposure potential, contaminated environmental media, and chemical contaminants present;	Little or no increased risk noted for maternal residence in census tract containing one or more waste sites; some association seen between a maternal residence within ¼ mile of a National Priority List site and neural tube defects (OR 2.1) and heart defects (OR 4.2), but the 95% CIs were compatible with the null.

Reference, Year, Country	Study design, Regional description	Congenital malformations included	Exposure description	Findings
			maternal exposure defined as residence in census tract and within 1 mile or less of one or more sites during periconceptional period	Positive associations were noted between a maternal residence within 1 mile of sites with some heavy metals, polycyclic hydrocarbons, and solvents and neural tube defects in offspring.
Marshall et al., 1997 USA	Population-based, case-control study 18 counties in New York State, 1983 – 1986 live births	Central nervous system and musculoskeletal system defects	Proximity and related exposure index of mother's address at delivery to waste sites with solvents, metals, and pesticides	Minimal or no association was noted between maternal residential proximity to waste sites (in general and those sites with solvents or metals) and central nervous system and musculoskeletal system defects. Adjustment for mother's race and education reduced crude ORs.
Dolk et al., 1998 Europe	Population-based case-control study Belgium, Denmark, France, Italy, UK Live births, fetal deaths, and pregnancy terminations	Non-chromosomal congenital anomalies	21 landfill sites with hazardous waste zone within 3 km radius of site was defined as "proximate zone"	Significant associations noted for residence within 3 km of site and neural tube defects (OR 1.9, 95% CI 1.2, 2.8), cardiac septal defects (OR 1.5, 95% CI 1.1, 3.2), and anomalies of the great arteries and veins (OR 1.8, 95% 1.0, 3.2). Elevated odds ratios were also found for tracheo-esophageal anomalies, hypospadias, and gastroschisis, although estimates were consistent with the null.
Eizaguirre-Garcia et al., 2000 United Kingdom	Population-based descriptive geographical study of birth defect cases and births during 1982 – 1989 in Glasgow and nearby areas	Congenital anomalies combined into one group	Residence in a circle within 10 km in radius around former site of factory and area contaminated by chromium; areas divided into 2 km area containing site and 8 concentric rings around it, each 1 km wide	Relative risk highest in an area 2-4 km away from pollutant (RR 1.5, 95% CI 1.2, 1.8); referent category 0-2 km from center of polluted area); investigators concluded that results did not point to possible teratogenic effect of chromium waste
Elliott et al., 2001 Great Britain	Population cohort including live births, stillbirths, congenital malformations including terminations; Great Britain, 1983-1998	All congenital anomalies combined; neural tube, cardiovascular, and abdominal wall defects;	Mother's address (unclear when ascertained or from what sources); distance to hazardous waste sites; within 2 km categorized as exposed	Unadjusted and adjusted relative risks close to 1.0 for all defects studied; modest association observed for surgical correction of gastroschisis and exomphalos (RR 1.2, 99% CI 1.1, 1.3) if mother lived within 2 km of site relative to living farther away

Table 1. (Continued)

Reference, Year, Country	Study design, Regional description	Congenital malformations included	Exposure description	Findings
		hypospadias and epispadias; surgical correction of hypospadias and epispadias; surgical correction of gastroschisis and exomphalos		
Dodds and Seviour, 2001 Canada	Population-based cohort study Nova Scotia, Canada Live births and stillbirths, 1988-1998	All major anomalies combined and nine anomaly sub-groups	Rates for malformations compared by maternal address at the time of delivery in Sydney (site of hazardous waste site), Nova Scotia, and Cape Breton County (excluding Sydney)	Residents in Sydney (with hazardous waste site) were 1.3 times more likely (95% CI 1.0, 1.5) to have births with a major congenital anomaly than Nova Scotia residents; also relative to Nova Scotia, rate ratio for NTDs elevated in births to Sydney residents (RR 1.8, 95% CI 1.1, 3.1)
Orr et al., 2002 USA	Population-based, case-control study; California (24 counties) births and fetal deaths, 1983 – 1988; focused on minority births	All birth defects combined; musculoskeletal, central nervous system, integumental, heart or circulatory, and oral cleft defects; chromosomal anomalies	Maternal address at child's birth (obtained from birth certificate) in census tract with one or more National Priority List (NPL) hazardous waste sites (n = 84 sites)	Modest effects observed for NTDs (OR 1.5, 95% CI 0.93, 2.6), anencephaly (OR 1.9, 95% CI 0.91, 3.8), and spina bifida (OR 1.3, 95% CI 0.61, 2.5) though estimates compatible with the null; associations also noted with trisomy 13 (OR 2.7, 95% CI 1.5, 4.6), trisomy 18 (OR 2.7, 95% CI 1.4, 5.1), and sex chromosome anomalies (OR 3.1, 95% CI 1.0, 9.6)
Vrijheid et al., 2002 Europe	Population-based case-control study Belgium, Denmark, France, Italy, UK live births, fetal deaths, and pregnancy terminations	Chromosomal congenital anomalies further classified as either Down's syndrome or non-Down's syndrome	23 landfill sites with hazardous waste zone within 3 km radius of site was defined as "proximate zone"	With adjustment for maternal age and socioeconomic status, women who lived within 3 km of hazardous waste site more likely to have a birth with a chromosomal anomaly than women who lived 3-7 km (OR 1.4, 95% CI 1.0, 2.0)

Reference, Year, Country	Study design, Regional description	Congenital malformations included	Exposure description	Findings
Morris et al., 2003 Great Britain	Population cohort including all births, stillbirths, and termination registries in Scotland between 1982 and 1997	All congenital anomalies combined; neural tube, cardiovascular, and abdominal wall defects; hypospadias and epispadias; surgical correction of hypospadias and epispadias; surgical correction of gastroschisis and exomphalos	Mother's address (unclear when ascertained or from what sources); distance to hazardous waste sites; within 2 km categorized as exposed	No statistical excess was found for all congenital anomalies combined (RR 0.96, 99% CI 0.89-1.02) or for any of the specific anomalies studied
Malik et al., 2004 USA	Population-based case-control study Dallas County, Texas live births 1979 - 1984	Live births diagnosed with congenital heart disease at any age	Mothers' residence at delivery within ¼ and 1 mile of hazardous waste site	Maternal residence within one mile of hazardous wastes site was slightly associated with congenital heart disease in offspring (OR 1.2, 95% CI 1.1, 1.4)
Yauck et al., 2004 USA	Population-based case-control study; Milwaukee, Wisconsin live births 1997 – 1999	Live births diagnosed with congenital heart defect (CHD) based on echocardiography, surgical findings, and autopsy reports	Mother's address at delivery within 1.32 miles of waste sites and industrial facilities with emissions of trichloroethylene (TCE)	Among older (> 37 years) mothers, CHD in offspring associated with a maternal residence within 1.32 miles of TCE-emitting sites (OR 3.2, 95% CI 1.2, 8.7); no relation found between living near these sites and CHD in offspring among younger women
Boyle et al., 2004 Great Britain	Population-based cohort and case-control studies; Eastern Region of Ireland births, 1986 - 1990	Births with congenital anomalies (all combined) detected by the regional congenital anomalies registry	Municipal landfill sites within 3 km (and other distances) of district electoral divisions; distance of case and control addresses from landfill sites	In both area-level analyses and the case-control study, congenital anomalies were not found to occur more commonly in proximity to municipal landfills
Palmer et al., 2005 Great Britain	Population-based cohort of live births in Wales 1983-1997	Congenital anomalies combined and sub-grouped as chromosomal anomalies, cardiovascular defects, and abdominal wall defects in live births	Expected rates of congenital anomalies in births to mothers (at time of delivery) living within 2 km of landfill sites, before and after opening of the sites, with referent group living at least 4 km away from these sites	Ratio of observed to expected rates of congenital malformations before landfills opened was less (0.87) than after their opening (1.2) giving a standardized risk ratio of 1.4 (95% CI 1.1, 1.7); although risk ratios for the subcategories of malformations also elevated, the 95% confidence intervals around these estimates included 1.0

Table 1. (Continued)

Reference, Year, Country	Study design, Regional description	Congenital malformations included	Exposure description	Findings
Kloppenborg et al., 2005 Denmark	Population-based cohort of live births in Denmark, 1997 - 2001	Congenital anomalies combined and sub-grouped as defects of the nervous or cardiovascular systems in live births	Distance of maternal residence from waste landfill sites in three buffer zones: 0-2, 2-4, and 4-6 km	No association noted between landfill location and congenital malformations combined or nervous system anomalies; noted small excess risk for anomalies of the cardiovascular system
Gilbreath and Kass, 2006 USA	Population-based retrospective cohort study of live births and fetal deaths in Alaska Native villages 1997 - 2001	Observable congenital anomalies as recorded on the birth records grouped into five categories including central nervous system, circulatory and respiratory, gastrointestinal, urogenital, and musculoskeletal or integumentary defects	Exposure variables obtained from hazard rankings of dumpsites; residence in village with open dumpsites ranked as lower or higher hazard	The 95% confidence limits for crude and adjusted rate ratios were consistent with the null, although the adjusted point estimates of the rate ratios (rate in villages with higher hazard dumpsites relative to rate in villages with lower hazard dumpsites) were positive for all congenital anomaly categories except gastrointestinal defects
Brender et al., 2006 USA	Population-based case-control study of live births and fetal deaths in Texas, 1996 - 2000	Live births and fetal deaths with cleft palate without cleft lip; cleft lip without or with cleft palate; isolated oral cleft (without any other major defect besides oral clefts)	Residence at delivery within 1 mile of NPL or state hazardous waste site; limited sample also available from the Texas portion of National Birth Defects Prevention Study (NBDPS) for residence during the periconceptional period (3 months before to 3 months after conception)	With adjustment for maternal race/ethnicity, education, and tobacco use, residence at delivery within 1 mile of waste site not associated with cleft palate (OR 0.60, 95% CI 0.31, 1.2) or cleft lip with or without cleft palate (OR 0.91, 95% CI 0.60, 1.4); in Texas NBDPS population for births during 1998 – 2000, women living within 1 mile of a hazardous waste site during the periconceptional period were not more likely to have offspring with oral clefts (OR 1.1, 95% CI 0.35, 3.2)
Suarez et al., 2007 USA	Population-based case-control study of live births and fetal deaths in Texas, 1996 - 2000	Live births and fetal deaths with neural tube defects	Residence at delivery within 1 mile of state or NPL hazardous waste site	With adjustment for maternal age, education, and race/ethnicity, no association noted between maternal residence near waste site and neural tube defects in offspring (OR 1.0, 95% CI 0.6, 1.7)

Reference, Year, Country	Study design, Regional description	Congenital malformations included	Exposure description	Findings
Kuehn et al., 2007 USA	Population-based case-control study of live births in Washington State, 1987 - 2001	Cases identified from linked birth-hospital discharge records and categorized into 11 groups based on system and type	Distance of maternal residence at delivery from hazardous waste sites; proximity defined as various distances up to 5 miles; waste sites categorized as high or low priority based on types of contaminants present and media contaminated	Relative to living > 5 miles from a site, living within 2 and 5 miles (OR 1.2), 1 and 2 miles (OR 1.3), 0.5 to 1 mile (1.3), and less than 0.5 miles (1.3) were significantly associated with increased risk of any congenital malformations in offspring; most associations with specific effect groups were modest except for birth defects involving skin if the mother lived within 1 mile of a site (OR 2.4, 95% CI 2.2, 2.7); associations for malformations stronger with sites in urban areas than in rural areas
Brender et al., 2008 USA	Population-based case-control study of live births and fetal deaths in Texas, 1996 - 2000	Live births and fetal deaths with chromosomal anomalies (combined) and categorized into nine categories based on BPA codes	Residence at delivery within 1 mile of state or NPL hazardous waste site	With adjustment for maternal age, education, and race/ethnicity, maternal residence not associated with chromosomal anomalies in offspring except Klinefelter variants among Hispanic births (OR 7.9, 95% CI 1.1, 42.4)

Abbreviations: OR, odds ratio; RR, relative risk or risk ratio; CI, confidence interval.

Table 2. Studies of maternal residence near industrial sources of pollution and congenital malformations

Reference, Year, Country	Study design, Regional description	Congenital malformations included	Exposure description	Findings
Wulff et al., 1996 Sweden	Population-based retrospective cohort study of 1973-1990 births in selected parishes of Sweden	All congenital malformations grouped into 13 categories; heart defects sub-categorized into 17 subcategories	Persons living in parishes within 20 km from a copper smelter; parent employed at smelter	Slightly larger proportion of malformations seen among exposed children in cohort than among reference population (RR 1.2, 95% CI 0.95, 1.4) with chromosomal anomalies more common among exposed group (RR 2.6, 95% CI 0.90, 6.7). Authors attributed chromosomal association to under-reporting in reference area and/or active surveillance in exposed area.
Marshall et al., 1997 USA	Population-based, case-control study in 18 counties in New York State, 1983 – 1986 live births	Central nervous system and musculoskeletal system defects	Proximity (within 1 mile) of maternal residence at delivery to industrial sources of air emissions; industrial sources identified from 1988 Toxic Release Inventory (TRI); general dispersion model used for solvent emissions	Central nervous system defects in offspring associated with maternal residence within 1 mile of TRI facility with air emissions of solvents (OR 1.3, 95% CI 1.0, 1.7) and metals (OR 1.4, 95% CI 1.1, 1.7). Minimal or no association seen for these potential maternal exposures and musculoskeletal system defects in offspring
Bhopal et al., 1998 United Kingdom	Population-based, retrospective cohort study of births, stillbirths, and terminations in Teesside and Sunderland, United Kingdom, 1986 - 1993	All congenital abnormalities (excluding isolated minor congenital abnormalities)	Residential proximity to major steel and petrochemical industries in Teesside divided into three zones based on distance with Sunderland serving as the reference population	No significant differences in congenital malformation rates (combined) found either across the Teesside zones or between these zones combined and Sunderland, the reference population (RR ranged from 0.87 – 1.2 with all 95% CI including the null).
Czeizel et al., 1999 Hungary	Retrospective cohort – cluster analysis of live births, stillbirths, and terminations in surrounding region of acrylonitrile factory in Nyergesujfalu, Hungary	Categorized into 32 isolated and 5 multiple congenital anomaly groups	Three concentric bands around the acrylonitrile factory (Nyergesujfalu with factory, within 5 km from epicenter, and 5 and 25 km from factory)	Decrease in risk of undescended testis with increasing distance from the acrylonitrile factory

Reference, Year, Country	Study design, Regional description	Congenital malformations included	Exposure description	Findings
Vinceti et al., 2001 Italy	Ecologic study of prevalence of birth defects during 1982-1986, 1987-1990, and 1991-1995 in Provinces of Reggio Emilia and Modena, northern Italy	All malformations combined and specific malformations divided into 18 groups	Prevalence of birth defects in Ceramic District (contaminated with lead) with the remainder of the two Provinces containing this District, but outside of the lead-contaminated area serving as the unexposed population	Relative to the unpolluted areas, excess risk of cardiovascular defects observed in lead-polluted area that decreased over time as pollution decreased (1982-1986: RR 2.6, 95% CI 1.7, 3.8; 1987-1990: RR 1.2, 95% CI 0.62, 2.1; 1991-1995: RR 0.97, 95% CI 0.57, 1.5); also found higher risks of oral clefts and musculoskeletal defects in lead-contaminated area with decreasing risk over time
Cresswell et al., 2003 United Kingdom	Ecologic study of prevalence of birth defects among livebirths in city of New Castle upon Tyne, United Kingdom, 1985-1999	Malformations among all livebirths, stillbirths, induced abortions, and fetal deaths after 14 weeks gestation categorized into chromosomal and non-chromosomal defects	Residence within 3 km of Byker waste combustion plant	Relative to living 3-7 km from plant, RR 1.11 (95% CI 0.96-1.3) for living within 3 km after site began operations; RR higher for non-chromosomal than for chromosomal defects among offspring of women living near waste combustion plant
Dummer et al, 2003 UK	Retrospective cohort study; Cumbria (northwest England), 1956-93	Deaths from congenital anomaly (ICD 740 – 749)	Distances of maternal address at child's birth (obtained from birth certificate) from incinerators and crematoriums	Risk of lethal congenital anomaly significantly increased ($p < 0.01$) with maternal address closer to incinerators (restricted to heart defects and spina bifida); increased risk of anencephaly and other congenital anomalies with maternal address near crematoriums
Cordier et al., 2004 France	Ecologic study of prevalence of birth defects in communities surrounding incinerators in southeast France, 1988-1997	Malformations among livebirths, stillbirths, and medical terminations divided into minor, chromosomal, monogenic, and other major anomalies; other major anomalies subdivided into 23 different subgroups	Dioxin concentration estimates in 194 communities with municipal waste incinerators; residence in these communities	With adjustment for population density and average family income, RR for other major anomalies was 1.1 (95% CI 0.98, 1.2) with living in exposed communities relative to unexposed communities; some association seen for facial clefts (RR 1.3, 95% CI 1.1, 1.6) and renal dysplasia (RR 1.6, 95% CI 1.1, 2.2); in exposed communities, dose-response trend of risk for obstructive uropathies seen with increasing exposure

Table 2. (Continued)

Reference, Year, Country	Study design, Regional description	Congenital malformations included	Exposure description	Findings
Tango et al., 2004 Japan	Retrospective cohort of births and fetal deaths in Japan, 1997-1998	Infant, neonatal, and fetal deaths due to congenital malformations (all combined)	Distance of street addresses (from vital records) from municipal solid waste incinerators divided into ten sub-areas delimited by ten circles of radii of 1,2,…, 10 km.	No significant excess noted in deaths due to congenital malformations with address within 2 km of municipal solid waste incinerator; statistical significant peak decline in risk of infant deaths due to congenital malformations (combined) with distance from the incinerators up to 10 km with peak at 1-2 km
Bentov et al., 2006 Israel	Ecologic study of 1995-2000 live births and stillbirths in Beer-Sheva subdistrict in Israel divided into Jewish and Bedouin localities	Major congenital malformations combined and subcategorized into anomalies of central nervous system, anomalies associated with chromosomal anomalies, and other major congenital malformations	Distance of localities from regional industrial park and predominant wind direction (17 chemical plants and one industrial toxic waste site)	Risk of congenital malformations among Bedouin populations higher in proximal than distal localities (RR 1.6, 95% CI 1.4, 1.8) especially risk of central nervous system defects (RR 2.3, 95% CI 1.4, 3.6); congenital malformations not associated with residential proximity to industrial park among births in Jewish localities
Brender et al., 2006 USA	Population-based case-control study of live births and fetal deaths in Texas, 1996 - 2000	Live births and fetal deaths with cleft palate without cleft lip; cleft lip without or with cleft palate; isolated oral cleft (without any other major defect besides oral clefts)	Residence at delivery within 1 mile of industries with reported air emissions of chemicals; limited sample also available from the Texas portion of National Birth Defects Prevention Study (NBDPS) for residence during the periconceptional period (3 months before to 3 months after conception)	Among women < 35 years, no association between residence within 1 mile of industrial facilities and oral clefts in offspring; among women ≥ 35 years, oral clefts in offspring associated with residence within 1 mile of industrial facilities (OR 2.4, 95% CI 1.3, 4.2) especially smelters (OR 15, 95% CI 2.8, 151)

Reference, Year, Country	Study design, Regional description	Congenital malformations included	Exposure description	Findings
Suarez et al., 2007 USA	Population-based case-control study of live births and fetal deaths in Texas, 1996 - 2000	Live births and fetal deaths with neural tube defects	Residence at delivery within 1 mile of industries with reported air emissions of chemicals	Modest risk seen for neural tube defects in offspring with maternal residence at delivery within 1 mile of industrial facility (OR 1.2, 95% CI 1.0, 1.5) with a stronger association among mothers 35 years and older (OR 2.7, 95% CI 1.4, 5.0) and among non-Hispanic white mothers (OR 1.8, 95% CI 1.1, 2.8)
Brender et al., 2008 USA	Population-based case-control study of live births and fetal deaths in Texas, 1996 - 2000	Live births and fetal deaths with chromosomal anomalies (combined) and categorized into nine categories based on BPA codes	Residence at delivery within 1 mile of industries with reported air emissions of chemicals	With adjustment for maternal age, race/ethnicity, and education, autosomal deletions in offspring (OR 1.5, 95% CI 1.0, 2.3) and Klinefelter syndrome (male births only, OR 2.9, 95% CI 1.1, 7.3) associated with maternal residence within 1 mile of industrial facility; among older women ($\geq$ 35 years), chromosomal anomalies in offspring associated with living near facilities with heavy metal or solvent emissions

Abbreviations: OR, odds ratio; RR, relative risk or risk ratio; CI, confidence interval.

Table 3. Relation Between Maternal Residential Proximity to Industrial Facilities with Air Emissions of Chemicals and Neural Tube Defects by Area of Residence, Texas, 1996 - 2000

Case-control status	Maternal residence within 1 mile of industrial facility					
	North-Central Texas			Other Regions of Texas		
	N (%)	Unadjusted OR[1] (95% CI)	Adjusted OR[2] (95% CI)	N (%)	Unadjusted OR[1] (95% CI)	Adjusted OR[2] (95% CI)
Controls	263 (18.1)	–		345 (11.8)	–	
Neural tube defects	64 (30.8)	2.0 (1.5, 2.8)	1.9 (1.3, 2.7)	51 (11.4)	0.95 (0.70, 1.3)	0.90 (0.66, 1.2)
Anencephaly	19 (34.5)	2.4 (1.3, 4.2)	2.3 (1.3, 4.3)	14 (11.3)	0.95 (0.54, 1.7)	0.95 (0.53, 1.7)
Spina bifida	38 (29.9)	1.9 (1.3, 2.9)	1.8 (1.2, 2.8)	32 (11.6)	0.97 (0.66, 1.4)	0.89 (0.60, 1.3)

[1]OR, odds ratio and CI, confidence interval. Referent groups are women who lived more than one mile from these facilities.
[2]Adjusted for maternal race/ethnicity and education.

In these studies, investigators obtained information about site location, conditions, and contaminants from a variety of governmental sources and entities tasked with monitoring and regulating landfills and/or hazardous waste sites. The definition of "residential proximity" varied by country. In the U.S., distances considered close to sites ranged from ¼ to 1 mile whereas in Europe, 2 or 3 km (1.2 miles or 1.9 miles respectively) were considered proximate. Although several studies developed measures to further classify these sites as "exposing" or "nonexposing" [Croen et al., 1997]; "high", "medium", or "low" probability of residential exposure [Marshall et al., 1997]; and "higher hazard" versus "lower hazard" [Gilbreath and Kass, 2006], all studies depended on some measure of residential proximity instead of biological markers of exposure or individual activity patterns as measures of potential exposure to site contaminants. Other studies used residence in a census tract with one more waste sites as a measure of potential exposure [Shaw et al., 1992; Orr et al., 2002], a method that potentially introduced further misclassification since census tracts vary in size.

With residential distance from waste sites as a proxy for maternal exposure to site contaminants, findings varied across studies. Several studies found positive associations between a maternal residence near waste sites and central nervous system defects [Geschwind et al., 1992; Croen et al., 1997 (neural tube defects); Dolk et al., 1998 (neural tube defects); Orr et al., 2002 (neural tube defects)], heart defects [Shaw et al., 1992; Croen et al., 1997; Dolk et al., 1998; Malik et al., 2004; Yauck et al., 2004 (among offspring of women 38 years or older)], and chromosomal anomalies [Geschwind et al., 1992 (sites with plastics); Orr et al., 2002; Vrijheid et al., 2002; Brender et al., 2008 (Klinefelter variants among Hispanic births)].

INDUSTRIAL FACILITIES AND BIRTH DEFECTS

Studies of the relation between potential maternal exposures to industrial emissions and birth defects have predominantly used proximity of residence as a proxy of exposure, although some studies have estimated air concentrations of various chemicals using a general dispersion model [Marshall et al., 1997] and a Gaussian plume model [Cordier et al., 2004]. Table 2 lists these studies by author and year of publication, country of study population, type of study design, malformations studied, methods of exposure assessment, and major findings.

Six studies focused on single facilities or industrial complexes including populations living around a copper smelter [Wulff et al., 1996], an industrial area with steel and petrochemical industries [Bhopal et al., 1998], an acrylonitrile factory [Czeizel et al., 1999], an industrial area contaminated with lead [Vinceti et al., 2001], a waste combustion plant [Cresswell et al., 2003], and a regional industrial park with chemical plants and a toxic waste site [Bentov et al., 2006]. Most residential associations with birth defects were modest with the exception of two studies.

In Italy [Vinceti et al., 2001], births with cardiovascular, musculoskeletal, and oral clefts were more likely to occur among women living in an industrial area contaminated with lead relative to women living away from this area; these risks decreased over time during the same periods that environmental lead contamination decreased. An ecologic study design was used, however, in which exposure was inferred by area of residence instead of distances from the industrial areas and contamination. Therefore, the positive associations found may be due to an ecologic bias. Ecologic biases occur when associations at an aggregate level do not represent exposures at the individual level among mothers of births with birth defects or normal outcomes [Morgenstern, 2008].

Bentov et al. [2006] also used an ecologic study design in Israel to examine the prevalence of birth defects in localities near a regional industrial park with emissions of aliphatic, aromatic, and polycyclic hydrocarbons and inorganic agents including heavy metals. Distance of localities from the industrial park was used as a proxy for exposure, and the localities were further classified as predominantly Jewish or Bedouin. Among Bedouin communities, risk of congenital malformations was higher among births to women living in localities close to the industrial park relative to women living localities farther away, especially central nervous system defects. These risks were not noted in the Jewish localities near the industrial area.

The remainder of studies examined industrial facilities, incinerators, or crematoriums in multiple locations within a state, region, or country. Central nervous system defects [Marshall et al., 1997] including neural tube defects [Suarez et al., 2007], oral clefts [Brender et al., 2006], and chromosomal anomalies [Brender et al., 2008] were more likely among births to women who lived within a mile of industrial facilities with reported air emissions of chemicals than among births to women who lived farther away. Some associations were also seen between a maternal residence near incinerators or crematoriums and neural tube defects [Dummer et al., 2003], heart defects [Dummer et al., 2003], oral clefts [Cordier et al., 2004], and renal dysplasia [Cordier et al., 2004]. Tango et al. [2004] found no excess of deaths due to congenital malformations among births to mothers who lived near municipal solid waste incinerators in Japan.

Study Issues and Limitations

Use of residential proximity as a proxy for exposure. In the studies discussed, most used some measure of residential proximity as a proxy for exposure to site contaminants and industrial emissions. However, proximity of residence should not be equated with exposure. For humans to be exposed to site or industrial contaminants, there must be a completed exposure pathway. The Agency for Toxic Substances and Disease Registry [2008] has identified five components that are required for a completed exposure pathway including:

1) A source of contamination (e.g., waste site) or release (e.g., stack and/or fugitive emissions from an industrial facility);
2) A release mechanism into the air, water, soil, or food chain;
3) A point of exposure (e.g., drinking water from a well, residential yard, ambient air, house dust);
4) An exposure route (e.g., ingestion, dermal contact, inhalation); and
5) A potentially exposed population (e.g., nearby residents).

Most studies established a source of contamination or emissions; identified a release mechanism to some degree for industrial facilities, but less so for hazardous waste sites; and identified potentially exposed populations (e.g., childbearing women who lived in close proximity to these sites and/or facilities). Points and routes of exposure, however, were inferred by residential proximity rather than established through biomarkers or environmental monitoring at personal residences. At least one study was able to establish historical human exposure from industrial activities in a residential area [Vinceti et al., 2001]. In an industrial area heavily polluted with lead, historical blood lead monitoring of residents indicated elevated blood lead levels compared with residents living outside the contaminated areas. However, blood lead levels of women giving birth to babies with or without major malformations were not reported.

Residential exposure to site contaminants and industrial emissions will also vary by climatic and topographic characteristics of geographic areas. These conditions were rarely considered in the studies discussed with the exception of those by Bentov et al. [2006] and Cordier et al. [2004].

Heterogeneity of potential exposures. While previous investigators have clearly established a source of contamination, in most cases, study populations were potentially exposed to a mixture of substances. Furthermore some women lived near more than one industrial facility. For example, 23% of childbearing women in a statewide sample of 1996 – 2000 births in Texas [Brender et al., 2006a, 2006b, 2008a; Suarez et al., 2007] lived within two miles of multiple industrial facilities with approximately 6% living within this distance of five or more such facilities. Focusing on populations that live near single sources of industrial pollution would reduce the problem of heterogeneity of exposure. However, investigations of single sites with limited contaminants often lead to inconclusive results from small sample size and low statistical power.

Availability of relevant maternal addresses. The morphogenesis of structural birth defects usually occur the first trimester of pregnancy while the period shortly before pregnancy (or even earlier for maternal meiosis I) or at the time of conception is the most critical in the generation of chromosomal anomalies. Therefore, the most appropriate residential exposure windows would be shortly before and at conception for chromosomal anomalies and the first trimester for non-chromosomal anomalies.

Many studies used address at delivery rather than the aforementioned exposure windows because of the ready availability of maternal addresses on birth and fetal death vital records. The estimated percent of women who change addresses between the time of conception and delivery has ranged between 12 and 33 percent depending on geographic location of study populations [Fell et al., 2004; Khoury et al., 1988; Shaw and Malcoe, 1992; Canfield et al., 2006]. While the percent of case- and control-mothers who moved have been similar [Shaw and Malcoe, 1992; Canfield et al., 2006], younger women have been more likely to move during pregnancy than older women.

Examining residential exposures at the time of delivery instead of around conception and the first trimester could lead to misclassification of environmental exposures. Because similar proportions of case- and control women change addresses during pregnancy, this misclassification would most likely be nondifferential and produce risk estimates closer to the null (1.00) than if maternal addresses during the critical periods of fetal development were consistently used.

Confounding factors. Maternal addresses are not randomly distributed, but depend on several factors such as socioeconomic status, parental occupation, and race/ethnicity. These same factors may also increase or decrease the likelihood of living near industries or hazardous waste sites. For instance, Perlin et al. [2001] noted that African American children under the age of six were more likely than white children to live in close proximity to industrial sources of air pollution in a U.S. study spanning three states (West Virginia, Louisiana, and Maryland). Woodruff et al. [2003] also noted racial disparities during pregnancy with respect to living in areas with higher air pollution; greater proportions of Hispanic, African American, and Asian/Pacific Islander women lived in such areas than white women. In this U.S. study, maternal education attainment was not associated with living in counties with the highest levels of air pollution with adjustment for maternal risk factors, region of country, and race/ethnicity.

Among comparison mothers in a case-control study of birth defects, Brender et al. [2006b] found both maternal race/ethnicity and education to be associated with living within a mile of a hazardous waste site or industrial facility with reported air emissions. With adjustment for maternal age, race/ethnicity, and education, women who were other than non-Hispanic white race/ethnicity and less educated (< 16 years) were more like to live near these sites and facilities than white, non-Hispanic women and women with 16 or more years of education.

Finally, there is some empirical evidence that occupation and location of residence are related. The percentages of employed persons in executive, managerial, or professional occupations were highly correlated with census and block group indicators in three cardiovascular study populations [Diez-Roux et al., 2001]. In two Texas case-control study populations of birth defects, women in production occupations were twice as likely to reside near industrial facilities with reported air releases than women in management/professional occupations and more likely to reside near such facilities than other occupations [Brender et al., 2008b].

Maternal race/ethnicity, education, and occupation are risk factors for several types of congenital malformations. Spurious associations might result between maternal exposures to industrial pollution and congenital malformations in offspring, if risk estimates are not adjusted for these potentially confounding factors. A confounding factor is causally related to the disease or condition under study (or serves as a proxy measure for unknown or unmeasured causes) and is associated with the exposure under study [Kelsey et al, 1996]. Since maternal race/ethnicity, education, and occupation are associated with both residential proximity to industrial facilities and selected birth defects, these maternal characteristics may be confounding factors in investigations of the relation between industrial pollution and birth defects.

Heterogeneity of effects by maternal characteristics and geographic location. Risk estimates for birth defects in relation to industrial pollution have varied by maternal characteristics and locality. For instance, Yauck et al. [2004] found that maternal residence near trichloroethylene-emitting sites was associated with congenital heart defects among offspring of older women (> 37 years) but not younger women (< 38 years). In Texas, risk estimates for several congenital anomalies were increased among offspring of older women (> 35 years) who lived within a mile of industrial facilities with air emissions including neural tube defects [(Suarez et al., 2007], oral clefts [Brender et al., 2006a], and chromosomal anomalies [Brender et al., 2008a]; minimal or no risk was noted for these congenital anomalies among offspring of younger mothers with this residential characteristic.

Besides age, variations of risk effects by maternal race/ethnicity have also been noted. Bedouin women in Israel who lived in localities near a regional industrial park were more likely to have babies with major congenital malformations than Bedouin women living in localities further away from this area [Bentov et al., 2006]. In contrast, the rates of major congenital malformations in Jewish women were similar whether they lived in localities proximate or distal from the industrial area. In Texas, an association between maternal residential proximity to industrial facilities and neural tube defects in offspring was noted among White, non-Hispanic women (OR 1.8, 95% CI 1.1, 2.8), but not among Hispanic women (OR 1.1, 95% CI 0.8, 1.4) [Suarez et al., 2007].

Risk estimates may also vary by geographic location. Heterogeneity of risk estimates across counties, regions, states, and countries might be due to differences in study design, case ascertainment or definition, exposure assessment and estimation, climatic factors, and site conditions. Findings may vary by geographic location even within a state or region. Table 3 shows risk estimates for neural tube defects in Texas by maternal residential proximity to industrial facilities.

The odds ratios for neural tube defects including anencephaly and spina bifida are elevated for births to women in north-central Texas who lived within one mile of industrial facilities relative to women living farther away from such facilities, but not for women who lived in other regions of the state with this residential characteristic. Risk estimates did not change appreciably with adjustment for maternal race/ethnicity and education.

Proximity Model for Exposure Assessment

One key issue in epidemiologic analyses of environmental health data is the assessment of exposure to a pollutant in question. Pathways of exposure include air, water, soil, and the food chain. We only discuss methods for exposure assessment related to air pollution in this and the following Section. Readers are directed to discussions by Stanek and Calabrese [2000], Martin and Earl [2002], Nieuwenhuijsen et al. [2006], and Paustenbach et al. [2006] for methods about exposure assessment related to soil contamination. Researchers interested in assessing exposure to water contamination may consult publications by Feijtel et al. [1998], Maslia et al. [2000], Harding and Walski [2000], Reif [2003], Keller [2006], Coffey et al. [2007], and Meliker et al. [2008] for more information. Islam et al. [2007] and Alonso et al. [2008] discuss assessment of food contamination from industrial pollution.

In this Section, we describe a proximity model and a Geographic Information System (GIS) based spatial search tool that uses distance as a surrogate for exposure in this section and briefly discuss more sophisticated models in the next Section. Researchers usually follow a 4-step procedure to perform exposure assessment in environmental epidemiology studies: 1) acquire environmental data; 2) prepare health outcome data; 3) link environmental and health outcome data; and 4) perform exposure assessment.

Acquire environmental data. When using proximity models for exposure assessment, we need to obtain two types of environmental data—spatial data (coordinates) of environmental sites in question and non-spatial data associated with these sites such as pollutants released from these sites. Non-spatial data is usually called attribute data. Spatial and attribute data should in two separate databases because each site under study may have an extensive list of attributes and pollutants. Having spatial and attribute data in a single database would generate many records with duplicating attributes about each site.

We use the preparation of data about superfund sites in Texas as an example to illustrate how to construct environmental GIS databases. After downloading data about the US National Priority List (NPL) superfund sites from the online Hazardous Substances Release/Health Effects Database (HazDat) at the Agency for Toxic Substances and Disease Registry (ATSDR) [ATSDR, 2005] and the state superfund sites from the Texas Commission on Environmental Quality (TCEQ) online superfund database [TCEQ 2005], we constructed databases containing the necessary data about the superfund sites in Texas for a project on the relation between residential proximity to waste sites and industrial facilities and selected congenital malformations [Brender et al. 2006b, 2008; Suarez et al. 2007].

Both point data showing the locations of superfund sites and polygons delineating the boundaries of these sites were obtained. The point locations were created by the longitude and latitude available in the ATSDR or TCEQ online databases. The polygons were created by manually digitizing the boundaries of the sites using Digital Orthophoto Quarter Quads (DOQQ) images with a 1-meter resolution from the Texas Natural Resources Information System (TNRIS). Additional information such as site status (active/deleted) and year of deletion (if applicable) associated with each site was also included in the databases. A separate database containing pollutants identified in each site was also constructed. Information about these pollutants was obtained either from the ATSDR databases or extracted from paper reports and microfilmed files available at TCEQ. Figure 1 is a map showing the locations of some superfund sites in Texas.

Figure 1. Selected superfund sites in exas (USA).

Prepare health outcome data. Similar to environmental data, health outcome data typically contain geographic locations of cases (persons with the adverse health outcome under study) and controls (persons without the adverse health outcome under study) and attribute data such as residence address, age, race/ethnicity, education, and other characteristics associated with a case or control. In most situations, attributes associated with a case or control can be obtained from existing health or medical records databases. It is more time consuming to obtain the coordinates associated with cases and controls because medical records typically only have street addresses and do not have coordinates. In the U.S., coordinates can be obtained from street addresses through a common and widely used procedure in Geographic Information Systems (GIS)—geocoding.

Geocoding uses known street networks as references and applies computational procedures to determine the corresponding coordinates of a street address [see, e.g., Karimi et al. 2004; Zhan et al. 2006a]. Other ways of obtaining geographic locations of cases or controls include field surveys using global positioning systems (GPS) devices or measurement from registered aerial photos or remote sensing images. After obtaining the geographic locations and required attributes of cases and controls for environmental epidemiologic studies, the data can be imported into GIS to construct GIS map layers and attribute databases. Geocoding should be conducted without knowledge of exposure or disease status to reduce the likelihood of information bias. Geocoding can be problematic for rural populations, but this problem is decreasing since emergency services are now encouraging the assignment of street addresses to these areas. The preparation of health outcome data can be considered complete once map layers and attribute databases are constructed within a GIS environment.

Link environmental and health outcome data. When the environmental databases and health outcome databases are in place, the next step is to link environmental data and health outcome data. When distance is used as surrogate for exposure, the basic procedure of linking environmental and health outcome data can be summarized into the six steps shown in Table 4. These steps can be automated in a GIS environment through the development of a program utilizing GIS functions.

Table 4. Steps in linking environmental and health data

Step	Summary of Procedures
1	Specify an exposure of interest—A specific substance (e.g., lead)
2	Search all environmental sites in the geographic area in question and select those sites with the exposure of interest specified in Step 1.
3	Specify a search threshold distance between the environmental site and the location of the maternal address of a case or control.
4	Identify the pool of cases and controls of interest.
5	Perform a spatial search to select the cases and controls from the pool identified in Step 4 that are within the threshold distance of an environmental site.
6	Generate an output file containing all possible combinations of the specified chemical, an environmental site, and a case or control for further analysis.

* Adapted from Zhan et al. 2006b.

Perform exposure assessment. In proximity models, exposure assessment is simply a search process of identifying a case or control within a given distance of an environmental site releasing a pollutant in question. Zhan et al. [2006b] described a spatial search tool called GIS-EpiLink that can be used to accomplish this task. GIS-EpiLink was used to perform exposure assessment in a reported study examining the relationship between proximity of maternal residences to environmental sites and congenital malformations in offspring [Brender et al. 2006b; Suarez et al. 2007]. Zhan et al. [2006b] provided an example showing the search results of cases and controls within a given distance of environmental sites releasing lead in Texas using GIS-EpiLink as a spatial search tool.

Other Methods of Exposure Assessment

In addition to proximity models for assessing exposure to air pollutants, other methods have also been developed. These methods include statistical interpolation models (e.g., Kriging), land use regression models, dispersion models, coupled emission-meteorological models, and hybrid models. Jerrett et al. [2005] provided a comprehensive review of these models and evaluated their usefulness in health related research. Readers are referred to this review article and the references cited in that article for detailed descriptions of these models. We only provide a brief description of these models in this chapter.

Statistical interpolation models use statistical procedures to estimate the intensity of a pollutant at a location based on observed intensity of the pollutant at monitoring locations where the intensity of the pollutant at a given time is known. Land use regression models use different types of land use land cover as independent variables and the intensity of a pollutant as a dependent variable. These models then use regression analysis to determine the intensity of the pollutant in question at different locations based on observed intensity of the pollutant at monitoring locations.

Dispersion models combine Gaussian plume equations and data about emissions, topography, and meteorological conditions to estimate the intensity of a pollutant at a given location. Coupled emission-meteorological models use both chemical and meteorological models to simulate the dynamics of pollutants in space and time. Hybrid models combine monitoring data from different sources and use one or some combination of the models mentioned above to determine the intensity of a pollutant at a given location.

We would like to mention two recent and promising models in this chapter: the Community Multi-scale Air Quality model (CMAQ) and the AMS/EPA Regulatory Model (AERMOD) [Stein et al. 2007; Marshall et al. 2008]. The CMAQ model is the result of a joint effort between the US National Oceanic and Atmospheric Administration (NOAA) and the US Environmental Protection Agency (EPA). This model is particularly suitable for air quality modeling at a regional scale. AERMOD was developed jointly by the American Meteorological Society (AMS) and the US EPA. AERMOD provides a modeling tool for estimating air quality at a spatial resolution that is higher than that of CMAQ. Stein et al. [2007] provides a study that combines the strength of both CMAQ and AERMOD. A combination of both models holds much promise for enhancing methods for assessing exposure to air pollution.

CONCLUSION

Risk estimates for birth defects in relation to maternal residential proximity and/or exposure to industrial pollution have varied across studies. Increased risks for central nervous system defects (including neural tube defects), congenital heart defects, and chromosomal anomalies in offspring were noted in several U.S. and European study populations that lived close to hazardous waste sites. Odds ratios and relative risks for other types of defects in relation to this residential characteristic tended to be close to the null (1.0). Several studies also noted a maternal residence near active sites with chemical emissions (industries, incinerators, crematoriums) to be associated with central nervous system defects, oral clefts, heart defects, or chromosomal anomalies in offspring.

Most studies used residential proximity as a proxy for exposure to industrial contaminants, which might have led to misclassification of exposures. On the other hand, that approach provides an objective method for addressing community perceptions and concerns regarding perceived excesses of births with congenital anomalies among women living near waste sites, industrial facilities, and other sources of chemical releases.

To more clearly understand the impact that industrial pollution has on risk of congenital malformations, better population-based studies are needed with sufficient sample size and variability of exposures. Although the relation between residential proximity to waste sites and other potential environmental hazards and health outcomes will continue to have relevance in public health investigations, methods of exposure assessment will need to be refined to determine if industrial pollution is a causal factor in the etiology of birth defects. Statistical interpolation models such as Kriging, land use regression models, dispersion models, coupled emission-meteorological models, and hybrid models will provide more precise estimates of ambient air exposures to pollutants on the individual level. As a matter of course, such studies should account for maternal race/ethnicity, education, and occupation that are potentially related to both industrial contaminant exposures and risk of birth defects. Activity patterns, vitamin use (i.e. folic acid supplementation), and home use of chemicals also need to be considered. Finally, it is worthwhile to remember that risk estimates of birth defects in relation to industrial pollution can vary by maternal and regional characteristics. Important and significant effects might be masked by combining heterogeneous populations and site conditions.

It is unlikely that public concern will abate about the impact that industrial pollution has on pregnancy outcomes. Given that some chemicals released into the environment are known teratogens in laboratory animal models and we do not understand the causes of the majority of human birth defects, potential risks of industrial pollution on fetal development need to be better characterized by well-designed epidemiologic studies.

ACKNOWLEDGMENT

Part of this chapter was written while F. Benjamin Zhan was visiting Wuhan University (China) as a Chang Jiang Scholar Guest Chair Professor, and support from the Chang Jiang Scholars Award Program is greatly appreciated.

REFERENCES

Agency for Toxic Substances and Disease Registry (ATSDR). HazDat Database [online]. Accessed June 22, 2005]. Available from: http://www.atsdr.cdc.gov/hazdat.html.

Agency for Toxic Substances and Disease Registry (ATSDR). Public Health Assessment Guidance Manual [Online]. [Accessed April 12, 2008]. Available from: http://www.atsdr.cdc.gov/HAC/PHAManual/ch2.html.

Alonso, E.; Tapie, N.; Budzinski, H.; Lemenach, K.; Peluhet, L.; Tarazona, J.V. (2008). A model for estimating the potential biomagnification of chemicals in a generic food web: preliminary development. *Environ Sci Pollut Res Int 15*, 31-40.

Bentov, Y.; Kordysh, E.; Hershkovitz, R.; Belmaker, I.; Polyakov, M.; Bilenko, N.; Sarov, B. (2006). Major congenital malformations and residential proximity to a regional industrial park including a national toxic waste site: An ecological study. *Environ Health* [online serial], Mar 29;5:8.

Bhopal, R.S.; Tate J.A.; Foy C.; Moffatt S.; Phillimore P.R. (1999). Residential proximity to industry and adverse birth outcomes. *Lancet, 354,* 920-921.

Boyle, E.; Johnson, H.; Kelly, A.; McDonnell, R. (2004). Congenital anomalies and proximity to landfill sites. *Ir. Med, J. 97*, 16-18.

Brender, J.D.; Zhan, F.B.; Suarez, L.; Langlois, P.; Moody, K. (2006a). Maternal residential proximity to waste sites and industrial facilities and oral clefts in offspring. *J Occup Environ Med, 48,* 565-572.

Brender, J.D.; Zhan, F.B.; Suarez, L.; Langlois, P.; Gilani, Z; Delima I.; Moody K. (2006b). Linking environmental hazards and birth defects data. *Int. J. Occup. Environ. Health, 12,* 126-133.

Brender, J.D.; Zhan, F.B.; Langlois, P.H.; Suarez, L.; Scheuerle, A. (2008a). Residential proximity to waste sites and industrial facilities and chromosomal anomalies in offspring. *Int. J. Hyg. Environ. Health, 211*, 50-58.

Brender, J.D.; Suarez, L.; Langlois, P.H.; Steck, M.; Zhan, F.B.; Moody, K. (2008b). Are maternal occupation and residential proximity to industrial sources of pollution related? *J. Occup. Environ. Med. (in press).*

Briggs, D. (2005). The role of GIS: Coping with space (and time) in air pollution exposure assessment. *J. Toxicol. Environ. Health A 68*, 1243-1261.

Canfield, M.A.; Ramadhani, T.A.; Langlois, P.H.; Waller, D.K. (2006). Residential mobility patterns and exposure misclassification in epidemiologic studies of birth defects. *J. Expo. Sci. Environ. Epidemiol., 16,* 538-543.

Coffey, R.; Cummins, E.; Cormican, M.; Flaherty, V.O.; Kelly, S. (2007). Microbial exposure assessment of waterborne pathogens. *Human and Ecological Risk Assessment, 13,* 1313-1351.

Cordier, S.; Chevrier, C.; Robert-Gnansia, E.; Lorente, C.; Brula, P.; Hours, M. (2004). Risk of congenital anomalies in the vicinity of municipal solid waste incinerators. *Occup. Environ. Med.*, 61, 8-15.

Cresswell, P.A.; Scott, J.E.S.; Pattenden, S.S.; Vrijheid, M. (2003). Risk of congenital anomalies near the Byker waste combustion plant. *J. Public Health Med., 25,* 237-242.

Croen, L.A.; Shaw, G.M.; Sanbonmatsu, L.; Selvin, S.; Buffler, P.A. (1997). Maternal residential proximity to hazardous waste sites and risk for selected congenital malformations. *Epidemiology, 8*, 347-354.

Czeizel, A.E.; Hegedus, S.; Timar, L. (1999). Congenital abnormalities and indicators of germinal mutations in the vicinity of an acrylonitrile producing factory. *Mutat Res., 427*, 105-123.

Diez-Roux, A.V.; Kiefe, C.I.; Jacobs, Jr., D.R.; Haan, M.; Jackson, S.A.; Nieto, F. J.; Paton, C.C.; Schulz, R. (2001). Area characteristics and individual-level socioeconomic position indicators in three population-based epidemiologic studies. *Ann. Epidemiol., 30*, 395-405.

Dodds, L., and Sevioiur, R. (2001). Congenital anomalies and other birth outcomes among infants born to women living near a hazardous waste site in Sydney, Nova Scotia. *Can. J. Public Health, 92,* 331-334.

Dolk, H.; Vrijheid, M.; Armstrong, B.; Abramsky, L.; Bianchi, F., Garne, E.; Nelen, V.; Robert, E.; Scott, J.E.S.; Stone, D.; Tenconi, R. (1998). Risk of congenital anomalies near hazardous-waste landfill sites in Europe: the EUROHAZCON study. *Lancet, 352,* 423-427.

Dummer, T.J.B.; Dickinson, H.O.; Parker, L. (2003). Adverse pregnancy outcomes around incinerators and crematoriums in Cumbria, north west England, 1956-93. *J. Epidemiol. Community Health, 57*, 456-461.

Eizaguirre-Garcia, D.; Rodriguez-Andres, C.; Watt, G.C. (2000). Congenital anomalies in Glasgow between 1982 and 1989 and chromium waste. *J. Public Health Med., 22*, 54-58.

Elliott, P.; Briggs, D.; Morris, S.; de Hoogh, C.; Hurt, C.; Jensen, T.K.; Maitland, I.; Richardson, S.; Wakefield, J.; Jarup, L. (2001). Risk of adverse birth outcomes in populations living near landfill sites. *BMJ, 323*, 363-368.

Feijtel, T.; Boeije, G.; Matthies, M.; Young, A.; Morris, G.; Gandolfi, C.; Hansen, B.; Fox, K.; Matthijs, E.; Koch, V.; Schroder, R.; Cassani, G.; Schowanek, D.; Rosenblom, J.; Holt, M. (1998). Development of a geography-referenced regional exposure assessment tool for European rivers - GREAT-ER. *Journal of Hazardous Materials 61,* 59-65.

Fell, D.B.; Dodds, L., King, W.D. (2004). Residential mobility during pregnancy. *Paediatr Perinat Epidemiol, 18*, 408-414.

Geschwind, S.A.; Stolwijk, J.A.; Bracken, M.; Fitzgerald, E.; Stark, A.; Olsen, C.; Melius, J. (1992). Risk of congenital malformations associated with proximity to hazardous waste sites. *Am. J. Epidemiol., 135*, 1197-1207.

Gilboa, S.M.; Mendola, P.; Olshan, A.F.; Langlois, P.H.; Savitz, D.A.; Loomis, D.; Herring, A.H.; Fixler, D.E. (2005. Relation between ambient air quality and selected birth defects, Seven County Study, Texas, 1997-2000. *Am. J. Epidemiology 162*, 238-252.

Gilbreath, S., and Kass, P.H. (2006). Fetal and neonatal deaths and congenital anomalies associated with open dumpsites in Alaska native villages. *Int. J. Circumpolar Health, 65*, 133-147.

Harding, B.L.; Walski, T.M. (2000). Long time-series stimulation of water quality in distribution systems. *Journal of Water Resources Planning and Management-ASCE 126,* 199-209.

Honein, M.A.; Paulozzi, L.J.; Cragan, J.D.; Correa, A. (1999). Evaluation of selected characteristics of pregnancy drug registries. *Teratology 60,* 356-364.

Islam, E., Yang, X.E.; He, Z.L.; Mahmood, Q. (2007). Assessing potential dietary toxicity of heavy metals in selected vegetables and food crops. *J. Zhejiang Univ. Sci. B 8*, 1 -13.

Jerrett, M; Arain, A; Kanaroglou, P; Beckerman, B; Potoglou, D; Sahsuvaroglu, T; Morrison, J; Giovis, C. (2005). A review and evaluation of intraurban air pollution exposure models. *J. Expo Anal. Environ. Epidemiol. 15*, 185-204.

Karimi, H.A.; Durcik, M.; Rasdorf, W. (2004). Evaluation of uncertainties associated with geocoding techniques. *Computer–Aided Civil and Infrastructure Engineering 19,* 170-185.

Keller, V. (2006). Risk assessment of "down-the-drain" chemicals: Search for a suitable model. *Sci. Total Environ 360*, 305-318.

Kelsey, J.L.; Whittemore, A.S.; Evans, A.S.; Thompson, W.D. (1996). *Methods in Observational Epidemiology* (second edition). New York: Oxford University Press.

Khoury, M.J.; Stewart, W.; Weinstein, A.; Panny, S.; Lindsay, P.; Eisenberg, M. (1988). Residential mobility during pregnancy: implications for environmental teratogenesis. *J. Clin. Epidemiol., 1988*, 15-20.

King, P.B.; Lie, R.T.; Irgens, L.M. (1996). Spina bifida and cleft lip among newborns of Norwegian women with epilepsy: Changes related to the use of anticonvulsants. *Am. J. Public Health 86*, 1454-1456.

Kloppenborg, S.; Brandt, U.K.; Gulis, G.; Ejstrud, B. (2005). Risk of congenital anomalies in the vicinity of waste landfills in Denmark; an epidemiological study using GIS. *Cent. Eur. J. Public Health*, 13, 137-143.

Kuehn, C.M.; Mueller, B.A.; Checkoway, H.; Williams, M. (2007). Risk of malformations associated with residential proximity to hazardous waste sites in Washington State. *Environ. Res., 103*, 405-412.

Langlois, P.H., and Scheuerle, A. (2007). Using registry data to suggest which birth defects may be more susceptible to artifactual clusters and trends. *Birth Defects Res. A Clin. Mol. Teratol, 79*, 798-805.

Malik, S; Schecter, A.; Caughy, M.; Fixler, D.E. (2004). Effect of proximity to hazardous waste sites on the development of congenital heart disease. *Arch. Environ. Health, 59*, 177-181.

Marshall, E.G.; Gensburg, L.J.; Deres, D.A.; Geary, N.S.; Cayo, M.R. (1997). Maternal residential exposure to hazardous wastes and risk of central nervous system and musculoskeletal birth defects. *Arch. Environ. Health, 52*, 416-425.

Marshall, J.D.; Nethery, E.; Brauer, M. (2008). Within-urban variability in ambient air pollution: Comparison of estimation methods. *Atmospheric Environment 42*, 1359-1369.

Martin, I., and Earl, N. (2002). Development of the Contaminated Land Exposure Assessment (CLEA) model used in the derivation of the UK Soil Guideline Values for human health. *Epidemiology 13*, S116-S117.

Maslia, M.L.; Sautner, J.B.; Aral, M.M.; Reyes, J.J.; Abraham, J.E.; Williams, R.C. (2000). Using water-distribution system modeling to assist epidemiologic investigations. *Journal of Water Resources Planning and Management-ASCE 126,* 180-198.

Meliker, J.R.; AvRuskin, G.A.; Slotnick, M.J.; Goovaerts, P.; Schottenfeld, D.; Jacquez, G.M.; Nriagu, J.O. (2008). Validity of spatial models of arsenic concentrations in private well water. *Environ. Res. 106*, 42-50.

Morgenstern, H. (2008). Ecologic studies. In: K.J. Rothman; S. Greenland.; T.L. Lash. (Eds.), *Modern epidemiology* (third edition, pp. 511-531). Philadelphia: Wolters Kluwer/Lippincott Williams and Wilkins.

Morris, S.E.; Thomson, A.O.; Jarup, L.; de Hoogh, C.; Briggs, D.J.; Elliott, P. No risk of adverse birth outcomes in populations living near special waste landfill sites in Scotland. *Scott. Med. J., 48,* 105-107.

Nelson, K, and Holmes, L.B. (1989). Malformations due to presumed spontaneous mutations in newborn infants. *New Engl. J. Med. 320,* 19-23.

Nieuwenhuijsen, M.; Paustenbach, D.; Duarte-Davidson, R. (2006). New developments in exposure assessment: The impact on the practice of health risk assessment and epidemiological studies. *Environment International 32*, 996-1009.

Nuckols, J.R.; Ward, M.H.; Jarup, L. (2004). Using geographic information systems for exposure assessment in environmental epidemiology studies. *Environ. Health Perspect 112*, 1007-1015.

Orr, M.; Bove, F.; Kaye, W.; Stone, M. (2002). Elevated birth defects in racial or ethnic minority children of women living near hazardous waste sites. *Int. J. Hyg Environ. Health, 205*, 19-27.

Palmer, S.R.; Dunstan, F.D.; Fielder, H.; Fone, D.L.; Higgs, G.; Senior, M.L. (2005). Risk of congenital anomalies after the opening of landfill sites. *Environ. Health Perspect, 113*, 1362-1365.

Paustenbach, D.J.; Fehling, K.; Scott, P.; Harris, M.; Kerger, B.D. (2006). Identifying soil cleanup criteria for dioxins in urban residential soils: How have 20 years of research and risk assessment experience affected the analysis? *J. Toxicol Environ. Health B Crit Rev, 9,* 87-145.

Perlin, S.A.; Wong, D.; Sexton, K. (2001). Residential proximity to industrial sources of air pollution: interrelationships among race, poverty, and age. *J. Air Waste Manag. Assoc, 51,* 406-421.

Petrini, J.; Damus, K.; Russell, R.; Poschman, K.; Davidoff, M.J.; Mattison, D. (2002). Contribution of birth defects to infant mortality in the United States. *Teratology 66 (Suppl 1*), 53-56.

Pew Charitable Trusts. (1999). Public opinion research on public health. Washington DC: Mellman Group Inc., and Public Opinion Strategies Inc.

Reif, J.S.; Burch, J.B.; Nuckols, J.R.; Metzger, L.; Ellington, D.; Anger, W.K. (2003). Neurobehavioral effects of exposure to trichloroethylene through a municipal water supply. *Environ. Res. 93,* 248-258.

Ritz, B.; Yu, F.; Fruin, S.; Chapa, G.; Shaw G.M.; Harris, J.A. (2002). Ambient air pollution and risk of birth defects in Southern California. *Am. J. Epidemiol. 155,* 17-25.

Schardein JL. (2000). Chemically induced birth defects (third edition). New York: Marcel Dekker, Inc.

Sever, L.E. (1997). Environmental contamination and health effects: What is the evidence? *Toxicol Ind Health 13*, 1-17.

Shaw, G.M.; Schulman, J.; Frisch, J.D.; Cummins, S.K.; Harris, J.A. Congenital malformations and birthweight in areas with potential environmental contamination. (1992). *Arch. Environ. Health, 47,* 147-154.

Shaw, G.M., and Malcoe, L.H. (1992). Residential mobility during pregnancy for mothers of infants with or without congenital cardiac anomalies: a reprint. *Arch. Environ. Health, 47*, 236-238.

Stanek, E.J. and Calabrese, E.J. (2000). Daily soil ingestion estimates for children at a superfund site. *Risk Anal., 20,* 627-635.

Stein, A.F.; Isakov, V.; Godowitch, J.; Draxler, R.R. (2007). A hybrid modeling approach to resolve pollutant concentrations in an urban area. *Atmospheric Environment, 41,* 9410-9426.

Suarez, L.; Brender, J.D.; Langlois, P.H.; Zhan, F.B.; Moody, K. (2007). Maternal exposures to hazardous waste sites and industrial facilities and risk of neural tube defects in offspring. *Ann. Epidemiol., 17,* 772-777.

Tango, T.; Fujita, T.; Tanihata, T.; Minowa, M.; Doi, Y.; Kato, N.; Kunikane, S.; Uchiyama, I.; Tanaka, M.; Uehata, T. (2004). Risk of adverse reproductive outcomes associated with proximity to municipal solid waste incinerators with high dioxin emission levels in Japan. *J. Epidemiol., 14,* 83-93.

Texas Commission on Environmental Quality (TCEQ). TCEQ Site Layers [online]. [Accessed June 22, 2005]. Available from: http://www.tceq.state.tx.us/gis/sites.html.

Vinceti, M.; Rovesti, S.; Bergomi, M.; Calzolari, E.; Candela, S.; Campagna, A.; Milan, M.; Vivoli, G. (2001). Risk of birth defects in a population exposed to environmental lead pollution. *Sci Total Environ, 278,* 23-30.

Vrijheid, M.; Dolk, H.; Armstrong, B.; Abramsky, L.; Bianchi, F.; Fazarinc, I.; Garne, E.; Ide, R.; Nelen, V.; Robert, E.; Scott, J.E.; Stone, D.; Tenconi, R. (2002). Chromosomal congenital anomalies and residence near hazardous waste landfill sites. *Lancet, 359,* 320-322.

Woodruff, T.J.; Parker, J.D.; Kyle, A.D.; Schoendorf, K.C. (2003). Disparities in exposure to air pollution during pregnancy. *Environ. Health Perspect, 111,* 942-946.

Wulff, M.; Hogberg, U.; Sandstrom-Holmgren, A. (1996). Congenital malformations in the vicinity of a smelter in Northern Sweden, 1973-1990. *Paediatr Perinat Epidemiol, 10,* 22-31.

Wyszynski, D.F., and Beaty, T.H. (1996). Review of the role of potential teratogens in the origin of human nonsyndromic oral clefts. *Teratology, 53,* 309-317.

Yauck, J.S., Malloy, M.E., Blair, K., Simpson, P.M., McCarver, D.G. (2004). Proximity of residence to trichloroethylene-emitting sites and increased risk of offspring congenital heart defects among older women. *Birth Defects Res. A Clin. Mol. Teratol, 70,* 808-814.

Yoon, P.W.; Olney, R.S.; Khoury, M.J.; Sappenfield, W.M.; Chavez, G.F.; Taylor, D. (1997). Contribution of birth defects and genetic diseases to pediatric hospitalizations. A population-based study. *Arch. Pediatr. Adolesc. Med., 151,* 1096-1103.

Yoon, P.W.; Rasmussen, S.A.; Lynberg, M.C.; Moore, C.A.; Anderka, M.; Carmichael S.L.; Costa, P.; Druschel, C.; Hobbs, C.A.; Romitti, P.A.; Langlois, P.H.; Edmonds, L.D. (2001). *Public Health Rep. 116 (Suppl 1),* 32-40.

Zhan, F.B.; Brender, J.D.; DeLima, I.; Suarez, L.; Langlois, P.H. (2006a). Match rate and positional accuracy of two geocoding methods for epidemiological research. *Ann. Epidemiol., 16,* 842-849.

Zhan, F.B.; Brender, J.D.; Han, Y.; Suarez, L.; Langlois, P.H. (2006b). GIS-EpiLink: a spatial search tool for linking environmental and health data. *J. Med. Syst. 30,* 405-412.

In: Industrial Pollution Including Oil
Editor: Harry Newbury and William De Lorne
ISBN 978-1-60456-917-9

Chapter 9

AOP AS AN EFFECTIVE TOOL FOR THE MINIMIZATION OF HAZARDOUS ORGANIC POLLUTANTS IN COLORED WASTEWATER; CHEMICAL AND PHOTOCHEMICAL PROCESSES

Natalija Koprivanac* and Hrvoje Kusic†
Faculty of Chemical Engineering and Technology,
University of Zagreb, Zagreb, Croatia

ABSTRACT

The manufacturing and the application of organic dyes involve the production and the handling of many organic compounds hazardous to human health. Many of these substances are considered toxic, even carcinogenic. Over the past couple of decades, manufacturers and users of dyes have faced increasingly stringent legal regulations promulgated to safeguard human health and the environment. So, there is a clear need to treat dye wastewater prior to discharge into the primary effluent. The limitation of traditional wastewater treatment technologies (biological and physical), such as low rate, disability to degrade many of recalcitrant organic dyes and the production of secondary waste which demands further treatment, can be overcome by the utilization of advanced oxidation processes (AOPs). These wastewater treatment methods are considered as low- or even non-waste generation technologies. AOPs are based on the production of very reactive species, such as hydroxyl radicals, able to decolorize and to reduce recalcitrant colored wastewater loads due to the high oxidation power and the lack of selectivity of OH radicals towards a broad range of organic pollutants present in wastewater. Generally, AOPs can be broadly classified concerning the way of OH radicals generation into chemical, photochemical, photocatalytic, mechanical and electrical technologies. This research deals with the application of chemical and photochemical AOPs for the minimization of recalcitrant colored pollutants present in wastewaters. The comparative study of several processes, Fe^0/H_2O_2, $UV/Fe^0/H_2O_2$, UV/O_3 and $UV/O_3/H_2O_2$, was performed through an experimental research of decolorization and mineralization of a

* nkopri@fkit.hr
† hkusic@fkit.hr

reactive azo dye C.I. Reactive Blue 137, as a model wastewater pollutant. Applied processes were optimized according to their process parameters, Fe^0 dosages, initial H_2O_2 dosages and initial pH values of treated solutions. The influence of initial organic dye concentration, as well as the addition of solid particles, synthetic zeolites, on the process effectiveness was also investigated. Studied AOPs were evaluated on the basis of their eco-effectiveness, by the means of color (A_{610}), aromaticity (UV_{280}), TOC and AOX value decrease, and their cost-effectiveness as well.

INTRODUCTION

Colored Wastewater

Dyes make our world beautiful, but they bring pollution. Synthetic dyes are present in many spheres of our everyday life and their application is continuously growing. Organic synthetic dyes have been widely used as colorants in different industries such as textile, paper, color photography, pharmaceutical, food, cosmetic, electric... [1]. More than 0.7 million tons of organic synthetic dyes are produced annually worldwide. In additon, over 10,000 different dyes and pigments have been applied in those industries. Studies indicated that approximately 15 % of produced synthetic dyes per year have been lost during manufacturing and processing operations that involve the production and handling with many organic compounds hazardous to human health. Wastewaters originated from dye production and application industries present a very serious environmental problem because of the aesthetic nature due to the fact that the coloration is visible even in a low dye concentration. Although the presence of color in water might not appear to have a negative effect other than aesthetic, this is not actually the case. The color can adsorb and reflect sunlight entering the water. Consequently, bacteria can not grow sufficiently in the water and that can have a disastrous effect on the food chain [2]. But even more important, many substances in such wastewaters are considered toxic, and even carcinogenic and mutagenic [3]. Nowadays, the mostly used group among all dyes are reactive synthetic dyes, mostly applied for dyeing of natural fibbers such as cotton and wool. Due to the high solubility and hydrolysis as a side effect of such types of dyes, they can often cause loaded colored effluents [4]. Almost 70 % of all reactive dyes are of the azo type, while other are mainly of anthraquinonic and phtalocyanine types [5]. Furthermore, commercial reactive azo dyes are intentionally designed to resist biodegradation, i.e. the wastewater originated from reactive dye processes is characterized by poor biodegradability, passing unaffected through conventional treatment systems and discharging into the environment [6-8]. Like other colored wastestreams, those from reactive azo dye manufacturing and utilization industries pose a major threat to the surrounding ecosystems due to the documented health hazards caused by the toxicity and the potentially carcinogenic nature of such organic pollutants [9]. The group of most dangerous contaminants in such wastewaters are aromatic amines which may cause methemoglobinemia, and they are generated as by-products of partial degradation of reactive dye wastewater treatment or as impurities present in commercial azo dyes due to their usage in production process. The amines oxidize the heme iron of hemoglobin from Fe(II) to Fe(III), blocking the oxygen binding. This results in characteristic symptoms like cyanosis of lips and nose, weakness and dizziness. [10]. A number of azo dyes have been shown to be animal carcinogenous, and analysis based on structure-activity relationships suggests that many more azo dyes are likely to have a carcinogenic potential. Reductive cleavage to carcinogenic aromatic amines, which are known to act via genotoxic mechanisms, is in all likelihood an important mechanism of azo dye carcinogenicity [7]. Since 20 years ago some intermediates for azo reactive dye manufacturing, such as: *o*-aminoazotoluene, *p*-aminoazobenzene, *o*-anisidine, 2-amino-4-nitrotoluene, benzidine, 2-naphthylamine, *o*-toluidine, etc, were become to EPA's "black list" of hazardous chemicals that resulted with their restriction. But today, many amines are still in use, resulting with their presence in wastewater originated from azo reactive dye manufacturing and application industries.

Wastewater Treatment Methods

Due to the above mentioned facts, manufacturers and users of reactive dyes have faced, over the past two decades, increasingly stringent legal regulations promulgated to safeguard human health and the environment [5]. To comply those requirements, the new technologies have been designed, specifically, to analyze and remove color and priority pollutants from wastewater effluents and the methodology of the cleaner production has been applied to circumvent pollution problems by eliminating their source [11].

Generally, wastewater treatment methods can be broadly classified into biological, physical (i.e. mechanical) and chemical [3, 12-14]. Biological treatment methods are widely applied for the treatment of municipal and industrial wastewaters. Although biological treatment methods have many advantages (economical acceptance, simplicity of treatment facility), they can not be successfully applied for the treatment of colored wastewater containing reactive dyes due to their poor biodegradability caused by recalcitrant nature [8, 14]. Reactive dyes are developed and designed to resist fading upon exposure to light, water and oxidizing agent [15, 16]. They are characterized by low biomass adsorbability and they are not degradable by aerobic conditions in conventional biological wastewater treatment facilities. Recently, some microorganism cultures, which are able to decolorize and mineralize some of reactive azo dyes, were developed [17-19]. However, biological processes by itself are often inappropriate due to the low rates of biological degradation. Physical, also known as mechanical, treatment methods such as: filtration, sedimentation, adsorption, coagulation/flocculation etc., like their name states, include only physical/mechanical processes, without chemical transformations of present contaminants, for wastewater treatment. Therefore, they generally transfer waste components from the liquid phase to a sludge phase. That makes them insufficiently effective and moreover, very expensive, due to the requested further treatment of generated secondary waste or due to the regeneration of inactive adsorbents. Although there are many examples in the literature about textile wastewater treatment by activated carbon adsorption with reported process effectiveness for wide range of textile dye types, the relatively low adsorption capacity of activated carbon for some of reactive dyes was shown [20]. Reactive dyes can be also removed from wastewaters by adsorption on various natural quaternized organic materials such as cellulose, sugarcane bagasse, rice husk, and coconut husk, alternatively cheap adsorbents. But here, too, no successful regeneration has been reported [21]. Reactive dye removal by adsorption onto mezoporous minerals such as sepiolites and zeolites was also studied, but successful removal was limited due to the low adsorption capacities of these natural minerals [20]. Removal by synthetic zeolites is possible but it depends on pore size and capacity level of applied zeolites [22-24]. Coagulation/flocculation process using Fe(III) and Al(III) salts, or polymeric flocculant Levafloc-R showed high efficiency for decolorization of wastewater containing reactive dyes [25-27]. The main problems of this relatively cheap and easily to maintain treatment method are related with the generation of large amounts of sludge as secondary waste that requires further treatment and disposal. Chemical treatment is based on the electron transfer, and most widespread application found processes using chlorine, ozone and potassium permanganate. The main disadvantage of chlorine usage for wastewater treatment is the production of chlorinated hydrocarbons that are in most cases even more toxic and carcinogenic than parent contaminants and therefore the usage of chlorine is restricted in many countries [28, 29]. Besides that, the usage of chlorine as an oxidant for wastewater treatment is characterized by its low selectivity and high consumption. The application of pure ozone oxidation is limited by high selectivity and slow kinetics. Although ozone was shown as very efficient for decolorization of colored wastewaters, it rarely produces complete mineralization to CO_2 and H_2O [30]. The usage of potassium permanganate as rather expensive oxidative agent is widespread due to the easiness of its application. But the main problem is caused due to the production and the precipitation of manganese oxide which than should be removed from the process equipment. From the previous statements concerning wastewater treatment methods it could be concluded that common treatment methods suffer from various limitations and that higher efficient treatment methods need to be found.

Advanced Oxidation Processes

Advances in chemical treatment of wastewater resulted with the development of a number of alternative chemical technologies; so-called Advanced Oxidation Processes (AOPs). In last two decades, the interest of not only world scientific community for the development and the application of AOPs as wastewater treatment methods was rapidly increased. The rapid increase of interest of research on AOPs resulted with numerous scientific publications, but also with their numerous applications on full-scale treatment facilities [31]. The main advantage of AOPs over other wastewater treatment processes; biological, physical-mechanical and classical chemical is their pronouncedly destructive nature that results with the mineralization of organic contaminants present in wastewater. AOPs are considered as low- or even none-waste generation technologies [32, 33], and generate short-lived chemical species with high oxidation power. At the first place, that are hydroxyl radicals, (HO•), which Glaze et al. [34] pointed out as main oxidation species in AOPs. Owing to high reactivity (Table 1) and unselectivity (Table 2) of HO•, AOPs are considered to be promising methods for the treatment of hazardous toxic organic pollutants in aqueous solutions [35]. AOPs generate OH• in sufficient quantities to oxidize the majority of organics present in the effluent water [36]. However, some of the simplest organic compound such as acetic, maleic and oxalic acid, as well as acetone, chloroform and tetrachloroethane can not be oxidized by OH radicals [37].

AOPs show high flexibility in their practical application due to the fact that it can be used either separately or in combination, or in combination with other classical wastewater treatment methods. Furthermore, another great advantage of AOPs over classical wastewater treatment methods is the possibility of conduction at ambient conditions, i.e. atmospheric pressure and room temperature.

Table 1. Standard redox potential of some oxidant species [35]

Oxidant	Redox potential, $E°$, V
Flour	3.03
Hydroxyl radical	2.80
Atomic oxygen	2.42
Ozone	2.07
Hydrogen peroxide	1.77
Permanganate ion	1.67
Chlorine	1.36
Chlorine dioxide	1.27

Table 2. Chemical species those are able to be oxidized by hydroxyl radicals [36]

Acids	formic, gluconic, lactic, malic, propionic, tartaric
Alcohols	benzyl, *tert*-butyl, ethanol, ethylene glycol, glycerol, isopropanol, methanol, propenediol
Aldehydes	acetaldehyde, benzaldehyde, formaldehyde, glyoxal, isobutyraldehyde, trichloroaldehyde
Aromates	benzene, chlorobenzene, chlorophenol, PCBs, phenol, hydroquinone, catechol, benzoquinone, *p*-nitrophenol, toluene, xylene, trinitrotoluene
Amines	aniline, cyclic amines, diethylamine, dimethylforamine, EDTA, propanediamine, *n*-propylamine
Dyes	azo, anthraqiunone, triphenylmethane
Ethers	tetrahydrofuran
Ketones	dihydroxyacetone, methylethylketone

Such so-called ambient AOPs can be used for the treatment of low loaded wastewaters (Figure 1.) [38]. For the treatment of highly loaded wastewaters, even up to value of COD (chemical oxygen demand) ecological parameter of 200 gL^{-1}, other processes generating also OH radicals can be used. Some authors consider those processes as AOPs, although these processes require pressure and temperature highly over ambient conditions, so they can be named as non-ambient AOPs. Such processes are wet air oxidation, supercritical wet air oxidation and other hydrothermal oxidation processes that requires pressures and temperatures even above 1 MPa and 150 °C, respectively [33, 35, 38].

Regarding to the value of TOC (total organic carbon) content, AOPs are suitable for the treatment of wastewaters loaded in range of 100-1000 mgC L^{-1} [39]. Ambient AOPs can be broadly classified concerning the way of OH radical generation. Generally, OH radical can be generated by chemical, electrical, mechanical or radiation energy. Therefore, AOPs can be classified into chemical and catalytic, photochemical and photocatalytic, mechanical and electrical processes (Figure 2.).

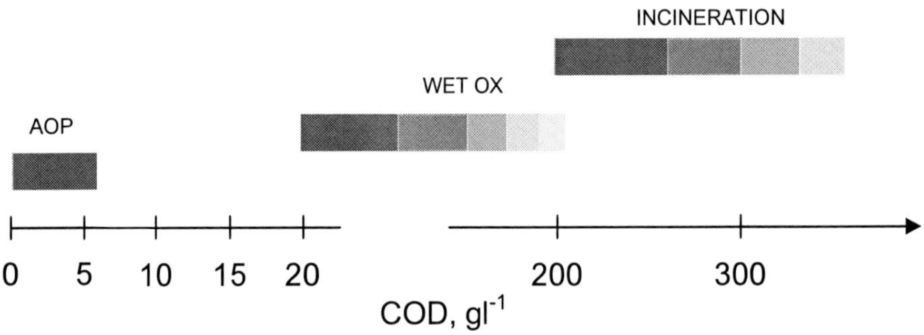

Figure 1. Suitability of water treatment technologies according to COD contents [38].

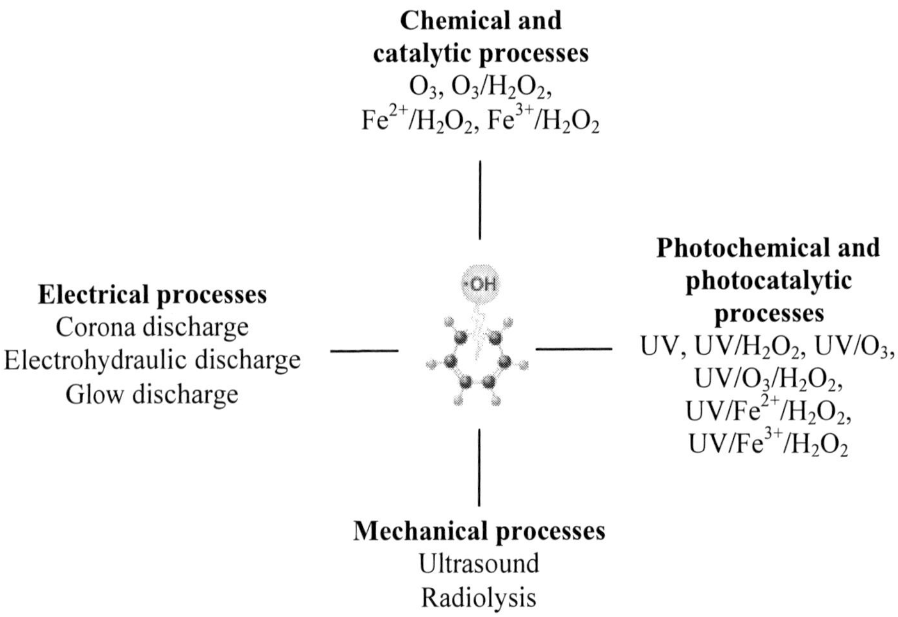

Figure 2. Schematic of advanced oxidation processes classification.

Chemical processes involve the application of ozone and/or hydrogen peroxide, while a subcategory of this type of AOPs can be named catalytic processes that involve usage of some powerful catalyst (e.g. iron or cupper ions) in combination with hydrogen peroxide to produce OH radicals, so-called Fenton type processes. Photochemical and photocatalytic processes involve application of UV or solar irradiation in combination with some powerful either oxidant (ozone and/or hydrogen peroxide) or photocatalyst (e.g. TiO_2, ZnO, etc.). OH radical can be also produce under influence of mechanical (e.g. ultrasound process, radiolysis) or electrical (e.g. electrohydraulic discharge and non-thermal plasma processes) energy [31-40].

Colored Wastewater Treatment by AOPs: A Review of Recent Studies

According to the number of AOPs investigations directed to the treatment of colored wastewater, the most studied processes are Fenton type processes, then UV-based technologies, ozone-based technologies, etc. This part of chapter contains the more detailed review of AOPs applied for the treatment of colored wastewater investigated in our laboratory, and the less detailed overview of other AOPs.

Fenton Type Processes

All processes that involve catalytic reaction between hydrogen peroxide and metal homogeneous ions, in the most cases Fe^{2+} ions, can be named as Fenton type processes. The reactivity of this system was first observed by Fenton in 1876 [41, 42]. Its utility was not recognized immediately, but 1930s when its mechanism based on OH radicals was proposed, the number of investigations concerning the application of Fenton reagent for the degradation of organic compounds dissolved in water is continuously growing. The application of Fenton process is based on the reaction that involves the oxidation of Fe^{2+} ions to Fe^{3+} ions with H_2O_2 and the subsequent generation of OH radicals (1):

$$Fe^{2+} + H_2O_2 \rightarrow Fe^{3+} + HO\bullet + HO^- \qquad (1).$$

Such generated OH radicals could then react with organic matter present in water. Although reaction shown by equation (1) is often referred as Fenton reaction [43], and presents the key step in the Fenton process, other important reactions also occur. The following set of reactions describes the occurrence of Fenton catalytic cycle:

$$Fe^{2+} + HO\bullet \rightarrow Fe^{3+} + HO^- \qquad (2)$$

$$Fe^{3+} + H_2O_2 \rightarrow Fe^{2+} + HO_2\bullet + H^+ \qquad (3)$$

$$Fe^{2+} + HO_2\bullet \rightarrow Fe^{3+} + HO_2^- \qquad (4)$$

$$Fe^{3+} + HO_2\bullet \rightarrow Fe^{2+} + O_2 + H^+ \qquad (5)$$

If the organic matter, i.e. organic pollutant, is present in the system, the next reactions are also involved in chain radical mechanism:

$$RH + HO\bullet \rightarrow H_2O + R\bullet \qquad (6)$$

$$R\bullet + Fe^{3+} \rightarrow Fe^{2+} + products \qquad (7)$$

By-products of above reaction (7) could be further degraded by radical mechanism to the complete mineralization. Besides Fenton process, Fe^{2+}/H_2O_2, that involves the application of ferrous salts, mostly ferrous sulphates, as a source of iron catalyst for Fenton reaction (1) [44-49], there is a number of studies that investigated the application of so-called Fenton "like" processes for the degradation of pollutants in wastewater. There are three types of Fenton "like" processes. The first group of processes considers the usage of ferric salts instead of ferrous salts as catalysts for the incitation of Fenton reaction (1) [48-51]. Next group of processes considers the usage of heterogeneous Fenton type catalysts such as iron powder, iron-oxides, iron-ligands, or iron ions doped in zeolites, pillared clays or resins, instead homogenous ferrous ions obtained from the dissolution of added ferrous salts [48, 49, 52-62]. Third group are processes that use other metal ions, e.g. cupper, manganese or cobalt, instead of ferrous ions in Fenton reaction [44, 63]. The primary benefits of Fenton type processes are their ability to convert a broad range of pollutants to harmless or biodegradable products and the fact that their relatively cheap reagents are safe to handle and are environmentally benign [43]. Even though these systems offer a cost effective source of HO•, their efficiency is limited with a couple of disadvantages: (a) the need for the removal of remaining iron ions after the treatment and (b) a limited yield of reaction process due to the formation of stable Fe^{3+}-complexes [37, 43]. Those limitations can be overcome by the usage of heterogeneous Fenton-type catalysts which can lower the final concentration of iron ions in the bulk after the treatment, while by the assistance of UV irradiation the formed Fe^{3+}-complexes can be destroyed, thus allowing Fe^{3+} ions to participate in the Fenton catalytic cycle [60, 64]. The oxidation power of Fenton type processes depends of several operating parameters such as iron concentration, source of iron catalysts, or catalyst if other metal ions are used instead of iron ions, H_2O_2 concentration, applied catalysts/hydrogen peroxide ratio (Fenton reagent), temperature, pH and treatment time [65]. According to the literature [65], a minimal threshold concentration of ferrous ions that allows the reaction to proceed within the reasonable period of time ranges between 3-15 mgL^{-1}, regardless of the concentration of organic pollutant. However, iron levels <25-50 mgL^{-1} can require excessive reaction times (10-24 hours). Typical Fenton catalyst/H_2O_2 ratios range between 1 : 5 to 1 : 25 [65, 66], while it is markedly proven that the optimal pH range for Fenton process is between 2.8 and 3.0 [43-54].

As it was mentioned above, Fenton type processes are the most studied processes for the treatment of colored wastewater. Numerous studies dealt with the application of Fenton process Fe^{2+}/H_2O_2 for the decolorization of colored wastewater including those that contain reactive azo dyes. Xu et al. [44] investigated the efficiency of Fe^{2+}/H_2O_2 for the decolorization and the mineralization of 20 organic dyes including 3 reactive dyes and they obtained rather high decolorization, >96 %, while mineralization extents after 120 min of treatment ranged between 72-79 % TOC removal. Lucas and Peres [45] achieved bleaching between 83 and 96 % depending on the initial dye concentration. Meric et al. [46] applied Fe^{2+}/H_2O_2 process for the decolorization and the detoxification of three reactive dyes with the initial dye concentrations of 100 and 200 mgL^{-1}. The overall decolorization obtained in that study was >98 %, while COD removal yielded between 92 and 95 %. In our previous study [49], Fenton process was applied for the degradation of two reactive dyes (azo and anthraquinonic type), and the bleaching >95 % was achieved in cases of both reactive dyes, while final extents of partial mineralization, 34.3 and 72.1 % of TOC removal, depended on dye structure. Neamtu et al. [51] obtained >99 % of color removal for both studied reactive dyes with the initial concentrations of 100 mgL^{-1}. Hsueh et al. [67] investigated the efficiency of Fe^{2+}/H_2O_2 using the low concentration of ferrous salts for the decolorization of reactive dye model solution, and they obtained similar results as in previous cases, rather high decolorization, >96 %, and <20 % of TOC removal after a one-hour treatment. From the above presented results of several studies applying Fe^{2+}/H_2O_2 process for the degradation of reactive dyes in wastewater, it can be observed that rather high decolorization was achieved in all cases, while according to the other monitored ecological parameters, i.e. TOC or COD, in all studies the mineralization of colored wastewater was only partial. Similar results were obtained by using ferric salts instead of ferrous salts [49, 67]. Several studies investigated application of heterogeneous catalysts for Fenton reaction and obtained similar results. Decolorization >98 % and mineralization extents of 45.3 and 46.9 % TOC removals for C.I. Reactive Blue 137 and C.I. Reactive Blue 49 model wastewaters, respectively, were obtained in our previous study by using iron powder instead of homogenous ferrous ions [49]. Tang and Chen [54] applied iron powder/H_2O_2 system for the treatment of colored wastewater containing C.I. Reactive Red 120, and obtained color removal >90 %. They pointed out iron powder/H_2O_2 system as effective for dye removal owing to the Fenton reaction that occurs in the bulk and on the iron powder surface. Several studies include the application of ferrous or ferric ions doped in zeolites or resins, as well as the usage of ferrioxalate, for the decolorization of model wastewaters containing reactive dyes [51, 60, 62, 69, 70]. By these heterogeneous Fenton type processes rather high dye degradation was estimated, and mostly owing to the occurrence of Fenton reaction on the surface of heterogeneous catalysts resulting with lower concentration of iron ions in the bulk after the treatment. But, in most of those studies the reactivity of Fenton type processes were improved by UV irradiation or by non ambient conditions (so-called wet oxidation process). Presence of UV light in Fenton type processes could give them several benefits. One of benefits is an additional source of OH radicals besides primary source throughout Fenton mechanism. Therefore, OH radicals could be generated in photo-Fenton processes from the photolysis of hydrogen peroxide, equation (8), and from the reduction of ferric ions to ferrous ions, equation (9). Furthermore, UV light could provide the avoidance of breaking the Fenton catalytic cycle due to the formation of stable Fe^{3+}-complexes between free Fe^{3+} ions and some aliphatic acids formed as by-products of dye degradation, shown by equation (10) [44, 45, 51, 62, 64, 68].

$$H_2O_2 \xrightarrow{h\upsilon} 2HO\bullet \tag{8}$$

$$FeOH^{2+} \xrightarrow{h\upsilon} Fe^{2+} + HO\bullet \tag{9}$$

$$Fe^{3+}(L^-) \xrightarrow{h\upsilon} Fe^{2+} + L\bullet \tag{10}$$

Also, the presence of UV irradiation could enable these processes to achieve the complete mineralization due to the degradation of some OH radical persistent by-products, such as oxalic and acetic acid, by UV light [64, 69-71]. Throughout all these studies it was found that the decolorization of reactive dye wastewater by Fenton type processes was rather high and fast. On the other hand, only partial degradation/mineralization of formed organic by-products after initial cleavage of dye chromophores indicate high susceptibility of dye chromophores to OH radicals in the comparison to the remained organic part of colored wastewater. However, Fenton type processes could be proposed as one of successful treatment methods for colored wastewaters containing reactive dyes.

UV-based Processes

Under the term "UV-based processes" are considered all processes that apply UV light either for the degradation of organic pollutants or for the initiation of oxidation mechanisms by the irradiation of some powerful oxidants or photo-catalysts, throughout then will be provided degradation of organic pollutants present in water. Hence, UV-based processes could be classified into photochemical processes, where UV light is used for the photolysis of some powerful oxidants such as H_2O_2 and/or O_3, and photocatalytic, where UV light is used for the irradiation of some powerful photocatalyst such as TiO_2, ZnO etc.

UV Photolysis

Since 1666 when Isaac Newton observed the diffraction of white light beam when passing through a prism, the investigations regarding UV light and its properties were began. At the begging of 19th century, the radiant energy beyond two ends of the spectra of visible lights was discovered. One of those is identified as infrared and another as ultraviolet region. Furthermore, it was shown that invisible chemically active irradiation beyond violet end of spectrum were the subject of laws of interference. All those observations, along with further investigations, indicated to the fact that irradiation with visible (VIS), infrared (IR) or ultraviolet (UV) light has characteristics of the same electromagnetic irradiation, but they differ regarding to its frequency, and, what was discovered latter, pertaining energy [35, 72].

Therefore, UV radiation comprises energies from about 300 kJ Einstein^{-1} (UV-A radiation, 1 Einstein = 1 mol of photons), up to 1200 kJ Einstein^{-1} (vacuum UV). Table 3. summarizes the wavelength and the pertaining energy of different types of irradiation. For oxidation purposes, UV-C radiation is the most often in use, although in the literature could be found the application of other types of UV radiation [68, 73-75]. The most common wavelength of UV-C radiation is 254 nm that could be achieved by low-pressure vapor mercury lamp invented by Hewitt at 1901 [35]. The primary usage of UV radiation in the past was for disinfection, but with the development of reaction mechanism, UV radiation nowadays found the usage for oxidation purpose as well. At the room temperature, most molecules reside in their lowest-energy electronic state, i.e. "ground state". When molecules were exposured to UV radiation, they transferred to state with higher energy, i.e. "excited sate".

Table 3. Radiation type and pertaining energy [35]

Radiation	Wavelength, nm	Energy range, kJ Einstein^{-1}
IC	>780	<155
VIS	780 – 400	155 – 300
UV-A	400 – 315	300 – 377
UV-B	315 – 280	377 – 425
UV-C	280 – 100	425 – 1198

The energy difference between the "ground" and the "excited" state depends on absorbed radiation $h\nu$, where h and ν are Planck constant and the frequency of absorbed radiation, respectively. The molecule in the "excited state" has very short lifetime (10^{-9} do 10^{-8} s), after which it returns to "ground state" by one of several mechanisms (fluorescence, phosphorescence...) or decomposes to yield a different molecule, resulting with photochemical reaction. Simple mechanism of direct photolysis is given below [35]:

$$M + h\upsilon \xrightarrow{k_a} M^* \quad (11)$$

$$M^* \xrightarrow{k_b} M \quad (12)$$

$$M^* \xrightarrow{k_c} produkti \quad (13)$$

UV radiation is always in use in combination with some powerful oxidant or photocatalyst. The efficiency of its separate use depends on the lot of limitation such as: (a) water solution should be treated in a way to achieve the highest possible UV transmission, i.e. turbidity should be as low as possible, (b) too high concentration of OH radicals could inhibit mineralization reaction of organic contaminant present in water, (c) water solution should be free of heavy metals and oils, and (d) the costs of UV radiation are higher than those for dark Fenton type processes. However, several authors investigated in their comparative studies the efficiency of UV radiation for the treatment of colored wastewater containing reactive dyes. Bali et al. [76] obtained only negligible decolorization, 2 %, of C.I. Reactive Black, initial concentration 100 mgL^{-1}, after two hours treatment by UV photolysis. Furthermore, Behnajady et al. [77] achieved after a one-hour treatment only 18.5 % of color removal with initial dye concentration of 30 mgL^{-1}. In our previous studies, in the case of C.I. Reactive Red 45 with initial concentration of 80 mgL^{-1} again poor color removal, 30 %, and the mineralization degree of 3 % of TOC removal were achieved [78]. Treating model dye solutions with lower initial dye concentrations, 20 mgL^{-1}, significantly higher color removals of 87 and 98 % were achieved, whilst only 17 and 6 % of dye solutions mineralized [52, 79]. Such low decolorization in the case of higher dye concentration could be the fact that although UV radiation itself has ability to destroy organic molecules [35], the efficiency of direct photolysis of organic dyes decay has been proved to be difficult and it depends on the dye's reactivity and photosensitivity, and more important, solution transmission is very low in the case of higher dye concentration [80]. Moreover, most of the commercially used dyes are usually designed to be light resistant.

Photochemical Processes

These processes combine UV radiation and the addition of some powerful oxidant, H_2O_2 and/or O_3, in the system. The application of UV/H_2O_2 process for water purification is widely investigated AOP due to the fact that involves the simplest way of OH radical generation, through direct photolysis of H_2O_2. For its successful application it is necessary to determine the type and the initial concentration of pollutants in water, as well as the presence of other organic and inorganic compounds, so-called "scavengers", that could inhibit or even stop the treatment process. This process is in use for the removal of micro- and macro-pollutants from drinking water, for the treatment of organic toxic compounds present in lower concentration in ground waters, for the treatment of smaller volume of highly recalcitrant pollutants in order to achieve their detoxification and faster degradation, for the control of exhaust gases in the case of volatile organic compounds (VOCs)... [81]. Quantity of energy that emits by direct photolysis of H_2O_2 under UV radiation is very high, and theoretically two OH radicals could be generated per absorbed energy quant, equation (8) [82]. In practice, the highest quantum yield for generation of OH radicals is 0.5 mol of H_2O_2 per Einstein [35, 83]. That value is called the basic energy of direct photolysis of H_2O_2. By using higher energy than basic, formed OH radicals can be recombined and in that way H_2O_2 is producing. According to the literature [84, 85], besides reaction (8), other important reactions should be also taken into account in UV/H_2O_2 process:

$$H_2O_2 \Leftrightarrow HO_2^- + H^+ \tag{14}$$

$$HO\bullet + H_2O_2 \rightarrow HO_2\bullet + H_2O \tag{15}$$

$$HO\bullet + HO_2^- \rightarrow HO_2\bullet + HO^- \tag{16}$$

$$HO_2\bullet + HO\bullet \rightarrow H_2O + O_2 \tag{17}$$

Above reactions present "scavenger" mechanism of H_2O_2 (14) or OH radicals (15-17) negatively influencing the overall process efficiency. Important process parameters of UV/H_2O_2, besides UV lamp characteristics and reactor performances, are pH of solution and the initial concentration of H_2O_2. One of limitations of UV/H_2O_2 process that should be taken into consideration is the presence of iron and potassium salts in treated water resulting with reduction of UV radiation, and that could be avoided by adjusting pH to the value where those salts can precipitate. Further limitation is related to the large quantities of suspended particles resulting with increased turbidity and that can be solved by the filtration as the pretreatment of such wastewaters. Besides H_2O_2, ozone is also very used as oxidant in photochemical processes. Moreover, ozone is even better oxidant than H_2O_2 due to the significantly higher value of molar absorption coefficient at 254 nm, typical value for UV-C radiation (3300 $M^{-1}cm^{-1}$>>>18.6 $M^{-1}cm^{-1}$). Also, the rate of ozone photolysis is almost 1000 times higher than that of H_2O_2 [86]. UV/O_3 process is based on the fact that by the decomposition of ozone under UV radiation two OH radicals are generated which rather form H_2O_2 than react with organic matter present in water, shown by equation (18):

$$O_3 + H_2O \xrightarrow{h\upsilon} H_2O_2 + O_2 \tag{18}$$

Furthermore, such formed H_2O_2 decomposes under UV radiation to OH radicals that react with organic matter present in water, equation (8). This oxidative system contains three components that could generate OH radicals and/or directly oxidize organic matter: UV radiation, H_2O_2 and O_3. Accordingly, there are several ways to generate OH radicals: (a) photolysis of formed H_2O_2 or (b) reaction between formed H_2O_2 and ozone, so-called peroxone reaction. Also, there are several mechanisms for the degradation of organic pollutants in water: (a) direct photolysis, (b) OH radical attack generated from different sources and (c) direct ozone attack [87]. Process efficiency is limited by concentration of ozone and other factors mentioned above in the case of UV/H_2O_2 process. There is also combination of these two binary systems (UV/H_2O_2 and UV/O_3) the $UV/O_3/H_2O_2$ process. This process could enable to achieve the complete mineralization of organics present in water due to the enhanced generation of OH radicals throughout added H_2O_2 in the system [36, 78].

In the literature could be found many studies concerning UV/H_2O_2 process and colored wastewaters, but application of UV/O_3 and $UV/O_3/H_2O_2$ processes are not so widely investigated, particularly for the treatment of wastewater containing reactive dyes. Mohey El-Dein et al. [88] achieved complete bleaching of C.I. Reactive Black 5 with initial dye concentration of 500 mgL^{-1} at pH 7 and 140 mM H_2O_2 solution after 150 minutes of treatment, while after 240 minutes mineralization reached almost 90 % of removed ecological parameter DOC (dissolved organic carbon). In our previous paper, time required for complete mineralization of C.I. Reactive Red 45 (80 mgL^{-1}) by UV/H_2O_2 process was 600 minutes, while complete bleaching was accomplished after 60 minutes of treatment [78]. Muruganandham and Swaminathan [89] achieved between 93.5 and 61.0 % of decolorization, whilst 62.9 and 24.6 % of aromatic products degraded after 150 minutes, applying UV/H_2O_2 process for the treatment of C.I. Reactive Orange 4 with various initial dye concentrations (0.1-0.5 mM). Bali et al. [76] decolorized C.I. Reactive Black (100 mgL^{-1}) model solution completely by UV/H_2O_2 process, pH 7 and initial H_2O_2 of 100 mM, after 60 minutes. In our previous study [79] where UV/H_2O_2 process was applied for the treatment of C.I. Reactive Blue 137 model solution with 20 and 80 mgL^{-1} initial dye concentrations, complete decolorization was achieved after 10 and 50 minutes, respectively. Colonna et al. [90] reported in their study of decolorization of several dyes, including one reactive, by UV/H_2O_2 process complete bleaching of C.I. Reactive Blue 19 after 110 minutes and that ratio $[H_2O_2]/[dye]$ plays important role in the oxidation efficiency of applied system. That was also confirmed by Ince [91] and he named that ratio "effective H_2O_2 level". Shu and Chang [92] reported significantly faster decolorization of studied dyes by UV/O_3 than by UV/H_2O_2. Similar results are obtained in our previous study [78], where significantly faster decolorization and correspondingly higher mineralization of C.I. Reactive Red 45 was obtained by UV/O_3 and $UV/O_3/H_2O_2$ than by UV/H_2O_2 process. It was established that mineralization by $UV/O_3/H_2O_2$ is four time faster than that by UV/H_2O_2 process. As the important process parameter of UV/O_3 and $UV/O_3/H_2O_2$ processes is pointed out pH of solution and it was established that optimal pH is at neutral or weak basic conditions [78, 93].

Photocatalytic Processes

Heterogeneous photocatalysis is based on the photonic excitation of a solid, which renders it more complex. The term photocatalysis may designate several phenomena that involve photons and catalyst, while this part of the chapter will consider only semiconductor photocatalysis. Photocatalytic activity of TiO_2 is based on its semiconductor properties. Radiation by photons, which have higher transfer energy, of such semiconductor leads to generation of electron-hole pairs [94]:

$$TiO_2 \xrightarrow{h\upsilon} h_{vb}^{\ +} + e_{cb}^{\ -} \tag{19}$$

Holes in valence band (h_{vb}^+) are strong oxidants, while electrons in conductance band (e_{cb}^-) act as reductants. Holes in valence band reacts with H_2O or hydroxyl ions on the surface and producing OH radicals:

$$H_2O + h_{vb}^+ \rightarrow HO\bullet + H^+ \qquad (20)$$

$$h_{vb}^+ + HO^- \rightarrow HO\bullet \qquad (21)$$

Electrons react with dissolved oxygen producing superoxid radical ion, $O_2\bullet^-$, or its protonated form, perhydroxyl radical, $HO_2\bullet$, equations (22) and (23):

$$O_2 + e_{cb}^- \rightarrow O_2\bullet^- \qquad (22)$$

$$O_2\bullet^- + 2H^+ \Leftrightarrow 2HO_2\bullet \qquad (23)$$

In water, two $HO_2\bullet$ can be recombined if their concentration allow them to react significantly yielding H_2O_2 and O_2. It follows photocatalytic reduction of H_2O_2 by scavenging an electron from the conduction band where OH radicals are generated [95]:

$$H_2O_2 + e_{cb}^- \rightarrow HO\bullet + HO^- \qquad (24)$$

The efficiency of TiO_2 photocatalysis could be improved by the addition of H_2O_2, but the addition of optimal H_2O_2 dosage should be taken into account. When it is in excess, it decrease process effectiveness. Photocatalysts, i.e. semiconductors, are comprised of microcrystalline or nanocrystalline particles and they are used in a form of thin film or as powder dispersion [96, 97]. Besides TiO_2 as the mostly known, used and studied photocatalyst, the usage of alternative photocatalysts, such as ZnO, CdS, SnO_2, is investigated as well [71, 94-101]. One of the main operating parameter of photocatalytic process is the dosage of photocatalyst.

In the studies concerning the application of photocatalytic processes for reactive dye wastewater degradation, TiO_2 is mostly used photocatalyst. Bizani et al. [102] investigated the efficiency of UV/TiO_2 process for the degradation of two commercial reactive dyes (initial concentrations 50 mgL^{-1}) and achieved the complete decolorization between 30 and 120 minutes of treatment depending on dye molecular structure and the type of TiO_2 photocatalyst. Goncalves et al. [97] investigated optimal pH range for reactive dye decolorization by UV/TiO_2 process and they reported the highest rate of C.I. Reactive Orange 4 decolorization at strong basic condition. Muruganandham and Swaminathan [103] investigated the application of other photocatalysts such as SnO_2, ZnO, CdS and Fe_2O_3 for decolorization of C.I. Reactive Yellow, and they reported that SnO_2, CdS and Fe_2O_3 have negligible activity on RY14 decolorization and degradation. In our previous paper [104], two effective photocatalysts, TiO_2 and ZnO, are compared in the case of C.I. Reactive Red 45 model solution decolorization and mineralization. The optimal conditions for both processes are established and it was found out that there are no significant differences between these two photocatalysts for decolorization of model solution at established optimal conditions for each process. Similar findings are reported in the study Akyol et al. [105] where complete color removal of Remazol Red was obtained after 35 minutes treatment by UV/TiO_2 and UV/ZnO processes.

Ozone-Based Processes

Ozone is main component in many oxidation processes assembled under the term ozonation processes. In these processes ozone is applied either alone (O_3 process) or with the addition of oxidant, e.g. H_2O_2 (O_3/H_2O_2 process), UV radiation (explained in above subchapter), catalyst, activated carbon, ultrasound etc. Ozone is inorganic molecule constituted by three atoms of oxygen. It is present in nature in upper atmosphere in the form of stratospheric layer around the earth, and it is formed by the photolysis of diatomic oxygen and further recombination of atomic and diatomic oxygen, shown by equations (25) and (26) [35]:

$$O_2 \xrightarrow{h\upsilon} 2O\bullet \tag{25}$$

$$O\bullet + O_2 \rightarrow O_3 \tag{26}$$

Ozone can be generated artificially in ozone generators. There is two ways of generating ozone by ozone generators: (a) the cleavage of oxygen molecules under the influence of strong electrical field and (b) the same mechanism like in nature, the photolysis of oxygen, but induced artificially. The mechanism of both ways can be described by above reactions (25) and (26). Ozone was discovered in 1840, while its molecular structure as triatomic oxygen was established 30 years latter. The first usage of ozone was reported at the half of 19th century where ozone was used as disinfectant in many water treatment plants and hospitals. Therefore, at 1856 the first usage of ozone as disinfectant for operating rooms in hospital over Europe was reported [106, 107]. First usage of ozone for the purification of drinking water in Monaco and Germany was performed at 1860 [107, 108]. Ozone application as oxidant for water purification was retained through 20th century, and its significant increase was noticed at 1970s when the production of trihalomethanes and other organohalogenated hydrocarbons were identified during the water treatment by chlorine [35]. Ozone is very reactive reagent with redox potential of 2.07 V (Table 2), either in liquid or gas phase. That high reactivity owes to electronic configuration; ozone can be presented as a hybrid in four different molecular resonance structures that give ozone characteristics of an electrophilic, dipolar or even nucleophilic agent. Hence, ozone molecule could react with organic compounds under two different mechanisms: direct and indirect. Direct mechanism involves organic compound degradation by molecular ozone, occurring at acidic pH range. The main reactions involved in direct mechanism are the reaction of addition to the unsaturated part of the hydrocarbon molecule and electron transfer. Rather high oxidation/reduction potential enables ozone to react with many organic, but also inorganic compounds [35].

$$O_3 + 2H^- + 2e^- \rightarrow O_2 + H_2O \quad (27)$$

Degradation of organics by indirect ozone mechanism occurs at basic conditions and involves generation of OH radicals and their further reaction with present organic compounds in water. Hydroxyl radicals are generated by the reaction of ozone with hydroxyl ions present in water [109]:

$$O_3 + H_2O \xrightarrow{OH^-} HO\bullet + O_2 + HO_2\bullet \quad (28)$$

Mechanism of ozone decomposition in water depends on the presence of chemical species that can initiate, promote and/or inhibit its decomposition. The most accepted ozone decomposition mechanism is given in Figure 3.

According to the presented mechanism of ozone decomposition in water, it can be seen that ozonation can classified in AOPs, when the degradation of organics occurs by indirect mechanism, and in classical chemical treatment methods, when direct mechanism is dominant in the degradation of organics in water. One of the limiting factors of ozonation influencing overall process effectiveness presents the efficiency of ozone mass transfer from gas to liquid phase. One of possible solutions gave Mathews et al. [110] by the application of *in situ* ozone generator with porous electrode. They obtained 90 % of C.I. Reactive Blue 19 decolorization in two times shorter treatment period than by the conventional ozone generator. Besides O_3 process, either direct or indirect, one of widely investigated process from the group of ozone-based processes is O_3/H_2O_2, so-called peroxone process or perozonation [111]. Efficiency of that process is based on the fact that the presence of H_2O_2 could enhance ozone reactivity in bulk phase. Moreover, some authors consider degradation of organics by direct mechanism negligible in the case of O_3/H_2O_2 process [111, 112]. Hydroxyl radicals are generated by reactions of ozone and perhydroxyl ion, HO_2^-, which is present in water by the dissociation of H_2O_2.

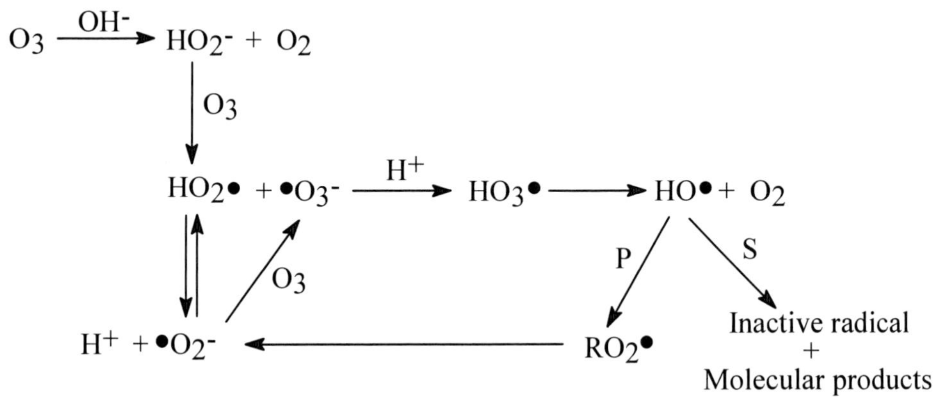

Figure 3. Scheme of ozone decomposition mechanism in water. P = promoter (e.g. ozone, methanol), S = scavenger or inhibitor (e.g. *t*-butanol, carbonate ion), I = initiator (e.g. hydroxyl ion, perhydroxyl ion) (adapted by Beltran [35]).

The simple overall mechanism of OH radical generation in peroxone process is shown by equation (29) [112]:

$$H_2O_2 + O_3 \rightarrow 2HO\bullet + 3O_2 \quad (29)$$

Degradation kinetic of organic pollutants by O_3 process, either direct or indirect mechanism, is reported as first order or "pseudo"-first order kinetic, while in the case of peroxone process, degradation kinetics follows second order due to the predominant degradation mechanism by OH radicals.

Many investigations are available in the case of ozonation as treatment methods for colored wastewaters containing reactive dyes. Chu and Ma [30] investigated direct and indirect O_3 process for the degradation of several dyes including three reactive. They reported dye decolorization by indirect mechanism significantly faster than that by direct mechanism. Wu and Wang [113] reported the complete decolorization of C.I. Reactive Black 5 (0.5 gL^{-1}) in 40 minutes of treatment by O_3 process with rate of ozone generation 26.1 $mgL^{-1}min^{-1}$, and their results revealed that rate of ozone transfer increased with increase in dye concentration, applied ozone dose and temperature. Peralta-Zamora et al. [114] obtained very fast complete decolorization of C.I. Reactive Blue 19, in <5 minutes, but degradation of dye monitored in UV region, i.e. degradation of aromatics formed after the initial cleavage of dye chromophores, was significantly slower. Furthermore, Arslan-Alaton et al. [115] investigated the application of O_3 process for the treatment of simulated dyebath effluent containing several different reactive dyes at different pH region: acidic, neutral and basic. Results of their study showed that decolorization of simulated dyebath is very fast at all three pH regions, yielding with 100 % color removal in couple of minutes, and that degradation of aromatics, monitored at 280 nm, was somewhat slower than decolorization, and that overall extent of aromatics degradation depends on operating pH value. Mineralization of solution was also monitored, and significantly lower extents of partial mineralization were achieved than those of decolorization and aromatic degradation. They obtained similar findings in previous study too [116]. Similar large difference between decolorization and mineralization extents were obtained in our previous comparative study of O_3, O_3/zeolite (NH_4ZSM5 or HY type) and O_3/H_2O_2 processes for degradation of C.I. Reactive Blue 137 model solution [117]. Model dye solution was completely decolorized at all initial dye concentrations (20-150 mgL^{-1}), while only partial mineralization was obtained in all cases. The somewhat higher mineralization extents were obtained by O_3/H_2O_2 than by O_3 process. Zeolites did not affected decolorization rate, only slightly due to changing of pH values of solution, but somewhat higher TOC removals were obtained in cases with zeolite addition than without. Explanation could be found in the adsorption of aromatic by-products with appropriate molecular structure to penetrate in zeolite pores.

Besides the improvement of O_3 process by H_2O_2 addition, some authors combined O_3 and ultrasound processes in order to improve degradation of reactive dye solution [118]. From the presented results it can be seen that decolorization of colored wastewater can be easily achieved by ozonation processes, but dye degradation in order to get complete mineralization is very hardly to achieve.

High Voltage Electrical Discharge Processes

Basic principles of processes including electrical discharge, so-called corona discharge, in liquid and gas phase, are that chemical and physical processes could be induced by strong electrical field, and therefore these processes could be consider as electrical AOPs. It is proved that by applying electrical discharge in liquid phase intensive UV irradiation and various active species (•OH, •H, •O, HO_2^-, $O_2^{\bullet -}$, H_2O_2, etc.) are considered to be produced [39, 40, 119-124]. By the presence of oxygen in the system, ozone and its ozone-related radical species can be formed. This process can be effective for the treatment of biological microorganisms and dissolved chemicals in liquid phase [125]. Application of the processes based on strong electric field in water or organic solution is investigated in the past due to importance in electrical transmission processes and they practical usage in biology, chemistry and electrochemistry. Recently, the application of reactors involving electrical discharge in liquid or above the surface, for the treatment of drinking and wastewaters is widely investigated. Nowadays, these reactors are in use, but only in laboratory scale. There are many types of corona reactors according to the configurations of their high voltage and ground electrodes. The most common reactor type has both electrodes, high voltage and ground, in liquid phase [40]. Generally, corona is type of electrical discharge that is formed when DC, AC or pulsed electric potential is applied between two non-uniform electrodes where a dielectric medium is placed between the electrodes. Corona onset begins when the discharge initiates in the dielectric medium. That manifests as thin violet layer which surrounds the high voltage electrode in liquid phase. That layer contains of very short streamers. Initial voltage of corona discharge depends on the type of applied electrodes and physical properties of dielectric material between the two electrodes. As the corona discharge voltage increases, the streamers become larger and corona discharge is formed. Further increasing of the voltage leads to the sparkover, i.e. the production of direct current channel that bridges two electrodes [119, 125]. As it was mentioned before, pulsed corona discharge is applied in gas and liquid phase. In gas phase corona, the mostly used electrode geometry is “plate-to-plane” and “wire-to-cylinder”, while in liquid phase the most studied geometries are “point-to-plane” and “plate-to-plane” (Figure 4.) [40].

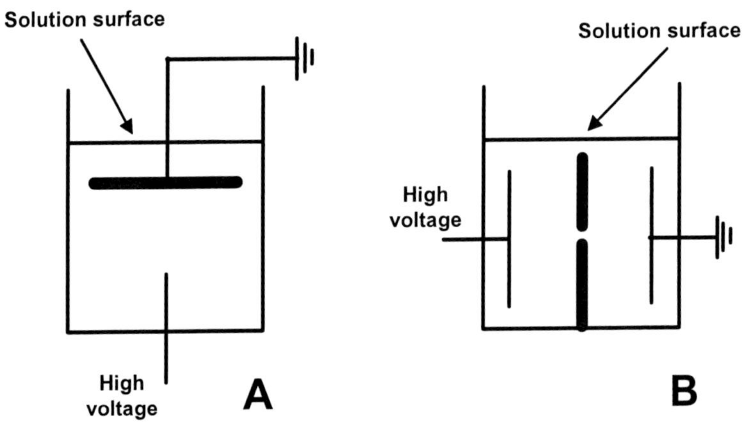

Figure 4. Most studied electrode geometry in liquid corona discharge reactors: (A) point-to-plane and (B) plate-to-plane.

Pulsed corona discharge produce repetitive (60 Hz), fast raising high voltage pulses with short lifetime (μs). Application of short high voltage pulses (200-1000 ns) consider presence of non-thermal conditions, that means electrons in plasma having a higher mean energy (T_e >> 1 eV), than the other constituents in liquid (T_{H2O} < 0.1 eV) [121, 126]. On this way, the energy loss for the migration of ions, chemical species with less mobility than electrons and which have no influence to generation of radicals, is minimized as high as possible [120]. Among all radical species generated in corona when electrons with high energy collide with water molecules, equations (30) and (31), OH radicals present main species responsible for degradation of organics.

$$H_2O + e^-* \rightarrow HO\bullet + H\bullet + e^- \quad (30)$$

$$H_2O + e^-* \rightarrow H_2O^+ + 2e^-* \quad (31)$$

$$H_2O + H_2O^+ \rightarrow H_3O^+ + HO\bullet \quad (32)$$

Such formed radical species could react between themselves resulting with the production of H_2O_2, O_2, or H_2.

$$H\bullet + H\bullet \rightarrow H_2 \quad (33)$$

$$HO\bullet + HO\bullet \rightarrow H_2O_2 \quad (34)$$

$$HO\bullet + H\bullet \rightarrow H_2O \quad (35)$$

Degradation of organics could be performed by: (a) direct reactions with highly reactive species (OH radicals), (b) indirect reactions over radicals formed from stabile molecules (H_2O_2), and (c) direct reactions with stabile molecules.

Hybrid Corona Reactors

These types of corona reactors involve simultaneous application of non-thermal plasma above liquid surface and direct high voltage electrical discharge in liquid phase. That is achieved by the combination of different electrode configuration and/or by changing electrode geometry of standard corona reactors which use "point-to-plane" geometry. There are two hybrid corona reactors, so-called hybrid-series and hybrid-parallel (Figure 5.). Hybrid-series involves "point-to-plane" geometry and configuration of electrodes where high voltage electrode is in liquid phase, and ground electrode is placed in gas phase just above the liquid solution. That enables the formation of conventional streamer like corona discharge in liquid and the formation of intensive plasma channels near to liquid surface. Hybrid-parallel corona reactor involves electrode geometries "point-to-plane" in liquid phase and "plane-to-plate" in gas phase. That reactor configuration combines mostly used electrode geometries in liquid and gas phase, i.e. on that way is combined classical corona discharge in liquid and gas.

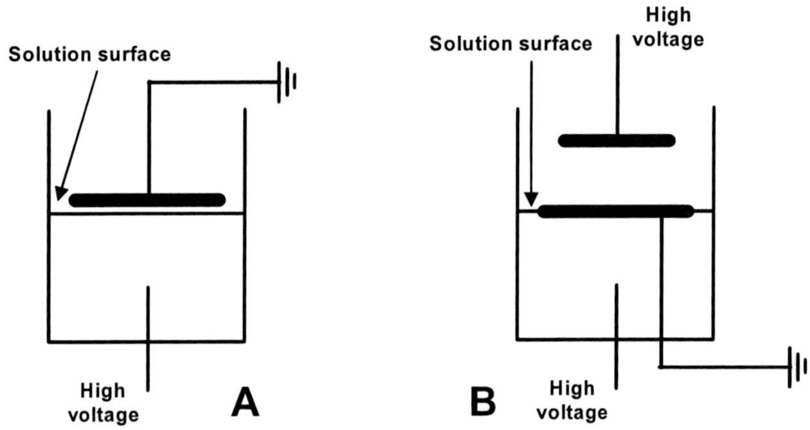

Figure 5. Configuration of hybrid corona reactors: (A) series and (B) parallel.

Both hybrid reactor types are very interested due to improvements concerning standard liquid corona type reactor (Figure 4.). It is reported that hybrid reactors, besides other earlier mentioned radical, ionic and molecular species, include generation of ozone in gas phase and its transfer to liquid phase, improving degradation of organics in liquid [127]. Besides explained corona discharge, there are other types of discharges in water such as glow, barrier, glidarc, but they are not discussed in this subchapter.

There are not too many studies directed to the investigation of corona discharge reactors application for the treatment of colored wastewater, particularly containing reactive dyes. Tri Sugiarto et al. [128] investigated the application of corona discharge reactor, "plane-to-plate" geometry, for decolorization of three dyes (initial dye concentrations of 10 mgL^{-1}). They reported complete decolorization at neutral pH after 120 minutes treatment and dependence of decolorization rate on pH value. They obtained the highest decolorization rate at basic conditions. Sunka et al. [129] investigated new type of corona discharge reactor, "pin hole" geometry, for decolorization of C.I. Reactive Blue 137 (50 mgL^{-1}). In first case where only corona discharge was used for decolorization of reactive dye solution, very low color removal was achieved after 30 minutes of treatment, while the addition of Fe ions in the system improved color removal and in 5 minutes of treatment the complete color removal was obtained. Similar findings were obtained in study by Loncaric Bozic et al. [130] treating C.I. Reactive Blue 49 in the same reactor type. One of our earliest paper related to corona discharge dealt with the application of corona discharge reactor with "point-to-plane" electrode geometry for decolorization of model solution of four different dyes (20 mgL^{-1}), including two of reactive type [131]. Decolorization extents ranged between 17-25.6 % after 60 minutes of treatment, while by the addition of Fe ions in the system, complete decolorization was achieved after 30 minutes. Our two new papers related to corona are directed to the application of hybrid corona reactors for decolorization of reactive dyes [132, 133]. C.I. Reactive Blue 137 (20 mgL^{-1}) was decolorized up to 72.9 and 85.0 % regarding reactor type. Both papers also investigated the influence of synthetic zeolites on decolorization by corona discharge and it was found that zeolites could improve color removal in hybrid-parallel corona reactor. Decolorization effectiveness was decreased by increasing initial dye concentration, resulting with color removal in the case of 80 mgL^{-1} initial dye concentrations between 34 and 48 % regarding dye structure, type of reactor, the addition of zeolites or not, and type of added zeolite. From the presented results of corona discharge processes for decolorization of colored wastewaters it can be seen that these type of processes could be efficient, but only when dye loading in wastewater is not so high.

Other AOPs

This part of subchapter contains less detailed review of ambient AOPs that are not studied in our laboratory, such as: ultrasound or ultrasonic irradiation, radiolysis of water and electrochemical processes.

Ultrasound

The term sonochemistry describes all chemical processes in which ultrasound irradiation is involved. The main premise of ultrasound process as wastewater treatment method is generation of free radicals, such as hydrogen and hydroxyl, upon the action of ultrasonic waves in liquid. The applied frequency ranges from 20-40 kHz; predominantly used in large scale applications, so-called "power ultrasound", up to 1 MHz; frequency used in so-called "diagnostic ultrasound" [134]. Ultrasound produces the chemical effects through several different physical mechanisms and the most important nonlinear acoustic process for sonochemistry is cavitation. Cavitation can be generated when large pressure differentials are applied in flowing liquid (hydrodynamical cavitation), or by means of an electrochemical transducer, piezoelectrical or magnetostrictive, in contact with the fluid (acoustical cavitation) [135]. Electrohydraulic cavitation in liquid causes bubbles. These bubbles start very small, and grow in diameter and implode under the periodic variations of the pressure field of the ultrasonic waves. The rapid implosion of the eventually instable gas bubbles causes adiabatic heating of the bubble vapor phase. It is assumed that creates drastic local conditions: temperatures of about 5500 °C inside the bubble and 2100 °C in the liquid that surrounds the cavity, and pressure up to 100 MPa within the collapsing cavity. Such conditions are limited to a very small region and the heat produced during cavitation is dissipated very quickly (heating and cooling rates greater than 109 K/s). As a result, the surrounding liquid remains at the ambient temperature. Organic compounds are highly degraded in such an environment and inorganic compounds can be oxidized or reduced. Water irradiation using ultrasound causes decomposition of the water molecules into extremely reactive radicals HO• and H•, shown by equation (36) [119, 136].

$$H_2O \xrightarrow{ultrasound} H\bullet + \bullet OH \tag{36}$$

Several authors investigated ultrasound as treatment process for colored wastewater degradation. Tezcanli-Guyer and Ince [137] investigated ultrasound, 520 kHz, for the treatment of four organic dyes including two of reactive type (initial concentrations 20-60 mgL^{-1}), and reported decolorization after two hours treatment between 81-99 % depending on dye structure, while degradation of formed aromatics after four hours treatment ranged between 11-93 %. Rehorek et al. [138] applied power ultrasound, 850 kHz, at different input powers for the treatment of several organic dyes. They obtained complete decolorization of all studied dyes in 1-15 hours of treatment regarding input power of ultrasound, and they reported that ultrasound is able to degrade organic dyes to non-toxic products.

Water Radiolysis

Irradiation of dilute aqueous solutions by high energy ionizing radiation, ranged from keV to MeV, results with the excitation and the ionization of water molecules leading to the production of radical species. The types of radiation may be classified by the source of production: (a) the decay of radioactive nuclei (α-, β- and γ-radiation), (b) beams of accelerated charged particles (electrons, protons and deutrones) and (c) short-wavelength electromagnetic radiation (X rays). OH radicals and hydrated electrons, e_{aq}^-, can be considered as predominant species formed by water radiolysis. Besides these two species, water radiolysis produces also hydrogen atoms, hydrogen peroxide (predominant specie in the presence of O_2), hydroxyl ions, etc. The mostly investigated radiation sources for degradation of organics are γ-radiation and electron beam [119, 139-141]. Dajka et al. [140] reported the role of e_{aq}^- in decolorization of C.I. Reactive Black. Zhang et al. [141] investigated γ-radiation for decolorization of three different dyes and reported that, depending on dye structure, different gamma irradiation dose are required to achieve the complete decolorization.

Electrochemical Processes

These processes can occur by a direct electron transfer reaction to (reduction) or from (oxidation) the present organic pollutant, or by a chemical reaction of the pollutant with previously electrogenerated species. The mechanism is generally viewed as a direct anodic oxidation of organic pollutant involving its reduction by direct electron transfer from organic molecule to the electrode to form a radical cation that readily deprotonates, equation (37):

$$RH \xrightarrow{electrolysis} R\bullet + H^+ + e^- \quad (37)$$

Nowadays, various types of electrodes are in use: Pt/IrO_2, Ti/Pt, SnO_2/Ti, $Ti/RuTiO_2$, etc [119, 142]. Catanho et al. [143] used $Ti/RuTiO_2$ electrode for decolorization of C.I. Reactive Red 198 and reported 75 % color removal after 2 hours treatment. Awad and Abo Galwa [144] investigated different conductive electrolytes for decolorization of colored wastewater and obtained the highest decolorization in the case of NaCl. Carneiro et al. [145] investigated different types of electrochemical processes for degradation of C.I. Reactive Blue 4 and reported the as the best process $Ti/SnO_2/SbO_x/RuO_2$ electrode in Na_2SO_4 0.2 M at pH 2.2 and potential of +2.4 V, where complete decolorization and 60 % mineralization were obtained.

Study of C.I. Reactive Blue 137 Wastewater: Treatment by Several AOPs

After a review of available and currently studied AOPs, two types of processes were chosen for our research directed to degradation of C.I. Reactive Blue 137 as pollutant in model wastewater. Due to the very high effectiveness and fastness of Fenton type processes, but also considering secondary treatment of solution due to iron ions and counter ions, Fenton type process using iron powder, Fe^0/H_2O_2, was one of studied treatment processes. It was concluded that the assistance of UV light to Fe^0/H_2O_2 could offer significant benefit to process effectiveness due to the achievement of complete mineralization, as it was mentioned above in "Fenton type processes" subchapter. Another group of chosen processes include those combining ozone and UV light, UV/O_3 and $UV/O_3/H_2O_2$, due to the conductance at neutral or at weak basic conditions that excludes after-treatment solution conditioning like in the case of Fenton type processes, and also due to constant ozone supply enabling to constant OH radical generation. The influence of synthetic zeolites on chosen AOPs was also investigated in the framework of this research. Synthetic zeolites were shown positive effect when added to some of AOPs for the treatment of colored wastewater in our previous studies [104, 117, 132, 133, 146, 147]. Generally, synthetic zeolites, are three-dimensional, microporous, crystalline solids with well defined structures that contain aluminum, silica, and oxygen in a regular framework. Zeolites have void pore space that can host cations, water or other molecules. They have the ability to act as catalysts for chemical reactions which take place within the internal cavities. Furthermore, synthetic zeolites may act as ion exchangers, because the loosely-bound nature of the extra-framework metal ions allows the exchange of other types of metals when present in aqueous solution. Shape-selective properties of zeolites are also the basis for their use in molecular adsorption [148-150].

Objectives

The aim of the study was to find the optimal treatment process for colored wastewater containing reactive dye C.I. Reactive Blue 137. The comparative study involved several processes, Fe^0/H_2O_2, $UV/Fe^0/H_2O_2$, UV/O_3 and $UV/O_3/H_2O_2$, and it was performed through an experimental research of decolorization and mineralization of a reactive azo dye C.I. Reactive Blue 137, as a model wastewater pollutant. Applied processes were optimized according to their process parameters, Fe^0 dosages, initial H_2O_2 dosages and initial pH values. The influence of initial organic dye concentration as well as the addition of solid particles, synthetic zeolites, on the process effectiveness was also investigated. Studied AOPs were evaluated on the basis of their eco-effectiveness, by the means of color (A_{610}), aromaticity (A_{280}), TOC and AOX value decrease, and their cost-effectiveness as well.

Materials and Methods

Synthetic wastewater containing organic model pollutant was prepared using two different concentrations, 20 and 80 mgL^{-1}, of the commercial dye Cibacron Marine P2R-01, C.I. Reactive Blue 137 (RB137) obtained from Ciba-Geigy, and deionezed water; pH 7, conductivity less than 1 $\mu S\ cm^{-1}$. Other chemicals (H_2O_2, 30 %; sulfuric acid, >95 %; and sodium hydroxide, p.a.) were supplied by Kemika, Croatia.

Series of experiments were conducted in order to find an optimal Fe^0/H_2O_2 ratio. At the beginning of the experiment, pH was adjusted at 3 using 25 % sulphuric acid, which was followed by the addition of iron powder and H_2O_2. Iron powder dosages were 28 and 56 mgL^{-1} that corresponded to the concentrations of 0.5 and 1.0 mM, respectively. Concentrations of hydrogen peroxide were varied to give molar ratios 1 : 5, 1 : 10, 1 : 20, 1 : 30, 1 : 40, and 1 : 50. Reaction mixture (V=250 mL) was continuously stirred at room temperature in an open batch system with magnetic stirring bar and was treated for two hours, while dye concentration and TOC values were measured at the end of the each experiment to established decolorization and mineralization extents.

All further experiments were carried out in a 0.8 L batch glass annular water-jacketed photoreactor, used in our previous studies [48, 52, 59, 78, 79, 87]. The value of incident photon flux at 254 nm, 7.36×10^{-6} Einstein s^{-1}, was calculated on the basis of hydrogen peroxide actinometry experiments [151]. In experiments that required the addition of iron powder, Fe^0/H_2O_2 and $UV/Fe^0/H_2O_2$, optimized dosages of iron powder and H_2O_2 were added accordingly to the results of experiments explained in previous paragraph (Fe^0/H_2O_2 optimization study part). In experiments where UV irradiation was required, $UV/Fe^0/H_2O_2$, UV/O_3 and $UV/H_2O_2/O_3$, the UV lamp was switched on. In the experiments involving ozone, UV/O_3 and $UV/H_2O_2/O_3$, ozone was introduced into the reactor. Ozone was generated from pure oxygen, >99.9 %, by introducing it into the ozone generator, MIC Systems Inc., Valdosta, Georgia, USA to produce ozone. The outlet gas mixture from the ozone generator, i.e. mixture of ozone and oxygen, was fed into the reactor through a sintered glass plate diffuser located in the bottom half of the reactor. The inlet gas flow into the ozone generator was fixed at 0.15 $Lmin^{-1}$. The rate of ozone generation in reactor inlet stream, 7.9×10^{-3} $gmin^{-1}$, was determined iodometrically [152]. In experiments of UV/O_3 and $UV/H_2O_2/O_3$ processes optimization, initial pH values and hydrogen peroxide concentration ranged from 3 to 11, and 0 to 10 mM, respectively, until the achievement of maximal mineralization extent. Initial pH was adjusted with the addition of NaOH or H_2SO_4. Added quantities of H_2SO_4, NaOH and H_2O_2 as an additional oxidizing agent, were negligible in comparison to the total volume of treated reaction mixture. The experiments were conducted by adjusting one variable, while others were held constant.

The total volume of the treated solution was 0.5 L in all cases, while the mixing of the solution was provided by both magnetic stirring and peristaltic pump at a flow rate of 0.1 $Lmin^{-1}$. Experiments were carried out at 25 (+/- 0.2) °C. The duration of each experiment was one hour. Samples were taken periodically from the reactor (2, 5, 10, 15, 20, 30, 40, 50 and 60 min) and thereafter immediately analyzed. In experiments where the influence of synthetic zeolites on process effectiveness was investigated, at the beginning of experiments different amounts of zeolites were added (0.4 gL^{-1} of NH4ZSM5 in the case of Fe^0/H_2O_2 process and 1.0 gL^{-1} in the cases of UV/O_3 and $UV/H_2O_2/O_3$ processes), accordingly to our previous studies [117, 146]. In these experiments, samples were centrifuged before the analysis in order to remove solid particles of zeolites from the solution. All experiments were repeated at least two times to give reproducibility of the experiments within 5 %, and averages are reported.

The initial pH values were measured by handylab pH/LF portable pH/conductivity-meter, Schott Instruments GmbH, Mainz, Germany. The decolorization and degradation of RB137 solution was monitored by a Perkin Elmer Lambda EZ 201 UV/VIS spectrophotometer, USA, by scanning complete UV and VIS spectra. The extent of RB137 mineralization was determined on the basis of measurements of total organic carbon (TOC), using TOC analyzer; TOC-V_{CPN} 5000 A, Shimadzu, Japan, and absorbable organic halides (AOX), performed by Organic Halide Analyzer, DX-2000, Dohrmann.

Results and Discussion

Fenton Type Processes

First group of investigated AOPs were addressed to Fenton type process. As it was mentioned earlier, the oxidation power of Fenton type processes, particularly those performed in the dark; strongly depends on process parameters such as iron catalyst concentration, concentration of H_2O_2, pH, temperature, and treatment time. Therefore, it was necessary to performe laboratory investigations, so-called "jar tests", of those parameters in order to find optimal for chosen system Fenton type process/organic pollutant [65]. In the research "jar tests" were directed to find optimal concentration of iron powder and optimal Fe^0/H_2O_2 ratio, while other process parameters were in the range reported as optimal in the literature: pH 3, room temperature, and a one-hour treatment time [44, 45, 49, 51, 52, 54, 66]. The results of Fe^0/H_2O_2 process parameters for the treatment of RB137 model solution with initial concentration of 20 mgL^{-1} are summarized in Figure 6.

It can be seen that through complete investigated range of Fe^0/H_2O_2 ratio at both iron powder dosages rather high decolorization of RB137 colored wastewater, >95 %, was achieved. Obtained decolorization extents were below visibility limit, <1 mgL^{-1} of remained dye in the solution [2]. During the same treatment period, only partial mineralization was obtained in all cases, with the highest 40.9 % of TOC removal in the case of $c(Fe^0)$=1.0 mM and 1 : 20 Fe^0/H_2O_2 ratio. It can be concluded that decolorization of RB137 was not significantly influenced by investigated process parameters, but at the same time, mineralization showed different results. Almost the same system behavior could be observed at both investigated iron powder dosages and the same Fe^0/H_2O_2 ratio. Mineralization of RB137 increased with increasing the Fe^0/H_2O_2 ratio, but only to the certain point.

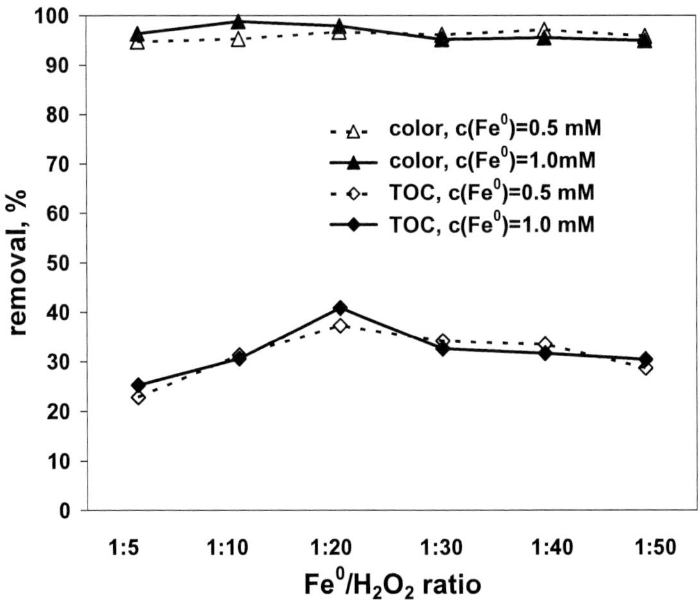

Figure 6. Influence of iron powder dosages and Fe^0/H_2O_2 ratio on color and TOC removal in Fenton "like" process for degradation of RB137 at pH 3.

That can be explained with the increasing of OH radical concentration in the bulk, resulting with an increase of mineralized part of RB137. According to the results of TOC removal, further increase over obtained optimal ratio showed negative effect to the process effectiveness. That can be explained with the fact that H_2O_2 when is in excess, act as OH radical scavenger, resulting with the lowering of RB137 mineralization [44, 45, 47-49, 65]. Besides that effect, the creation of an inert oxidative film on the surface of iron powder at higher concentration of H_2O_2 should also be taken into account. That inert film could disable the leaching of Fe^{2+} ions from iron powder surface in the bulk, as well as the occurrence of Fenton reaction on the iron powder surface [54].

After determining the optimal iron powder dosage and Fe^0/H_2O_2 ratio, the kinetic of RB137 degradation with Fe^0/H_2O_2 and $UV/Fe^0/H_2O_2$ processes was investigated. In Figure 7. the results of RB137 degradation with both processes are presented. By monitoring complete UV/VIS spectra, it was tried to get an overview on the degradation of complete dye molecular structure.

By full black line the chromatogram of RB137 through complete UV/VIS spectra was presented. Three characteristic peaks can be observed. First one absorbing in the VIS region between 480-680 nm with maximum wavelength at 610 nm corresponds to RB137 chromophore. Next two peaks absorb in UV region, first between 260-350 nm with maximum wavelength at 280 nm corresponding to the absorbance of aromatic part of dye molecule and second between 200-260 with maximum wavelength at 230 nm corresponding to triazine ring in dye molecular structure [45, 72, 153, 154]. Comparing degradation of dye chromophores, i.e. the disappearance of peak absorbing at λ_{max}=610 nm, it can be observed that decolorization is much slower with Fe^0/H_2O_2 than $UV/Fe^0/H_2O_2$ process. Therefore, complete decolorization of RB137 by $UV/Fe^0/H_2O_2$ process was achieved after 5 minutes of treatment, while in the case of Fe^0/H_2O_2 process, the complete bleaching, i.e. zero absorbance of RB137 model solution at 610 nm was observed barely after 20 minutes treatment process. The band of RB137 in UV region with two peaks at 230 and 280 nm was interfered through almost complete treatment time by absorbance of added H_2O_2 [35, 155]. It can be observed that the monitoring of dye degradation in UV region, as well as degradation of aromatic structures and triazine ring was partially disabled due to the interference by H_2O_2. But, it can be seen that new peak, produced by the interference of triazine ring peak with the absorbance of added H_2O_2, is lowering much faster in the case of $UV/Fe^0/H_2O_2$ than Fe^0/H_2O_2 process. That might be due to the much faster consumption of added H_2O_2, but also due to much faster degradation of triazine rings, by process combined with UV light. That is in accordance with the literature where is reported that $UV/Fe^0/H_2O_2$ process consumed H_2O_2 much faster than Fe^0/H_2O_2 process [48]. Furthermore, triazine rings have very high molar absorption coefficient at 230 nm, ≈40000 $M^{-1}cm^{-1}$, that makes them very unstable under UV-C irradiation [72]. Degradation of aromatic structures at λ_{max}=280 nm is also much faster by $UV/Fe^0/H_2O_2$ than by Fe^0/H_2O_2 process. Although that peak was also interfered by the addition of H_2O_2, in the case of $UV/Fe^0/H_2O_2$ process a significant decrease of that peak during the treatment period from 2.-20. minute could be observed, meaning that aromatic structures in RB137 molecules were degraded. After 30. minute of treatment time, the appearance and the constant increasing of new peak in UV region with maximum wavelength at 290 nm can be observed. That peak could correspond to the formation of organic acids as products of degradation of aromatic compounds in model solution that absorb in UV region at

220 and 290 nm [71, 153]. That is also confirmed by the increasing of peak at 220 nm in treatment period between 40. and 60. minute.

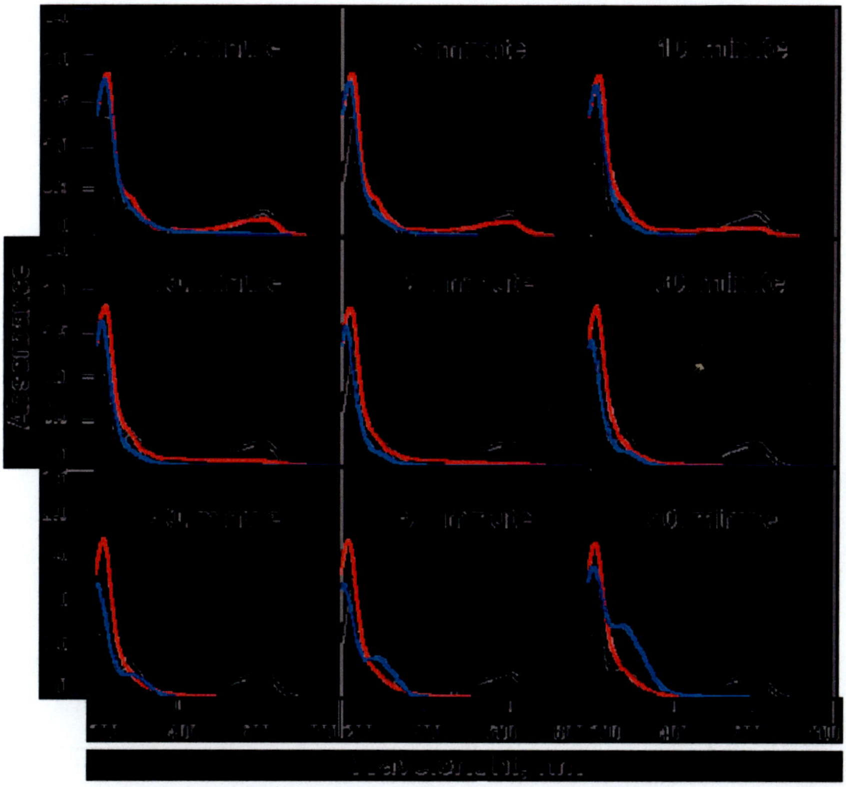

Figure 7. Degradation of RB137 with initial concentration 20 mgL^{-1} ( ) by Fe^0/H_2O_2 ( ) and $UV/Fe^0/H_2O_2$ ( ) processes monitoring through complete UV/VIS spectra

The higher efficiency of $UV/Fe^0/H_2O_2$ process for degradation of RB137 colored wastewater in comparison Fe^0/H_2O_2 process at the same process parameters can be also observed from the results of TOC removal during the treatment by those two processes (Figure 8.). In the first treatment period, until 15. minute, that difference is not so pronounced. But after 15. minute, in the case of Fe^0/H_2O_2 process strong inhibition in mineralization of RB137 model solution can be observed. On the other hand, the inhibition of $UV/Fe^0/H_2O_2$ process occurred after 40. minute of process. Inhibition and possible complete stopping of dark Fenton type process could be due to the complete consumption of H_2O_2, or due to the breaking of Fenton catalytic cycle by the formation of Fe^{3+}-complexes, mostly carboxylates [64]. Inhibition of photo-Fenton type processes could be caused by complete consumption of H_2O_2 and in that case degradation of organics will be continued only by their direct photolysis. Further reason for the inhibition of photo-Fenton type process might be due to the formation of organics acids resistant to degradation by OH radicals, such as acetic and oxalic [37], which can be degraded under UV light but with lower rate than other organic content by OH radicals. In this case it is more likely that the inhibition of both processes was not caused by the complete consumption of H_2O_2, due to the results of rather high remained

H_2O_2 concentration in bulk, >20 %, obtained in our previous study at the same process conditions, but with other organic pollutant [48].

Hence, it can be assumed that the inhibition of dark process, Fe^0/H_2O_2, occurred due to the formation of Fe^{3+}-complexes resulting with a lack of OH radicals in bulk produced through bulk Fenton reaction, while the mineralization efficiency of $UV/Fe^0/H_2O_2$ process was inhibited by the formation of large amounts of OH radical attack resistant organics acids. That is in accordance with the observation from Figure 7., where the appearance of new peaks occurred after 30. minute of treatment process.

The possible improvement of dye degradation efficiency of dark Fenton type process was tried to get by the addition of synthetic zeolite NH_4ZSM5. In our previous study [146], that type of zeolite showed positive effect on TOC removal by Fe^0/H_2O_2 process when added in lower dosages, <0.5 gL^{-1}. In Table 4 the comparison of Fe^0/H_2O_2, $Fe^0/H_2O_2/NH_4ZSM5$ and $UV/Fe^0/H_2O_2$ processes concerning the final decolorization, degradation of chloro-triazine rings and mineralization of RB137 model solution (20 mgL^{-1}) after a one-hour treatment is presented. It can be observed that all three compared processes showed very high decolorization efficiency, >99 % color removal in all cases. It should be noted that so-called "blank-tests" were performed in order to investigate adsorption of dye molecules on zeolites or possible penetration in zeolites pores. No changes of dye spectra monitored through complete UV/VIS region were obtained, indicating no dye adsorption by NH_4ZSM5 zeolite. Comparing mineralization efficiency of organic content by all three investigated processes, it can be observed that partial mineralization extents were obtained in all cases.

The increase of TOC removal for 10.3 % by $Fe^0/H_2O_2/NH_4ZSM5$ process in comparison to Fe^0/H_2O_2 could be contributed to the addition of solid particles. But, it is more likely that synthetic zeolite acted as adsorbents for molecule of appropriate size to be adsorbed/penetrated on/in zeolite. No effect of zeolite addition to other monitored parameters, color and AOX removal, and the findings of our previous study where scanned IR spectrograms of zeolites after the treatment with $Fe^0/H_2O_2/NH_4ZSM5$ showed new peaks which corresponded to possible dye degradation by-products [146], indicating to above mentioned assumption. Moreover, Kawai and Tsutsumi [157] reported, and the same was confirmed in our previous studies [59, 158], that NH_4ZSM5 zeolite could adsorb phenol. If phenol could be adsorbed, it can be assumed that other mono-substituted benzenes could also be removed from the aqueous solution by adsorption on/in NH_4ZSM5 zeolite. Furthermore, Feng et al. [156] proposed degradation mechanism of reactive dye with similar structure like RB137. They reported that after the initial cleavage of azo bonds, in the bulk are remained mono- and di-substituted benzenes and naphtholes and triazine rings. Moreover, Stylidi et al. [71] detected phenol and other mono-substituted benzenes as by-products of azo dye degradation by OH radicals.

Table 4. Degradation efficiency of RB137 model solution (20 mgL^{-1}) by Fe^0/H_2O_2, $Fe^0/H_2O_2/NH_4ZSM5$ and $UV/Fe^0/H_2O_2$ processes

Process	Fe^0/H_2O_2	$Fe^0/H_2O_2/NH_4ZSM5$	$UV/Fe^0/H_2O_2$
color removal, %	100	99.2	100
TOC removal, %	40.8	51.5	83.5
AOX removal, %	42.2	42.5	90.5

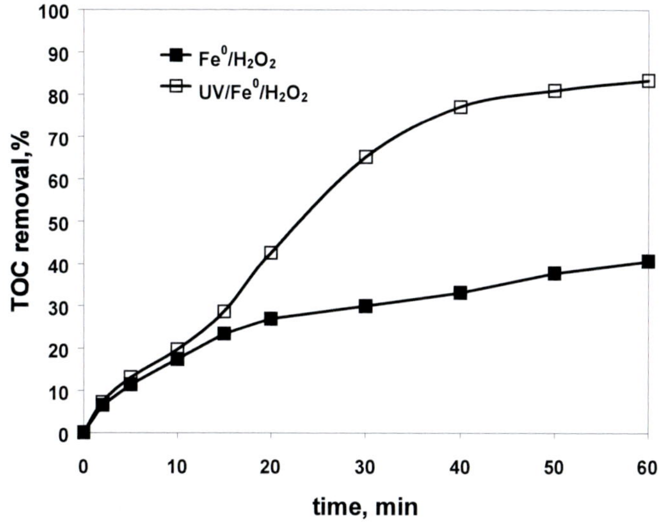

Figure 8. Mineralization of RB137 with initial concentration 20 mgL^{-1} by Fe^0/H_2O_2 and $UV/Fe^0/H_2O_2$ processes monitoring decrease of TOC ecological parameter.

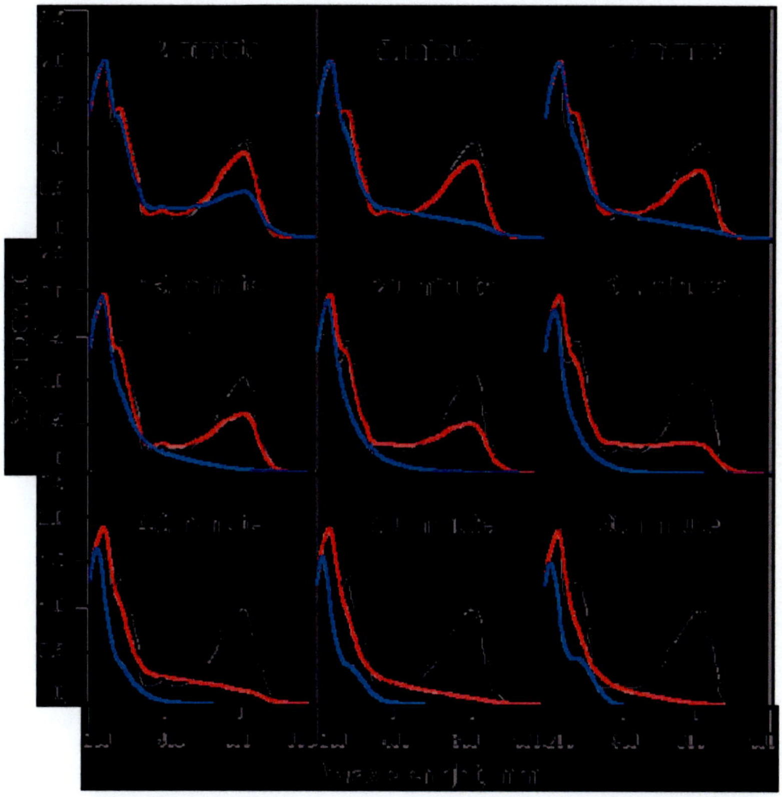

Figure 9. Degradation of RB137 with initial concentration 80 mgL^{-1} (▬) by Fe^0/H_2O_2 (▬) and $UV/Fe^0/H_2O_2$ (▬) processes monitoring through complete UV/VIS spectra.

In the favor of adsorption of only mono-substituted benzenes speaks also the fact that AOX removal was almost the same in the case with and without zeolite addition, indicating inability of NH_4ZSM5 zeolite for adsorption of triazine rings. The highest removal of all parameters; color, TOC and AOX was obtained when Fe^0/H_2O_2 process is combined with UV irradiation, resulting with 105 and 115 % improvement in TOC and AOX removal efficiency. The efficiency of Fe^0/H_2O_2 and $UV/Fe^0/H_2O_2$ processes was also tested on colored wastewater containing 80 mgL^{-1} of RB137 (Figure 9.).

Again, three specific peaks can be observed: dye cromophore peak at 610 nm, aromatic peak at 280 nm and triazine ring peak at 230 nm. The peaks in UV region were also interfered by absorbance of added H_2O_2, but not in such high manner like in the case with the initial dye concentration of 20 mgL^{-1}. $UV/Fe^0/H_2O_2$ process again showed rather high decolorization efficiency of colored wastewater; the complete bleaching was achieved between 20. and 30. minute of treatment, while in the case of Fe^0/H_2O_2 process complete bleaching was not achieved even after 60 minutes treatment. From Figure 9. a rather fast degradation of aromatics, i.e. disappearance of peak at 280 nm, can be observed even more clearly than in case of lower initial dye concentration due weaker interference of that peak by H_2O_2 absorbance. On the other hand, in the case of Fe^0/H_2O_2 process the lowering of that peak was much slower than in the case of $UV/Fe^0/H_2O_2$ process. Also, an appearance of new peak with maximum wavelength at 290, corresponding to the formation of aliphatic by-products, was observed after 40. minute of treatment in the case $UV/Fe^0/H_2O_2$ process.

Mineralization of RB137 solution (80 mgL^{-1}) by those two processes is showed in Figure 10. Again, the pronounced difference in process effectiveness can be observed after 10. minute of treatment, where the inhibition of Fe^0/H_2O_2 process can be observed.

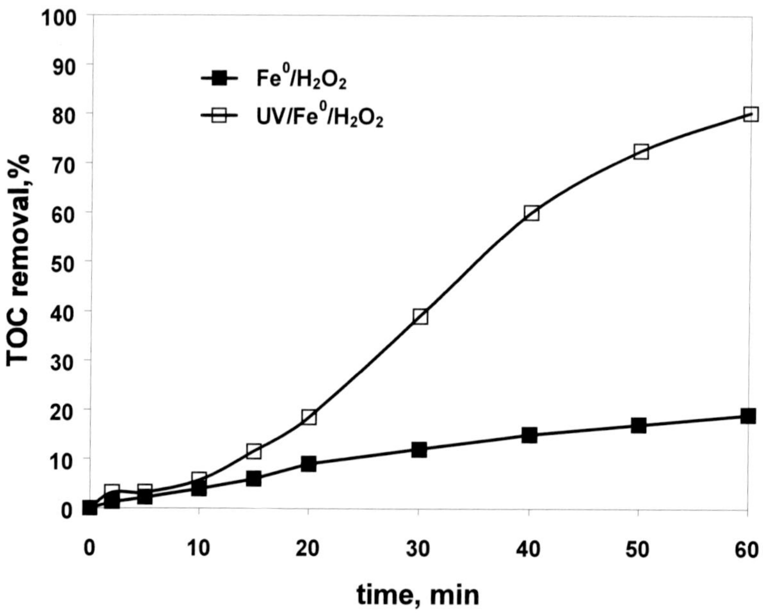

Figure 10. Mineralization of RB137 with initial concentration 80 mgL^{-1} by Fe^0/H_2O_2 and $UV/Fe^0/H_2O_2$ processes monitoring decrease of TOC ecological parameter.

In the case of UV/Fe^0/H_2O_2 process strong inhibition like in the case of lower initial dye concentration is not observed, but it is obvious that the rate of mineralization process is lowering by increasing of mineralization degree of model solution. It should be noted that final achieved mineralization extents; 19.1 and 80.3 % of TOC removal by Fe^0/H_2O_2 and UV/Fe^0/H_2O_2 process, respectively, are not four times lower than those obtained in the case of lower dye concentration, 40.8 and 83.5 %. Moreover, overall removed organic content expressed in mgC per L are much higher in the case of higher dye concentration: 4.9 >> 2.6 in the case of Fe^0/H_2O_2 and 20.4 >> 5.3 in the case of UV/Fe^0/H_2O_2 process. Such a large difference in the overall amount of mineralized dye might be due to the fact that besides OH radicals some other radical species, produced by the destruction of chromophoric parts and aromatic rings of the dye molecules and dye by-products, may be involved in further chain reactions of the dye degradation mechanism. Namely, with higher dye concentration, after initial cleavage of the chromophoric parts of the dye molecules, more radical species originated from chromophores and aromatic by-products destruction are present in the bulk and may participate in further dye degradation/mineralization. A similar effect is known from the literature. Chen and Pignatello [47] reported the formation and the involvement of organic radical species in phenol degradation mechanism by OH radicals. Furthermore, the simple mechanism proposed for the UV radiation chemistry is very similar. When UV radiation is incident upon the dye molecule, the radiation can be adsorbed, promoting the molecule to an excited state which can be involved in further degradation of dye molecules via radical mechanism [35, 159].

***UV-based Process*es**

Next group of investigated AOPs were photochemical processes: UV/O_3 and UV/O_3/H_2O_2. Like in Fenton type processes, the efficiency of UV-based processes strongly depends on process parameters such as: light intensity, reactor configuration, pH, type of oxidant, concentration of oxidant. Therefore, it was necessary to find optimal parameters for those which are changeable in our case; pH and initial H_2O_2 concentration (when added), while other process parameters, such as ozone feed concentration, light intensity, were held constant. In Figure 11. final color removals, TOC removals after 30 and 60 minutes treatment, and final pH values after the treatment of RB137 model wastewater (20 mgL^{-1}) by UV/O_3 process through initial pH range between 3 and 11 are presented. According to the complete color removal, achieved in <10 minutes of treatment, through entire pH investigated, it can be concluded that bleaching is not influenced by initial pH value. On the other hand, presented results of TOC removal after 30 and 60 minutes treatment were influenced by initial pH value, particularly at strong basic conditions.

The optimal pH area for the mineralization of RB137 is found to be between 6-9, i.e. at neutral and weak basic conditions, with the highest TOC removal, 88.7 %, achieved after 60 minutes at pH 8. It can be seen that the mineralization process is much faster in first 30 minutes of treatment process than in next 30 minutes. That can be attributed to the formation of large quantities of organic acids, including those resistant to OH radical degradation, in the last 30 minutes of treatment with significantly lower mineralization rate. The decrease in pH value after the treatment in comparison to initial value corresponds to the formation of acid by-products, mostly aliphatic [92].

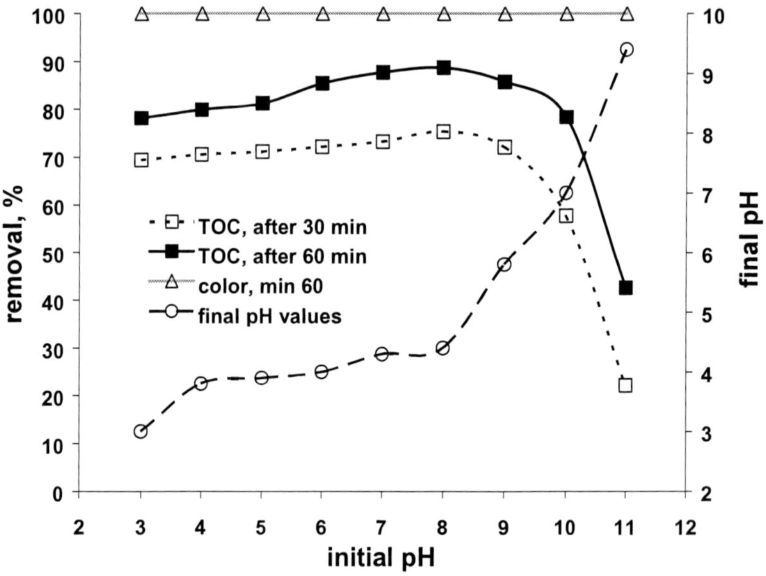

Figure 11. Influence of the initial pH value on the decolorization and mineralization efficiency, and final pH values after a one-hour treatment by UV/O_3 process.

After optimizing UV/O_3 process regarding initial pH value, the kinetic of RB137 model solution (20 mgL^{-1}) decolorization and mineralization, as well changes of pH value during the treatment by UV/O_3 process (pH 8) were monitored (Figure 12.A). It can be seen that complete bleaching was achieved even after 5 minutes of treatment, while the rate of RB137 mineralization was decreased when treatment process progressed. That is particularly pronounced in last 30 minutes of treatment time. As it was mentioned above, the reason for such system behavior could be the formation of large quantities of organic acids, mostly aliphatic, as by-products after initial cleavage of azo bonds in dye molecules, and subsequent formation of aromatic by-products, and thereafter their degradation [71, 160]. Some of formed acids are recalcitrant to degradation through OH radial degradation mechanism [37]. However, they can be destroyed by UV irradiation, but slower than other organic content by OH radicals.

Additional reason for the slowing down of mineralization process could be found in the fact that mineralization efficiency of UV/O_3 process is slightly lower in strong acidic pH region than at neutral and weak basic pH (Figure 11.). Formation of acid by-products is also confirmed by large drop in pH value. As treatment process progressed, pH value started slightly to increase, that can be attributed to the mineralization of acid by-products. In Figure 12.B are presented results of experiments with the addition of synthetic zeolites NH_4ZSM5 and HY to UV/O_3 process. These two types of zeolites were added to the oxidative system combining UV irradiation and ozone due to the observed positive effect of those zeolite types on decolorization of model wastewaters containing reactive dyes in hybrid-parallel corona reactor, which involves, besides classical corona discharge in liquid and gas phase, UV irradiation in liquid, as well as ozone formation in gas and its transfer to liquid phase [132, 133]. It should be noted, that like in the case of NH_4ZSM5 zeolite, series of experiments for the determining adsorption characteristics of HY zeolite for RB137 removal from aqueous solution were conducted showing no dye adsorption/penetration on/in zeolite. From Figure 12.B it can be seen that the addition of synthetic zeolites showed slight negative influence to the rate of RB137 decolorization and mineralization. That can be explained by the fact that efficiency of processes based on UV irradiation strongly depends on solution turbidity, i.e. solution transparency and clearness [35, 59].

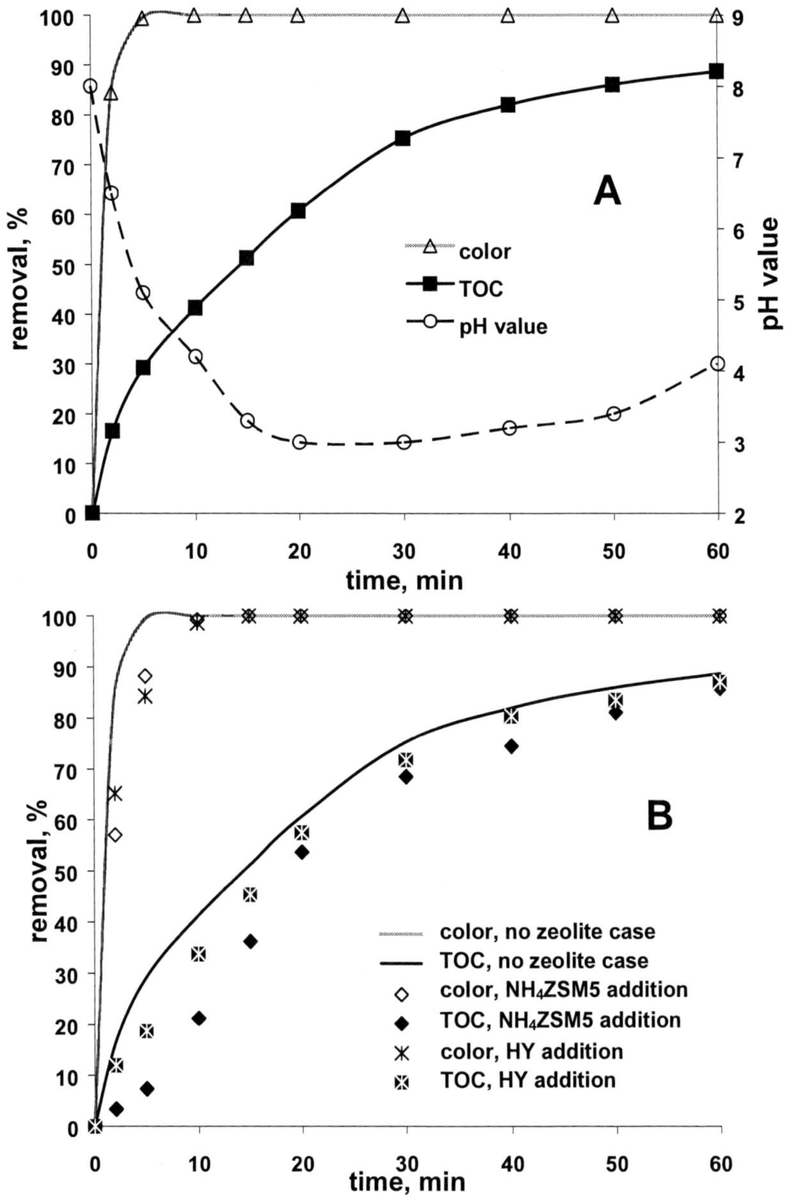

Figure 12. RB137 (20 mgL^{-1}) decolorization and mineralization efficiency as well as changes of pH values during the treatment (A), and influence of NH_4ZSM5 and HY zeolites on RB137 degradation efficiency (B) by UV/O_3 process (pH 8).

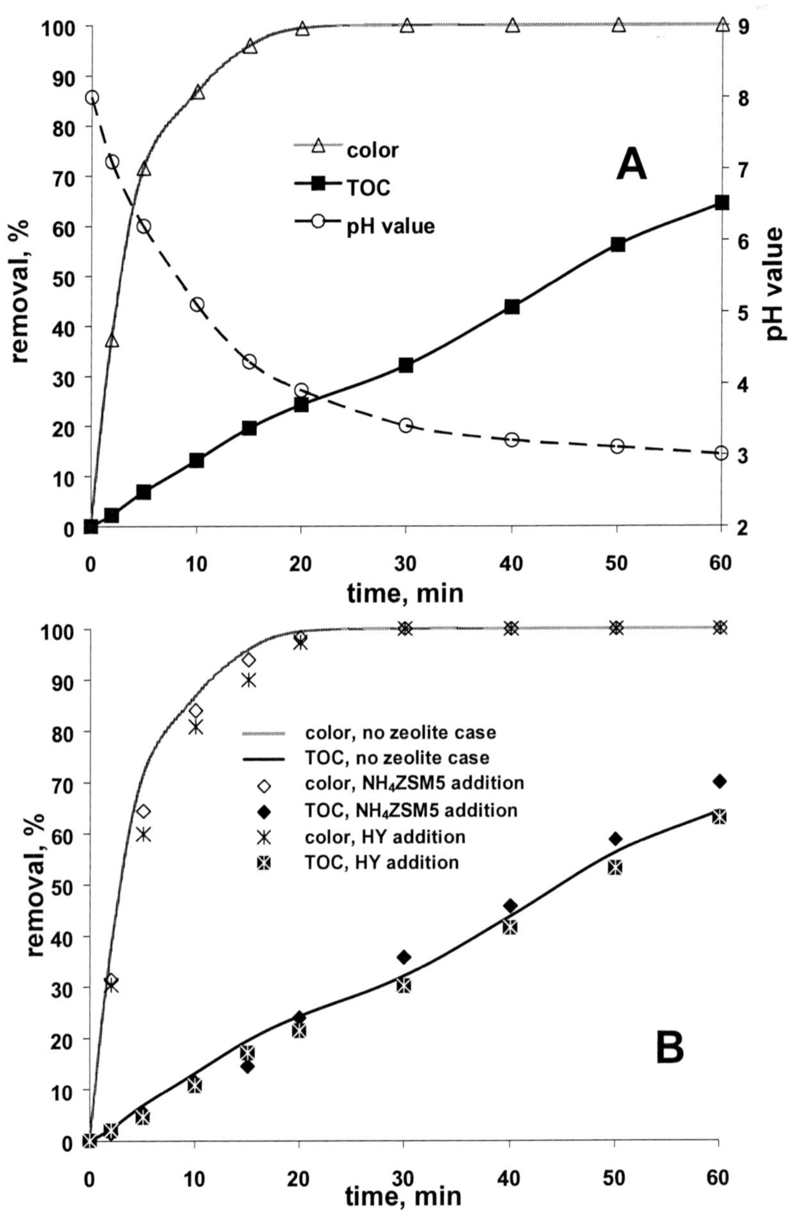

Figure 13. RB137 (80 mgL^{-1}) decolorization and mineralization efficiency as well as changes of pH values during the treatment (A), and influence of NH_4ZSM5 and HY zeolites on RB137 degradation efficiency (B) by UV/O_3 process (pH 8).

The efficiency of UV/O_3 process (pH 8) for decolorization and mineralization of RB137 was investigated at higher dye loading of model wastewater, 80 mgL^{-1} (Figure 13.). It can be seen that complete bleaching was achieved after 20 minutes of treatment, while final extent of partial mineralization after a one-hour treatment is 64.4 % TOC removal. Trend of the curve that connects points of TOC removal extents measured during the treatment is almost linear, i.e. very different than that in the case of lower initial dye concentration. Although drop in pH value indicate the formation of acidic by-products, the mineralization of RB137 model solution was not slowed, indicating that the most of formed by-products are not resistant to OH radical attack. But, with further process progressing over 60 minutes, the slowing down of mineralization process could be expected, due to the fact that Hoigne and Bader [162] pointed out that organic acids with low reaction rate constants under UV irradiation and OH radical resistant, such as acetic and oxalic, always accumulate as final products when any type of reactive organic aqueous solutes are degraded by ozone or OH radicals in water. Hence, the overall oxidation of organic material to CO_2 and H_2O will be delayed whenever the oxidations lead to oxalic or acetic acid as intermediates, which take part at acids conditions. Again, like in the case of lower dye loading, the influence of synthetic zeolites to UV/O_3 process efficiency was investigated (Figure 13.B). It can be seen that zeolite addition caused slight delay in decolorization of RB137 model wastewater (80 mgL^{-1}). That can be explained by earlier mentioned increase in the solution turbidity, that decrease UV irradiation efficiency for ozone photolysis resulting with slower generation of OH radicals.

On the other hand, the addition of NH_4ZSM5 zeolite showed slight positive effect to TOC removal in treatment period after 30. minute, but that was not the case with HY zeolite. If addition of zeolites negatively influence decolorization, it can be assumed that negatively effects the mineralization too, due to the same reason; increase in system turbidity. But, such different system behavior concerning the addition of different type of zeolite could be explained with the adsorption/penetration of organic molecules formed as dye degradation by-products with appropriate size on/in zeolite NH_4ZSM5. Namely, these two zeolites are different according to pore size and structure, Si/Al ratio, surface area, etc. ZSM5 type zeolite has a pentasyl structure which is characterized by parallel channels with pores sizes 0.53-0.56 nm with crossed channels with pores sizes 0.51-0.55 nm, while Y zeolite has a basic cubic structure with pores sizes 0.74 nm [148-150]. As it was proposed above in the case of $Fe^0/H_2O_2/NH_4$ZSM5, NH_4ZSM5 zeolite showed ability to adsorb mono-substituted benzenes. According to the results presented in Figure 13.B, where TOC removal raised over that obtained in the non-zeolite case, it can be assumed that at particular treatment time, formed mono-substituted benzenes as by-products of dye degradation adsorbed on zeolites or even penetrated in zeolite pores. Such system behavior in process using ozone is already observed in our previous study [117].

In the same manner like UV/O_3 process, UV/O_3/H_2O_2 process was optimized regarding initial pH value, but also regarding to the other important changeable process parameters; initial H_2O_2 concentration. The results of optimization study are presented in Figure 14. Firstly, UV/O_3/H_2O_2 process was optimized concerning initial H_2O_2 concentration by changing that parameter and keeping others constant, including initial pH too, at zero pH charge of RB137 model wastewater (pH 6). It can be seen that the complete color removal was achieved through entire range of initial H_2O_2 investigated (0.25-10 mM). Moreover, that is achieved in very short treatment period, <10 minutes. Hence, it can be concluded that initial H_2O_2 concentration does not influence to much on decolorization efficiency of UV/O_3/H_2O_2 process. But, as it can be seen, initial H_2O_2 concentration strongly influence the mineralization efficiency. By increasing the initial H_2O_2 concentration, RB137 mineralization efficiency of UV/O_3/H_2O_2 process also increased, but only to the certain level, where further increase of H_2O_2 concentration negatively influences to the system effectiveness. Explanation for such system behavior is given as follows: slight enhancement of mineralization extents in the first part of the range of initial H_2O_2 concentration studied (0.25-1.0 mM), could be contributed to an increase of OH radical concentration in the bulk according to H_2O_2

photolysis mechanism, shown by equation (8). The decrease of RB137 degradation efficiency by $UV/O_3/H_2O_2$ process with further increasing of the initial H_2O_2 concentration could be the result of H_2O_2 bulk concentration higher than 1.0 mM. When hydrogen peroxide is in excess, it behaves as a scavenger, consequently lowering the hydroxyl radicals concentration and decreasing the overall mineralization efficiency of applied process [78, 87]. Similar system behavior was observed during the optimization of H_2O_2 concentration in Fenton type process (Figure 6.). By comparing mineralization extents of RB137 model wastewater by UV/O_3 at pH 6 (Figure 11.) and by $UV/O_3/H_2O_2$ at the same pH with optimal concentration of H_2O_2 (Figure 14.A), it can be seen that almost 20 % higher TOC removal was obtained by $UV/O_3/H_2O_2$ process.

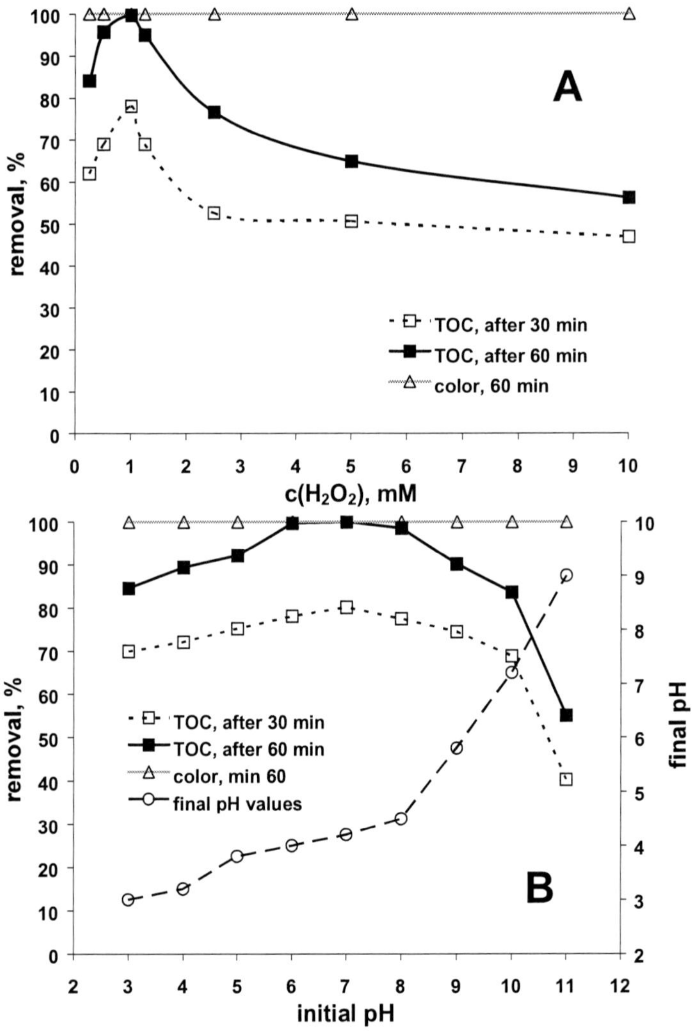

Figure 14. Influence of the initial H_2O_2 concentration (A), and initial pH value (B) on the decolorization and mineralization efficiency, and final pH values after a one-hour treatment by $UV/O_3/H_2O_2$ process.

In Figure 14.B the optimization of $UV/O_3/H_2O_2$ process regarding to the initial pH value is presented. That was performed in the way that pH values were changed from 3-11, while initial H_2O_2 concentration was held constant at 1.0 mM, found as optimal in the previous set of experiments (Figure 14.A). Again, complete bleaching was achieved in very short treatment period through complete range of pH investigated. According to mineralization extents achieved after 30. and 60. of minute treatment, the optimal pH region for degradation of RB137 by $UV/O_3/H_2O_2$ process was established. It can be observed similar system behavior like in the case of UV/O_3 process (Figure 11.); a significant decrease in process effectiveness at strong basic conditions, and the highest mineralization efficiency at neutral and weak basic pH region. But it should be pointed out that the addition of H_2O_2 to UV/O_3 system improved mineralization efficiency through complete investigated pH range. Moreover, the complete mineralization was achieved after a one-hour treatment in the pH region between 6 and 8. Due to such system behavior, optimal pH was found out according to the mineralization extents obtained after 30 minutes treatment time. On that way, it was found the highest mineralization of RB137, 80.1 % TOC removal, by $UV/O_3/H_2O_2$ process ($c(H_2O_2)$=1.0 mM) at pH 7. It should be noted that the trend of curve that connects points of final pH values measured after a one-hour treatment by $UV/O_3/H_2O_2$ process (Figure 14.B) is very similar to that obtained by UV/O_3 (Figure 11.).

Degradation kinetics of RB137 (20 mgL^{-1}) by $UV/O_3/H_2O_2$ process at optimal process conditions; pH 7 and $c(H_2O_2)$=1.0 mM, is showed in Figure 15.A. It can be seen that complete bleaching was achieved after 5 minutes of treatment. The same system behavior, like in the case of the treatment of RB137 (20 mgL^{-1}) by UV/O_3 process, slowing down of mineralization process with the progress of treatment process can be observed. However, complete mineralization was achieved. Decrease in the rate of mineralization is due to the earlier mentioned reason, production of large quantities of organic acids with lower rate of degradation. The drop in pH during the treatment time corresponds to the formation of acid by-products, while the achievement of complete mineralization resulted with reaching the initial pH 7 at the end of the process. Like in the case of UV/O_3/zeolite system for degradation of RB137 (20 mgL^{-1}) (Figure 12.B), the same system behavior could be observed in the case of $UV/O_3/H_2O_2$/zeolite (Figure 15.B). Moreover, negative influence of synthetic zeolites NH_4ZSM5 and HY addition is even more pronounced to RB137 decolorization and mineralization efficiency.

The $UV/O_3/H_2O_2$ process efficiency was tested for decolorization and mineralization of RB137 model wastewater with higher dye loading (Figure 16.). In all three cases: non-zeolite (Figure 16.A), with the addition of NH_4ZSM5 and with addition of HY zeolite (Figure 16.B) the very similar system behavior like that showed in the case of UV/O_3 process (Figure 13.) can be observed.

Complete bleaching with non-zeolite case was achieved after 20 minutes of treatment, trend of the curve which connects points of measured TOC removals is almost linear, zeolite addition caused delay in decolorization of RB137, while NH_4ZSM5 zeolite addition showed positive effect to TOC removal after 30. minute of process. But it should be pointed out that final achieved mineralization, 69.3 % TOC removal, by $UV/O_3/H_2O_2$ process is somewhat higher than that obtained by UV/O_3 process, 64.4 % TOC removal.

AOPs Cost Estimation

The efficiency of studied AOPs in laboratory scale: Fe^0/H_2O_2, $UV/Fe^0/H_2O_2$, UV/O_3 and $UV/O_3/H_2O_2$, for the treatment of RB137 model wastewater was estimated from the point of view of required operating costs. The evaluation was performed on the basis of the obtained results in the case of higher dye loading in colored model wastewater. It should be noted that evaluation of only operating costs was performed. The overall costs are represented by the sum of the capital costs, the operating costs and the maintenance.

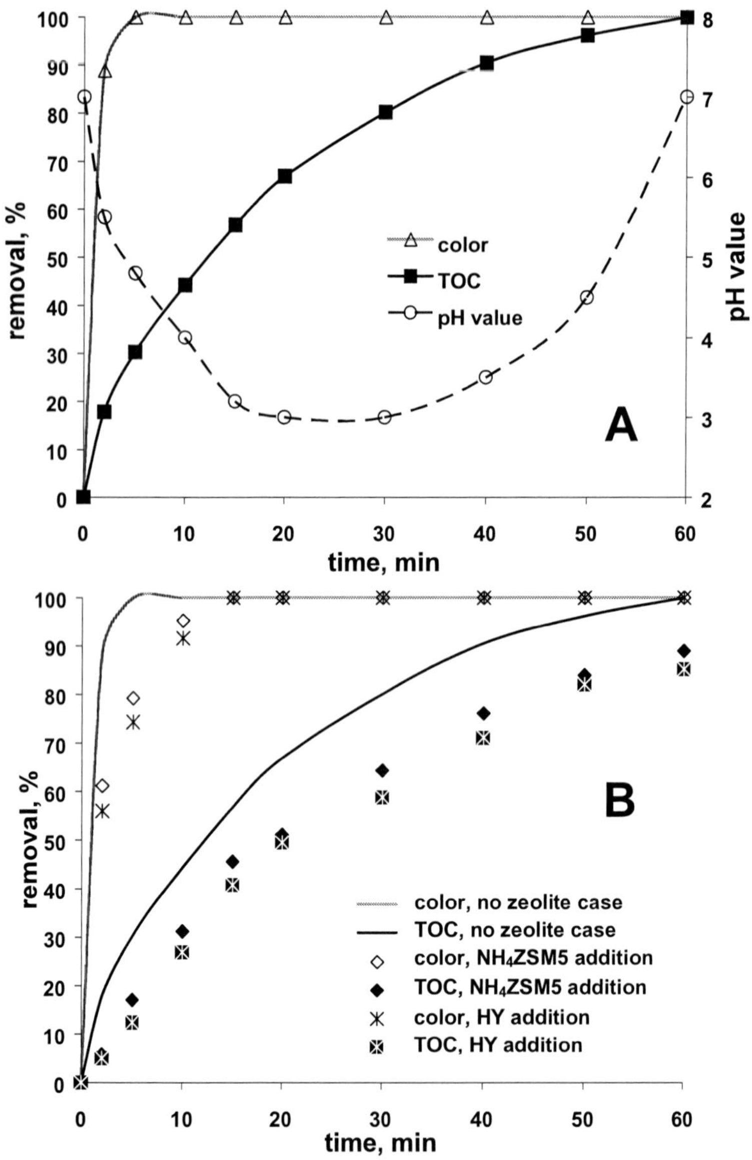

Figure 15. RB137 (20 mgL^{-1}) decolorization and mineralization efficiency as well as changes of pH values during the treatment (A), and influence of NH_4ZSM5 and HY zeolites on RB137 degradation efficiency (B) by $UV/O_3/H_2O_2$ process (pH 7 and $c(H_2O_2)$=1.0 mM).

For a full-scale system these costs will, besides the nature and the concentration of pollutants, strongly depend on the flow rate of the effluent and the configuration of the reactor [38]. In order to evaluate the operating costs for proposed processes, the costs of used reagents valid for Croatian market were taken into account (Table 5).

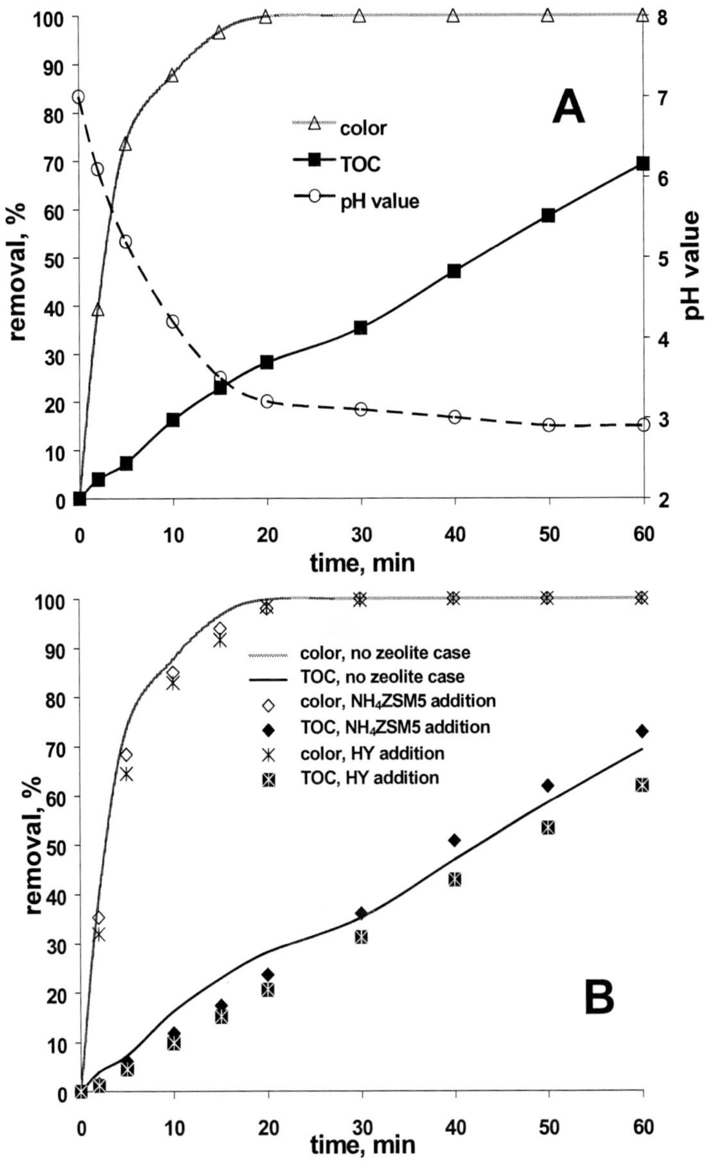

Figure 16. RB137 (80 mgL^{-1}) decolorization and mineralization efficiency as well as changes of pH values during the treatment (A), and influence of NH_4ZSM5 and HY zeolites on RB137 degradation efficiency (B) by $UV/O_3/H_2O_2$ process (pH 7 and $c(H_2O_2)$=1.0 mM)

Table 5. Cost of reagents used in applied processes

Reagent	Basis	Cost (US$)
Fe^0	100 g	13.62
H_2O_2, 30 %	1 L	37.59
O_2, >99.5 %	100 L	0.23
electricity	1 kWh	0.16

Note: costs are given in US$ and the values are valid for Croatian market.

The main estimation results are values of ratio *mineralization extent/process operating costs* (ME/POC). Besides costs estimation on the basis of mineralized organic content, the cost for decolorization below visibility limit, <1 mgL^{-1} that corresponds to decolorization below 1.25 % in our case, are also showed.

From the point of view of the most suitable processes for decolorization of RB137 model wastewater below visibility limit, it can be seen that processes combining ozone and UV irradiation, UV/O_3 and $UV/O_3/H_2O_2$, are double cheaper than processes using iron powder, and more important, decolorization is even faster. Moreover, by Fe^0/H_2O_2 process the complete decolorization was not achieved even after a one-hour treatment.

From the ratio ME/POC point of view, it can be seen that processes using UV irradiation are much more efficient, with ME/POC ratios ranged between 3.20-3.45 US$ g^{-1}, than that one conducted in dark. Moreover, UV irradiation could provide complete mineralization, which is not the case when only degradation over OH radicals is used. However, processes combining UV and ozone seem to be more appropriate due to that successful maintenance of $UV/Fe^0/H_2O_2$ process required strong acid conditions, pH 3. That could increase operating costs of treatment process, but also includes post-treatment as necessary due to the strong conditioning to the neutral pH, which also causes additional costs. Furthermore, post-treatment after the treatment by $UV/Fe^0/H_2O_2$ process includes removal of non-reacted iron powder and remaining iron ions from the bulk. On the contrary, UV/O_3 and $UV/O_3/H_2O_2$ processes were maintained at neutral or weak basic conditions, and after the complete mineralization, the pH values will be the same as at the beginning of treatment process. That excludes additional costs for the strong conditioning of solution before and after the treatment. The cost of processes using UV irradiation could be drastically reduced if solar light is used instead artificial UV.

Table 6. Cost evaluation for all applied processes (processes conducted in laboratory scale)

Process	time required for decolorization <1mgL^{-1} (min)	operational cost for decolorization <1mgL^{-1} (US$)	final TOC removal (%)	costs (US$ g^{-1})
Fe^0/H_2O_2	>60	0.046*	19.0	9.37
$UV/Fe^0/H_2O_2$	<30	0.055	80.3	3.20
UV/O_3	<20	0.025	64.4	3.45
$UV/H_2O_2/O_3$	<20	0.028	69.3	3.31

* Indicates treatment time over 60 minutes.

CONCLUSION

This chapter is dedicated to the investigation of highly efficient and destructive methods for the minimization of hazardous liquid waste containing reactive dyes as parent molecules and other contaminants present in such wastewaters. Besides an overview of recent publications concerning worldwide literature about AOPs for colored wastewaters treatment given in this chapter, a large part of our scientific investigations up to now, related to this subject, is presented too. The chapter describes different AOPs and the possibility of their usage for the treatment of colored wastewater containing reactive dyes in order to achieve colorless and non-toxic wastewater effluents at first, and, if it is possible, to reach significant minimization of overall organic content. In the research part of this chapter, several AOPs; Fe^0/H_2O_2, $UV/Fe^0/H_2O_2$, UV/O_3 and $UV/O_3/H_2O_2$, are investigated on degradation of C.I. Reactive Blue 137, azo reactive dye, as model pollutant in colored wastewater. It was shown that studied AOPs are very efficient for the bleaching of RB137 model solution. That is also reported through the review part of this chapter, where chemical and photochemical processes shown as the most efficient among all AOPs. Those findings are very important due to the fact that the present color in wastewaters could cause serious problems to aquatic life in recipient waters. Moreover, after decolorization less complex structures are formed that can be degraded by microorganisms, which is not the case when reactive dyes are present in water as parent molecules. Although AOPs seem to be much more expensive to start than common commercial treatment methods such as filtration, adsorption and biodegradation, they are highly efficient for the degradation of recalcitrant pollutants, such as reactive dyes. Moreover, evaluated values of ratio *mineralization extent/process operating costs* (ME/POC) for $UV/Fe^0/H_2O_2$, UV/O_3 and $UV/O_3/H_2O_2$ processes in the case of RB137 model solution treatment ranged between 3.20-3.45 US$ per gram of mineralized organic content. That seems reasonable if complete mineralization considers, i.e. complete removal of toxic organic substances from wastewater by destructive mechanism of AOPs.

ACKNOWLEDGMENT

We would like to acknowledge financial support from the Ministry of Science, Education and Sport, Republic of Croatia, Project # 0125-018. We are gratefully acknowledged for the support from the National Science Foundation (USA) (INT-0086351) for founding the ozone generator and synthetic zeolites.

REFERENCE

[1] Zheng, Z; Levin, RE; Pinkham, JL; Shetty, K. Decolorization of polymeric dyes by a novel penicillium isolate. *Process Biochemistry,* 1999 34, 31-37.

[2] Zollinger, H. *Colour Chemistry – Synthesis, Properties and Applications of Organic Dyes and Pigments.* New York: VHC Publishers; 1987.

[3] Riefe, A; Freeman, HS. *Environmental Chemistry of Dyes and Pigments.* New York: John Wiley and Sons; 1996.

[4] Phillips, D. Environmentally friendly, productive and reliable: priorities for cotton dyes and processes. *Journal of Society of Dyers and Colourists,* 1996 112, 183-186.

[5] Hunger, K. *Industrial Dyes; Chemistry, Properties, Application.* Weinheim: Wiley-VCH; 2002.

[6] Allen, SJ; Khader, KYH; Bino, M. Electrooxidation of dyestuff in waste waters. *Journal of Chemical and Biochemical Technology,* 1999 62, 111-117.

[7] Wang, C; Yediler, A; Lienert, D; Wang, Z; Kettrup, A. Toxicity evaluation of reactive dyestuffs, auxiliaries and selected effluents in textile finishing industry to luminescent bacteria Vibrio fischeri. *Chemosphere*, 2002 46 (2), 339-344.

[8] Gottlieb, A; Shaw, C; Smith, A; Wheatley, A; Forsythe, S. The toxicity of textile reactive azo dyes after hydrolysis and decolourisation. *Journal of Biotechnology*, 2003 101(1), 49-56.

[9] Nillson, R; Nordlinder, R; Wass, U. Asthma, rhinitis, and dermatitis in workers exposed to reactive dyes. *British Journal of Industrial Medicine*, 1993 50, 65-70.

[10] Arpe, H-J. Ullmann's Encyclopedia of Industrial Chemistry: Volume A2, Amines, Aliphatic to Antibiotics. 5th Edition, Weinheim: Wiley-VCH; 1985.

[11] Kelshaw, P. The future for West European dyestuff manufacturers. *Journal of Society of Dyers and Colourists*, 1998 114, 35-37.

[12] Sincero, AP; Sincero, GA. *Physical-Chemical Treatment of Water and Wastewater*. New York: CRC Press, IWA Publishing; 2003.

[13] Droste, RJ. *Theory and Practice of Water and Wastewater Treatment*. New York: John Wiley and Sons; 1997.

[14] Forgacs, E; Cserhati, T; Oros, G. Removal of synthetic dyes from wastewaters: a review. *Environmental International*, 2004 30, 953-971.

[15] Pagga, U; Brown, D. The degradation of dyestuffs: part ii behavior of dyestuffs in aerobic biodegradation tests. *Chemosphere*, 1986 15, 479–491.

[16] Seshadri, S; Bishop, PI; Agha, AM. Anaerobic/aerobic treatment of selected azo dyes in wastewater. *Waste Management*, 1994 15, 127–137.

[17] Lourenço, ND; Novais, JM; Pinheiro, HM. Effect of some operational parameters on textile dye biodegradation in a sequential batch reactor. *Journal of Biotechnology*, 2001 89, 163-174.

[18] Georgiou, D; Hatiras, J; Aivasidis, A. Microbial immobilization in a two-stage fixed-bed-reactor pilot plant for on-site anaerobic decolorization of textile wastewater. *Enzyme and Microbial Technology*, 2005 37 (6) 597-605.

[19] Ambrósio, ST; Campos-Takaki, GM. Decolorization of reactive azo dyes by Cunninghamella elegans UCP 542 under co-metabolic conditions. *Bioresource Technology*, 2004 91, 69-75.

[20] Ozdemir, O; Armagan, B. Turan, M; Celik, MS. Comparison of adsorption characteristics of azo-reactive dyes on mezoporous minerals. *Dyes and Pigments*, 2004 62, 49-60.

[21] Karcher, S; Kornmüller, A; Jekel, M. Screening of commercial sorbents for the removal of reactive dyes. *Dyes and Pigments*, 2001 51, 111-125.

[22] Wang, S; Li, H; Xu, L. Application of zeolite MCM-22 for basic dye removal from wastewater. *Journal of Colloid and Interface Science*, 2006 295, 71-78.

[23] Benkli, YE; Can, MF; Turan, M; Çelik, MS. Modification of organo-zeolite surface for removal of reactive azo dyes in fixed-bed reactors. *Water Research*, 2005 39, 487-493.

[24] Armaan, B; Turan, M; Cęlik, MS. Equilibrium studies on the adsorption of reactive azodyes into zeolite. *Desalination*, 2004 170, 33-39.

[25] Papić, S; Koprivanac, N; Lončarić Božić, A. Removal of reactive dyes from wastewater using Fe(III) coagulant. *Journal of Society of Dyers and Colourist*, 2000 116, 352-358.

[26] Koprivanac, N, Lončarić Božić, A, Papić, S. Cleaner production processes in the synthesis of blue anthraquinone reactive dye. *Dyes and Pigments,* 2000 44, 33-40.
[27] Papić, S; Koprivanac, N; Lončarić Božić, A; Meteš A. Removal of some reactive dyes from synthetic wastewater by combined Al(III) coagulation/carbon adsorption process. *Dyes and Pigments,* 2004 62, 293-300.
[28] White, GC. Handbook of Chlorination and Alternative Disinfectants. New York: John Wiley & Sons; 1999.
[29] La Grega, MD; Buckingham, PL; Evans, JC. *Hazardous Waste Management.* New York: McGraw-Hill; 1994.
[30] Chu, W; Ma, C-W. Quantitative prediction of direct and indirect dye ozonation kinetics. *Water Research,* 2000 34 (12), 3153-3160.
[31] Parsons, SA; Williams, M. Introduction. In: Parsons SA, editor. *Advanced oxidation processes for water and wastewater treatment.* London: IWA Publishing; 2004; 1-6.
[32] Gogate, PR; Pandit, AB. A review of imperative technologies for wastewater treatment I: oxidation technologies at ambient conditions. *Advances in Environmental Research,* 2004 8, 501-551.
[33] Gogate, PR; Pandit, AB. A review of imperative technologies for wastewater treatment II: hybrid methods. *Advances in Environmental Research,* 2004 8, 553-597.
[34] Glaze, WH; Kang, JW; Chapin, DH. The chemistry of water treatment processes involving ozone, hydrogen peroxide and ultraviolet irradiation. *Ozone Science and Engineering,* 1987 9, 335–352.
[35] Beltran, FJ. Ozone-UV radiation-hydrogen peroxide oxidation technologies. In: Tarr MA, editor. *Chemical degradation methods for wastes and pollutants - environmental and industrial applications*. New York: Marcel Dekker; 2003; 1-75.
[36] Rodriguez, M. Fenton and UV-VIS based advanced oxidation processes in wastewater treatment: Degradation, mineralization and biodegradability enhancement. PhD Thesis. Barcelona: University of Barcelona, 2003.
[37] Bigda, RJ. Consider Fenton's chemistry for wastewater treatment. *Chemical Engineering Progressing,* 1995 91, 62-66.
[38] Andreozzi, R; Vaprio, V; Insola, A; Marotta, R. Advanced oxidation processes (AOP) for water purification and recovery. *Catalysis Today,* 1999 53, 51-59.
[39] Grymonpre, DR; Finney, WC; Clark, RJ; Locke BR. Hybrid gas-liquid electrical discharge reactors for organic compound degradation, *Industrial and Engineering Chemistry Research,* 2004 43, 1975-1989.
[40] Locke, BR; Sato, M; Sunka, P; Hoffmann, MR, Chang; JS. Electrohydraulic discharge and nonthermal plasma for water treatment. *Industrial and Engineering Chemistry Research,* 2006 45 (3), 882-905.
[41] Fenton, HJH. On a new reaction of tartaric acid. *Chemical News,* 1876 33, 190.
[42] Fenton, HJH. Oxidation of tartaric acid in presence of iron. *Journal of Chemical Society,* 1894 65, 899-910.
[43] Tarr, MA. Fenton and modified Fenton methods for pollutant degradation. In: Tarr MA, editor. *Chemical degradation methods for wastes and pollutants - environmental and industrial applications.* New York: Marcel Dekker; 2003; 1-77.
[44] Xu, X-R; Li, H-B; Wang, W-H; Gu, J-D. Degradation of dyes in aqueous solutions by the Fenton process. *Chemosphere,* 2004 57, 595-600.

[45] Lucas, MS; Peres JA. Decolorization of the azo dye Reactive Black 5 by Fenton and photo-Fenton oxidation. *Dyes and Pigments,* 2006 71, 236-244.

[46] Meric, S; Selcuk, H; Gallo, M; Belgiorno V. Decolourisation and detoxifying Remazol Red dye and mixture using Fenton's reagent. *Desalination*, 2005 173, 239-248

[47] Chen, R; Pignatello, JJ. Role of quinone intermediates as electron shuttles in Fenton and photoassisted Fenton oxidations of aromatic compounds. *Environmental Science Technology,* 1997 31, 2399-2406.

[48] Kušić, H; Koprivanac, N; Lončarić Božić, A; Selanec, I. Photo-assisted Fenton type processes for the degradation of phenol: a kinetic study. *Journal of Hazardous Materials,* 2006 *in press.*

[49] Kusic, H; Loncaric Bozic, A; Koprivanac, N. Fenton type processes for minimization of organic content in coloured wastewaters: Part I: processes optimization. *Dyes and Pigments,* 2006 *in press.*

[50] Gallard, H; De Laat, J. Kinetics of oxidation of chlorobenzenes and phenyl-ureas by Fe(II)/H_2O_2 and Fe(III)/H_2O_2. Evidence of reduction and oxidation reactions of intermediates by Fe(II) and Fe(III). *Chemosphere*, 2001 42, 405-413.

[51] Neamtu, M; Yediler, A; Siminiceanu, I; Kettrup, A. Oxidation of commercial reactive azo dye aqueous solutions by the photo-Fenton and Fenton-like processes. *Journal of Photochemistry and Photobiology A: Chemistry,* 2003 161, 87–93.

[52] Kusic, H, Koprivanac, N; Srsan, L. Azo dye degradation using Fenton type processes assisted by UV irradiation: A kinetic study. *Journal of Photochemistry and Photobiology A: Chemistry,* 2006 181 (2-3), 195-202.

[53] Bergendahl, JA; Thies, TP. Fenton's oxidation of MTBE with zero-valent iron. *Water Research,* 2004 38, 327-334.

[54] Tang, WZ; Chen, RZ. Decolorization kinetics and mechanism of commercial dyes by H_2O_2/iron powder system. *Chemosphere*, 1996 32 (5), 947-958.

[55] Costa, RCC; Lelis, MFF; Oliveira, LCA; Fabris, JD; Ardisson, JD; Rios, RRVA; Silva, CN; Lago, RM. Novel active heterogeneous Fenton system based on $Fe_{3-x}M_xO_4$ (Fe, Co, Mn, Ni): The role of M^{2+} species on the reactivity towards H_2O_2 reactions. *Journal of Hazardous Materials,* 2006 129, 171-178.

[56] Moura, FCC; Araujo, MH; Costa, RCC; Fabris, JD; Ardisson, JD; Macedo, WAA; Lago, RM. Efficient use of Fe metal as an electron transfer agent in a heterogeneous Fenton system based on Fe^0/Fe_3O_4 composites. *Chemosphere*, 2005 60, 1118-1123.

[57] He, J; Ma, W; Song, W; Zhao, J; Qian, X; Zhang, S; Yu JC. Photoreaction of aromatic compounds at α-FeOOH/H_2O interface in the presence of H_2O_2: evidence for organic-goethite surface complex formation. *Water Research,* 2005 39, 119-128.

[58] Phu, NH; Hoa, TTK; Tan, NV; Thang HV; Ha, PL. Characterization and activity of Fe-ZSM-5 catalysts for the total oxidation of phenol in aqueous solutions. *Applied Catalysis, B: Environmental, 2001* 34, 267-275.

[59] Kušić, H; Koprivanac, N; Selanec, I. Fe-exchanged zeolite as the effective heterogeneous Fenton-type catalyst for the organic pollutant minimization: UV irradiation assistance, *Chemosphere*, 2006 *in press*

[60] Catrinescu, C; Neamtu, M; Yediler, A; Macoveanu, M; Kettrup, A. Catalytic wet peroxide oxidation of an azo dye, Reactive Yellow 84, over Fe-exchanged ultrastable Y zeolite. *Environmental Engineering and Management Journal,* 2002 1, 177-186.

[61] Catrinescu, C; Teodosiu, C, Macoveanu, M; Miehe-Brendlé, J; Le Dred, R. Catalytic wet peroxide oxidation of phenol over Fe-exchanged pillared beidellite. *Water Research,* 2003 37, 1154-1160.

[62] Ly, X; Xu, Y; Ly, K; Zhang, G. Photo-assisted degradation of anionic and cationic dyes over iron(III)-loaded resin in the presence of hydrogen peroxide. *Journal of Photochemistry and Photobiology, A: Chemistry,* 2005 173, 121-127.

[63] Verma, P; Baldrian, P; Nerud, F. Decolorization of structurally different synthetic dyes using cobalt(II)/ascorbic acid/hydrogen peroxide system. *Chemosphere*, 2003 50, 975-979

[64] Kavitha, V; Palanivelu, K. The role of ferrous ion in Fenton and photo-Fenton processes for the degradation of phenol. *Chemosphere*, 2004 55, 1235-1243.

[65] www.h2o2.com

[66] Arslan, I; Balcioglu, IA. Degradation of Remazol Black B dye and its simulated dyebath wastewater by advanced oxidation processes in heterogeneous and homogeneous media. *Coloration Technology,* 2001 117, 38-42.

[67] Hsueh, CL; Huang, YH; Wang, CC; Chen; CY. Degradation of azo dyes using low iron concentration of Fenton and Fenton-like system. *Chemosphere*, 2005 58 (10), 1409-1414.

[68] Carneiro, PA; Pupo Nogueira, RF; Zanoni, MVB. Homogeneous photodegradation of C.I. Reactive Blue 4 using a photo-Fenton process under artificial and solar irradiation. *Dyes and Pigments*, 2006 *in press*.

[69] Hsueh, C-L; Huanga, Y-H; Wang, C-C; Chena, C-Y. Photoassisted Fenton degradation of nonbiodegradable azo-dye (Reactive Black 5) over a novel supported iron oxide catalyst at neutral pH. *Journal of Molecular Catalysis A: Chemical,* 2006 245, 78–86.

[70] Muruganandham, M; Swaminathan, M. Decolourisation of Reactive Orange 4 by Fenton and photo-Fenton oxidation technology. Dyes and Pigments, 2004 63, 315-321.

[71] Stylidi, M; Kondarides, DI; Verykios, XE. Pathways of solar light-induced photocatalytic degradation of azo dyes in aqueous TiO_2 suspensions. *Applied Catalysis B: Environmental,* 2003 40, 271-286.

[72] Stefan, MI. UV photolysis: background. In: Parsons SA, editor. *Advanced oxidation processes for water and wastewater treatment.* London: IWA Publishing; 2004; 7-48.

[73] Legrini, O; Oliveros, E; Braun, AM. Photochemical process for water treatment. *Chemical Review,* 1993 93 (2), 671-698.

[74] Arslan, I; Balcioglu, IA; Bahnemann, D. Advanced chemical oxidation of reactive dyes in simulated dyehouse effluents by ferrioxalate–Fenton/UV-A and TiO_2/UV-A processes. *Water Research,* 2001 37, 3061-3069.

[75] Muruganandham, M; Swaminathan, M. Solar driven decolourisation of Reactive Yellow 14 by advanced oxidation processes in heterogeneous and homogeneous media. *Dyes and Pigments,* 2007 72 (2), 137-143.

[76] Bali, U; Catalkaya, E; Sengul F. Photodegradation of Reactive Black 5, Direct Red 28 and Direct Yellow 12 using UV, UV/H_2O_2 and UV/H_2O_2/Fe^{2+}: a comparative study. *Journal of Hazardous Materials,* 2004 B114, 159–166.

[77] Behnajady, MA; Modirshahla, N; Fathi H. Kinetics of decolorization of an azo dye in UV alone and UV/H_2O_2 processes. *Journal of Hazardous Materials,* 2006 133 (1-3), 226-232.

[78] Peternel, I; Koprivanac, N; Kusic, H. UV-based processes for reactive azo dye mineralization. *Water Research,* 2006 40 (3), 525-532.
[79] Kušić, H; Koprivanac, N; Lončarić Božić, A; Papić, S; Peternel, I; Vujević, D. Reactive dye degradation by photochemical AOP – development of a kinetic model for UV/H_2O_2 process. *Chemical and Biochemical Engineering Quarterly,* 2006 *in press.*
[80] Neppolian, B; Choi, HC; Sakthivel, S; Arabindoo, B; Murugesan, V. Solar/UV-induced photocatalytic degradation of three commercial textile dyes. *Journal of Hazardous Materials,* 2002 B89, 303-317.
[81] Tuhkanen, T. UV/H_2O_2 process. In: Parsons SA, editor. *Advanced oxidation processes for water and wastewater treatment.* London: IWA Publishing; 2004; 86-110.
[82] Hunt, JP; Taube, H. The photochemical decomposition of hydrogen peroxide, Quantum yields, transfer and fractionation effects. *Journal of American Chemical Society,* 1952 74, 5999-6002.
[83] Eckenfelder, WW. *Industrial Water Pollution Control.* 2nd Edition. New York: McGraw-Hill; 1989.
[84] Crittenden, C; Hu, S; Hand, DW; Green, SA. A kinetic model for H_2O_2/UV process in a completely mixed batch reactor. *Water Research,* 1999 33 (10), 2315-2328.
[85] De, AK; Chaudhuri, B; Bhattacharjee, S; Dutta, BK. Estimation of $OH^{\bullet}$ radical reaction rate constants for phenol and chlorinated phenols using UV/H_2O_2 photo-oxidation. *Journal of Hazardous Materials,* 1999 64 (1), 91-104.
[86] Guittonneau, S; De Laat, J; Duguet, JP; Bonnel, C; Dore, M. Oxidation of parachloronitrobenzene in dilute Aqueous solution by O_3 + UV and H_2O_2 + UV: A comparative study, *Ozone Science and Engineering,* 1990 12, 73-94.
[87] Kusic, H; Koprivanac, N; Loncaric Bozic, A. Minimization of organic pollutant content in aqueous solution by means of AOPs: UV- and ozone–based technologies. *Chemical Engineering Journal,* 2006 *in press*
[88] Mohey El-Dein, A; Libra, JA; Wiesmann U. Mechanism and kinetic model for the decolorization of the azo dye Reactive Black 5 by hydrogen peroxide and UV radiation. *Chemosphere*, 2003 52, 1069–1077.
[89] Muruganandham, M; Swaminathan M. Photochemical oxidation of reactive azo dye with UV–H_2O_2 process. *Dyes and Pigments,* 2004 62, 269–275.
[90] Colonna, GM; Caronna, T; Marcandalli, B. Oxidative degradation of dyes by ultraviolet radiation in the presence of hydrogen peroxide. *Dyes and Pigments,* 1999 41, 211-220.
[91] Ince, N. Critical effect of hydrogen peroxide in photochemical dye degradation. *Water Research,* 1999 33, 1080-1084.
[92] Shu, H; Chang, M. Decolorization effects of six azo dyes by O_3, UV/O_3 and UV/H_2O_2 processes. *Dyes and Pigments*, 2005 65, 25-31.
[93] Wu, C-H; Chang, C-L. Decolorization of Reactive Red 2 by advanced oxidation processes: Comparative studies of homogeneous and heterogeneous systems. *Journal of Hazardous Materials,* 2006 B128, 265–272.
[94] Mills, A; Lee, S-K. Semiconductor photocatalysis. In: Parsons SA, editor. *Advanced oxidation processes for water and wastewater treatment.* London: IWA Publishing; 2004; 137-166.
[95] Pichat, P. Photocatalytic degradation of pollutants in water and air:basic concepts and applications. In: Tarr MA, editor. *Chemical degradation methods for wastes and*

pollutants - environmental and industrial applications. New York: Marcel Dekker; 2003; 77-119.

[96] Carneiro, PA; Osugi ME; Sene, JJ; Anderson, MA; Boldrin Zanoni, MV. Evaluation of color removal and degradation of a reactive textile azo dye on nanoporous TiO_2 thin-film electrodes. *Electrochimica Acta,* 2004 49, 3807–3820.

[97] Goncalves, MST; Pinto, EMS; Nkeonye, P; Oliveira-Campos, AMF. Degradation of C.I. Reactive Orange 4 and its simulated dyebath wastewater by heterogeneous photocatalysis. *Dyes and Pigments,* 2005 64, 135-139.

[98] Reutergradh, LB; Iangpashuk, M. Photocatalytic decolorization of reactive azo dye: a Comparison between TiO_2 and CdS photocatalysis, *Chemosphere*, 1997 35, 585-596.

[99] Daneshvar, N; Salari, D; Khataee, AR. Photocatalytic degradation of azo dye Acid Red 14 in water on ZnO as an alternative catalyst to TiO_2. *Journal Photochemistry and Photobiology A: Chemistry,* 2004 162, 317-322.

[100] Sakthivel, S; Neppolian, B; Shankar, MV; Arabindoo, B; Palanichamy, M, Murugesan, V. Solar photocatalytic degradation of azo dye: Comparison of photocatalytic efficiency of ZnO and TiO_2. *Solar Energy Materials and Solar Cells,* 2003 77, 65-82.

[101] Vinogdopal, K; Wynkoop, DE; Kamat, PV. Environmental chemistry on semiconductor surfaces: Photosensitized degradation of a textile azo dye, Acid Orange 7, on the TiO_2 particles using visible light. *Environmental Science and Technology,* 1996 30, 1660-1666.

[102] Bizani, E; Fytianos, K; Poulios I; Tsiridis V. Photocatalytic decolorization and degradation of dye solutions and wastewaters in the presence of titanium dioxide. *Journal of Hazardous Materials,* 2006 *in press*.

[103] Muruganandham, M; Swaminathan M. TiO_2–UV photocatalytic oxidation of Reactive Yellow 14: Effect of operational parameters. *Journal of Hazardous Materials*, 2006 *in press*.

[104] Peternel, I; Koprivanac, N; Loncaric Bozic, A; Kusic, H. Comparative study of UV/TiO_2, UV/ZnO and UV/Fenton processes for the organic reactive dye in aqueous solution. *Waste Management,* 2006 *submitted*

[105] Akyol A; Yatmaz, HC; Bayramoglu, M. Photocatalytic decolorization of Remazol Red RR in aqueous ZnO suspensions. *Applied Catalysis B: Environmental,* 2004 54, 19–24.

[106] www.medicina.hr

[107] www.tifm.com/uvbozone

[108] Langlais, B; Reckhow, DA; Brink, DR. *Ozone in water treatment; application and engineering.* Boca Raton: Lewis Publishers; 1991.

[109] Tomiyasu, H; Fukutomi, H; Gordon, G. Kinetics and mechanism of ozone decomposition in basic aqueous solution. *Inorganic Chemistry,* 1985 24 (19), 2962-2966.

[110] Mathews, AP; Panda, KK; Ananthi, S; Padmanabhan, K. Mass transfer and oxidation kinetics in an in situ ozone generator. *Water Science and Technology,* 2004 49, 13-18.

[111] Arslan-Alaton, I; Dogruel, S; Baykal, E; Gerone, G. Combined chemical and biological oxidation of penicillin formulation effluent. *Journal of Environmental Management,* 2004 73 (2), 155-163.

[112] Glaze, W; Kang, J. Advanced oxidation processes. Test of a kinetic model for the oxidation of organic compounds with ozone and hydrogen peroxide in a semi batch reactor, *Industrial and Engineering Chemistry Research,* 1989 28, 1580-1587.

[113] Wu, J; Wang, T. Ozonation of aqueous azo dye in semi-batch reactor. *Water Research,* 2001 35 (4), 1093-1099.

[114] Peralta–Zamora, P; Kunz, A; Gomes de Moreas, S; Pelegrini, R; de Campos Moleiro, P; Reyes, J; Durán, N. Degradation of reactive dyes I. a comparative study of ozonation, enzymatic and photochemical processes. *Chemosphere*, 1999 38 (4), 835-852.

[115] Arslan-Alaton, I; Bacioglu, IA; Bahnemann, DW. Advanced oxidation of a reactive dyebath effluent: comparison of O_3, H_2O_2/UV-C and TiO_2/UV-A processes. *Water Research,* 2002 36, 1143-1154.

[116] Arslan, I; Bacioglu, IA. Effect of common reactive dye auxiliaries on the ozonation of dyehouse effluents contaning vinylsulphone and aminochlorotriazine dyes. *Desalination,* 2000 130, 61-71.

[117] Kušić, H; Lončarić Božić, A; Koprivanac, N; Peternel, I; Vujević, D; Papić, S. Comparative study of several ozone based processes for reactive dye degradation. In: Kruithof J, editor. Proceedings of 17th world Congress & exhibition: Ozone & Related Oxidants, Innovative & Current Technologies. *Strasbourg: International Ozone Association;* 2005; VII.3.24.1-10.

[118] Ince, NH; Tezcanli, G. Reactive dyestuff degradation by combined sonolysis and ozonation. *Dyes and Pigments,* 2001 49 (3), 145-153.

[119] Lukes, P. *Water treatment by pulsed streamer corona discharge*. PhD Thesis. Prague: Institute of Plasma Physics AS CR; 2001.

[120] Joshi, AA; Locke, BR; Arce, P; Finney, WC. Formation of hydroxyl radicals, hydrogen peroxide and aqueous electrons by pulsed streamer corona discharge in aqueous solution. *Journal of Hazardous Materials,* 1995 41, 3-30.

[121] Sunka, P; Babicky, V; Clupek, M; Lukes, P; Simek, M; Schmidt, J; Cernak, M. Generation of chemically active species by electrical discharges in water. *Plasma Sources Science and Technology,* 1999 8 (2), 258-265.

[122] Clements, JS; Sato, M; Davis, RH. Preliminary investigation of prebreakdown phenomena and chemical reactions using a pulsed high-voltage discharge in water. IEEE Trans. *Industrial Applications,* 1987 IA-23, 224-235.

[123] Sato, M; Sun, B; Ohshima, T; Sagi, Y. Characteristics of active species and removal of organic compounds by a pulsed corona discharge in water. *Journal of Advanced Oxidation Technologies,* 1999 4, 339-342.

[124] Sun, B; Sato, M; Clements, JS. Optical study of active species produced by a pulsed streamer corona discharge in water. *Journal of Electrostatics,* 1997 39, 189-202.

[125] Grymonpre, DR. *An experimental and theoretical analysis of phenol degradation by pulsed corona discharge*. PhD Thesis. Tallahassee: Florida State University; 2001.

[126] van Craeynest, K; van Langenhove, H; Stuetz, RM. AOPs for VOCs and odour treatment: Non-thermal palsma. In: Parsons SA, editor. *Advanced oxidation processes for water and wastewater treatment.* London: IWA Publishing; 2004; 275-301.

[127] Lukes, P; Appleton, AT; Locke, BR. Hydrogen peroxide and ozone formation in hybrid gas-liquid electrical discharge reactors. *IEEE Trans. Industrial. Applications,* 2004 40 (1), 60–67.

[128] Sugiarto, AT; Ohshima, T; Sato, M. Advanced oxidation processes using pulsed streamer corona discharge in water. *Thin Solid Films*, 2002 407, 174-178.

[129] Sunka, P; Babicky, V; Clupek, M; Fuciman, M; Lukes, P; Simek, M; Benes, J; Locke, BR; Majcherova, Z. Potential applications of pulse electrical discharges in water, *Acta Physica Slovaca,* 2004 54 (2), 135-145.

[130] Loncaric Bozic, A; Koprivanac, N; Sunka, P, Clupek, M; Babicky, V. Organic synthetic dye degradation by modified pinhole discharge. *Czechoslovak Journal of Physics,* 2004 54C, C958-C963.

[131] Koprivanac, N; Kusic, H; Vujevic, D; Peterenel, I; Locke, BR. Influence of iron on degradation of organic dyes in corona. *Journal of Hazardous Materials*, 2005 B117, 113-119.

[132] Kusic, H; Koprivanac, N; Peternel, I; Locke BR. Hybrid gas/liquid electrical discharge reactors with zeolites for colored wastewater degradation. *Journal of Advanced Oxidation Technologies, 2005* 8 (2), 172-181.

[133] Peternel, I; Kusic, H; Koprivanac, N; Locke, BR. The roles of ozone and zeolite on reactive dye degradation in electrical discharge reactors. *Environmental Technology,* 2006 27 (5) 545-558.

[134] Mason, TJ; Petrier, C. Ultrasound processes. In: Parsons SA, editor. *Advanced oxidation processes for water and wastewater treatment.* London: IWA Publishing; 2004; 185-208.

[135] Destaillats, H; Hoffmann, MR; Wallace, HC. Sonochemical degradation of polutants. In: Tarr MA, editor. *Chemical degradation methods for wastes and pollutants - environmental and industrial applications.* New York: Marcel Dekker; 2003; 201-233.

[136] Vajnhandl, S; Majcen Le Marechal, A. Ultrasound in textile dyeing and the decolouration/mineralization of textile dyes. *Dyes and Pigments,* 2005 65, 89-101.

[137] Tezcanli-Guyer, G. Ince, NH. Degradation and toxicity reduction of textile dyestuff by ultrasound. *Ultrasonics Sonochemistry,* 2003 10, 235–240.

[138] Rehorek, A; Tauber M; Gubitz G. Application of power ultrasound for azo dye degradation. *Ultrasonics Sonochemistry,* 2004 11, 177–182.

[139] Cooper, WJ; Gehringer, P; Pikaev, AK; Kurcuz, CN; Mincher, BJ. Radiation processes. In: Parsons SA, editor. *Advanced oxidation processes for water and wastewater treatment.* London: IWA Publishing; 2004; 209-248.

[140] Dajka, K; Takacs, E; Solpan, D; Wojnarovits, L; Guven, O. High-energy irradiation treatment of aqueous solutions of C.I. Reactive Black 5 azo dye: pulse radiolysis experiments. *Radiation Physics and Chemistry,* 2003 67, 535–538.

[141] Zhang, S-J; Yu, H-Q; Li, Q-R. Radiolytic degradation of Acid Orange 7: A mechanistic study. *Chemosphere*, 2005 61, 1003–1011.

[142] Brillas, E; Cabot, P-L; Casado, J. Electrchemical methods for degradation of organic pollutants in aqueous media. In: Tarr MA, editor. *Chemical degradation methods for wastes and pollutants - environmental and industrial applications.* New York: Marcel Dekker; 2003; 235-304.

[143] Catanho, M; Malpass, GRP; Motheo AJ. Photoelectrochemical treatment of the dye reactive red 198 using DSA electrodes. *Applied Catalysis B: Environmental,* 2006 62, 193–200.

[144] Awad, HS; Abo Galwa N. Electrochemical degradation of Acid Blue and Basic Brown dyes on Pb/PbO_2 electrode in the presence of different conductive electrolyte and effect of various operating factors. *Chemosphere*, 2005 61, 1327–1335.

[145] Carneiro, PA; Osugi, ME; Fugivara, CS; Boralle, N; Furlan, M; Zanoni, MVB. Evaluation of different electrochemical methods on the oxidation and degradation of Reactive Blue 4 in aqueous solution. *Chemosphere*, 2005 59, 431–439.

[146] Kusic, H; Lonacaric Bozic, A; Koprivanac, N; Papic, S. Fenton type processes for minimization of organic content in coloured wastewaters: Part II: combination with zeolites. Dyes and Pigments, 2006 *in press*

[147] Papić, S; Koprivanac, N; Lončarić Božić, A; Vujević, D; Dragičević Kučar, S; Kušić, H; Peternel, I. AOPs in azo dye wastewater treatment. Water Environment Research, 2006 78 (6), 572-579.

[148] Tomlinson, AAG. *Modern zeolites structure and function in detergents and petrochemicals.* Zurich; Trans Tech Publications; 1998.

[149] www.bza.org/zeolites

[150] www.zeolyst.com

[151] Nicole, I; De Laat, J; Dore, M, Duguet, JP; Bonnel, C. Utilisation du rayonnement ultraviolet dans le traitement des eaux: mesure du flux photonique par actinometrie chimique au peroxyde d'hydrogene: Use of U.V. radiation in water treatment: measurement of photonic flux by hydrogen peroxide actinometry. *Water Research,* 1990 24 (2), 157-168.

[152] APHA Standard methods for the examination of water and wastewater treatment. 20th Edition. Washington DC: American Public Health Association; 1998.

[153] Pretsch, E; Seibl, J; Simon, W. *Tabellen zur strukturaufklarung organischer verbindungen mit spektroskopischen methoden.* Berlin: Springer Verlag; 1981.

[154] Alaton, IA; Balcioglu, IA; Bahnemann, D. Advanced oxidation of reactive dyebath effluent: comparison of O_3, H_2O_2/UV-A and TiO_2/UV-A processes. *Water Research,* 2002 36, 1143-1154.

[155] Gutowska, A; Kaluzna-Czaplinska, J; Jozwiak WK. Degradation mechanism of Reactive Orange 113 dye by H_2O_2/Fe^{2+} and ozone in aqueous solution. *Dyes and Pigments,* 2006 *in press.*

[156] Feng, W; Nansheng, D; Helin H. Degradation mechanism of azo dye C. I. reactive red 2 by iron powder reduction and photooxidation in aqueous solutions. *Chemosphere,* 2000 41, 1233-1238.

[157] Kawai, T; Tsutsumi, K. Adsorption characteristic of surfactants and phenol on modified zeolites from their aqueous solutions. *Colloid Polymeric Science,* 1995 273, 787-792.

[158] Kušić, H; Koprivanac, N; Locke, BR. Decomposition of phenol by hybrid gas/liquid electrical discharge reactors with zeolite catalysts. *Journal of Hazardous Materials,* 2005 125, 190-200.

[159] Daneshvar, N; Salari, D; Khataee AR. Photocatalytic degradation of azo dye acid red 14 in water: investigation of the effect of operational parameters. *Journal of Photochemistry and Photobiology A: Chemistry,* 2003 157, 111-116.

[160] Bilgi, S; Demir, C. Identification of photooxidation degradation products of C.I. Reactive Orange 16 dye by gas chromatography - mass spectrometry. *Dyes and Pigments,* 2005 66, 69-76.

[161] Hoigné, J; Bader, H. Rate constants of reactions of ozone with organic and inorganic compounds in water—II: Dissociating organic compounds. *Water Research,* 1983 17 (2), 185-194.

In: Industrial Pollution Including Oil
Editor: Harry Newbury and William De Lorne
ISBN 978-1-60456-917-9

Short Communication

EFFECTS OF OIL POLLUTION ON NICKEL AND VANADIUM CONCENTRATIONS IN SOIL AND IN THE TISSUES OF THE DESERT LIZARD *ACANTHODACTYLUS SCUTELLATUS* AT KUWAIT'S GREATER AL-BURGAN OIL FIELD

M. A. Al-Hashem[1*] and P. F. Brain[2]

[1]Department of Biological Sciences, Kuwait University, 5969, Safat, State of Kuwait
[2]Department of Biological Sciences, School of the Environment and Society, Swansea University, Swansea, SA2 8PP, UK

ABSTRACT

Contamination of soils by heavy metals from oil spills is a serious problem. The environment of the Arabian Gulf region was exposed to such factors by the 1991 oil spill and is now a focus of study. Sites with apparently different levels of pollution, namely tar mat, soot and clear sites were located at the Greater Al-Burgan oil field. Eight field soil samples were taken from each of the types along with control samples from outside the area in 2001. Ten *Acanthodactylus scutellatus* lizards (5 of each sex) were collected from each field study site, weighed, humanely killed and stored in jars in a freezer at - 20°C until analysis. All samples (both soil and lizard) were analyzed using the Inductively Coupled Plasma/Atomic Emission Spectrophotometry methodology. The mean nickel and vanadium concentrations were highest in the tar mat soils and lowest in the control samples. The mean nickel concentrations in whole body lizard tissue were high in reptiles from both the clear and tar mat sites but lowest at the control sites. Mean concentrations of vanadium were low in lizard tissues and did not significantly differ in animals taken from the different sites. The results confirm that the 1991 oil spill was still resulting in

* Tel: ++9654819371, Fax: ++9654847054; E-mail: alkawther5@yahoo.com

elevated levels of both heavy metals in contaminated areas by 2001. It also confirms that *Acanthodactylus scutellatus* is a useful bioindicator species in this desert environment for nickel but not vanadium.

Keywords: *Acanthodactylus scutellatus*; Kuwait; oil spill; heavy metals; nickel; vanadium; ICP/AES.

INTRODUCTION

The impact of oil well fires at Burgan and Ahmadi in the second Gulf war of 1990 on the surrounding desert ecosystems have been well documented (Al-Hassan, 1992; Omar *et al.*, 2000) and are described in detail elsewhere (Al-Hashem, 2006; Al-Hashem *et al.* 2007). The fires, resulting from the destruction of oil wells throughout Kuwait, produced a substantial cloud of oily pollutants. The country (and then progressively the whole region) was covered by an oily film from the growing black cloud. Dark clouds blocked the sunlight for months, resulting in air temperatures decreasing by an average of 10° C and even the sea temperature declined by several degrees. The temperatures around the blazing oil wells in Kuwait were very high. Their effects on soil reached some 300 m around the well centre (Al-Sayegh, 2000).

The oil fires caused the release of particles, organic and inorganic gases, hydrocarbons and oil droplets (Al-Hassan, 1992). Oil spills, aerosol deposits, and seawater use have all had adverse effects on the desert ecosystem. The explosion of oil wells in Burgan and Ahmadi produced enormous volumes of soot and unburned oil in the form of oil-mist that was carried to distant areas. In areas covered by oil-soot, a thin black crust of about 2-5 mm of slightly compacted superficial soil was formed on the surface. Unburned oil escaping from exploded oil wells accumulated in low-lying areas to form oil-logged soil where hydrocarbons penetrated 15 cm or more in the soil profile (Omar *et al.*, 2000).

The term 'heavy metals' refers to metallic elements including antimony, arsenic, barium, beryllium, boron, cadmium, chromium, cobalt, copper, lead, manganese, mercury, molybdenum, nickel, selenium, silver, thallium, tin, vanadium and zinc. Such metals only cause problems in organisms under defined circumstances. They have to come in contact with the organism to be of any biological consequence and be in a form in which they will be able to enter the organism (Adriano, 2001). This last-mentioned property is termed 'bioavailability' and refers to the property of the metal being in a condition where it can be taken up by the organism and react with its metabolic machinery (Campbell, 1995). Information is only provided here on the two metals (vanadium and nickel) which were found in samples in the present study.

Vanadium is a shiny, silvery metal with few naturally-occurring isotopes. It appears to be required in comparatively small quantities for normal growth and differentiation in all organisms but whether the element is essential and what its precise function is remains obscure (Golden and Golden, 1981; Nielsen, 1984). Vanadium is found in many foods, in deed, there are significant amounts in milk, sea-foods, cereals, and vegetables. The element has a natural affinity for fats and oils (cooking oils often have high concentrations of this metal). Although man-made emissions of vanadium into the environment have been increasing by burning of fossil fuels, the environmental behaviour of this element has rarely been studied. This paucity of studies is related to the absence of any obvious environmental problem caused by this element and, unlike mercury the lack of sensitive analytical methods for its analysis. There has been even less information available regarding vanadium's effects on behaviour and distribution in wild animals (Sperling *et al.*, 2000). Saeki *et al.* (1999) observed high concentrations of vanadium in the liver, hair and bone of Northern fur seals (*Callorhinus ursinus*). This increase was caused by an elevated retention in the nuclear and mitochondrial fraction of the cells in these tissues.

Nickel is a silvery-white, lustrous, malleable and ductile metal that resists corrosion and is soluble in all acids except concentrated nitric (Pais *et al.*, 1997; Adriano, 2001). Although nickel is toxic to organisms at elevated concentrations, trace amounts of this element are required for several biological processes.

Tochimoto *et al.* (2003) found that nickel concentration in larvae of aquatic insect (*Stenopsyche marmorata*) is mainly taken from their food source. Herkovits *et al.* (2000) concluded that the exposure threshold for nickel to exert acute lethal effects on *Bufo arenarum* embryos is 96 h of treatment.

Contamination from heavy metals is a serious problem in most countries of the world. The environment of the Arabian Gulf region has been a subject of study in this respect in recent years due to the 1991 oil spill. Direct and indirect methods have been developed to detect heavy metals quantitatively and qualitatively and to evaluate the level of pollution in this area. Direct methods are applied to soil and water; indirect methods use plants and animals as bioindicators. Reptiles have rarely been used as bioindicators of pollution. It can be argued, however, that lizards would be good indicators of the quality of terrestrial habitats because of their relatively restricted mobility (Lambert, 1993). As invertebrates form a substantial part of the diet of most lizards, they are likely to concentrate metals that have been incorporated in the bodies of insects by eating plants. Heavy metals thus accumulate through the ingestion of contaminated food, and also by the incidental ingestion of soil (Loumbourdis, 1997). In the few studies using reptiles (especially turtles, e.g. Albers *et al.*, 1986) accumulation of high concentrations of these metals, confirms that reptiles are likely to be good pollution bioindicators.

Very few studies analyzed heavy metal concentrations in lizards. Beck (1956) included copper concentrations found in lizard livers. Lance *et al.* (1995) analyzed blood plasma in 16 different species of lizards to determine normal levels of zinc that were present. There are also heavy metal studies involving lizards from Brazil (Schmidt, 1984), India (Kaur, 1988) and Puerto Rico (Burger *et al.*, 1992). Loumbourdis (1997) analyzed aluminum, barium, cadmium, cesium, chromium, cobalt, copper, lead, manganese, molybdenum, nickel, rubidium, strontium and zinc concentrations in the whole body (minus liver and digestive tract) and the liver of the lizard *A. stellio stellio* in a high altitude urban site and a low-altitude agricultural area in Greece. This study was ground-breaking in that so many heavy metals were analyzed and lizards were used as the focus animals.

The Greater Al-Burgan oil field has an area of 349.65 km^2 and lies 20 kilometers to the south of Kuwait City. The land is solely used for oil and gas production and is fenced to prevent unauthorized entry. Types of contaminated soils were categorized according to the ground observations. The types used were designated as tar mat, soot and clear. The tar mat areas had a soil surface that had been solidified by oil, forming a layer about 1 cm thick that could be peeled off the underlying clean soil. The soot areas were found within the upper layer of soil and could be a 1-8 mm in depth. Contamination is, however, sometimes continuous and at other times discontinuous. The clear sites had no visual evidence of soil pollution. Two sites for each category of contaminated soils were located. The control area (Sulaibiya) is an Agriculture Research Station at Kabd which was established in 1975. The research station covers a total area of 20 km^2, being 4 km (east to west) x 5 km (north to south). Sulaibiya station is fenced and permission is required to enter the site.

Materials and Methods

Eight soil samples (initial weight for each is 500g) from each of the two control, clear, soot and tar mat sites were taken from the field, air dried and sieved using 2mm mesh and kept in sealed bottles. 1g of each sample was weighed and transferred into 100 ml conical flask before adding 30 ml of 37% concentrated (12N) hydrochloric acid (Fluka Wacker Ghemie GmbH, Munich, Germany). The sample was digested at a temperature near 200°C where it did not boil vigorously and froth was not formed. Digestion continued for 2 – 3 hours until the volume was reduced to 3 ml. The resultant solution was cooled and then filtered through Whatman No. 1 filter paper (Fluka Wacker Ghemie GmbH, Munich, Germany) into 25 ml volumetric flask (while transferring from conical to volumetric flask, the sides of the conical flask were washed down using distilled water which diluted the acid). The elute was analyzed using the Inductively Coupled Plasma/Atomic Emission Spectrophotometry (ICP/AES) methodology.

Ten lizards (5 of each sex) from each study site were collected from the field, weighed, humanely killed and kept in jars in a freezer at - 20°C. They were then freeze-dried and weighed. The samples were ground and kept in sealed bottles for further analysis.

One gram of the ground sample was taken in 100 ml conical flask. 20 ml of 70% analytical grade concentrated (16N) nitric acid (Fluka Wacker Ghemie GmbH, Munich, Germany) and 10 ml analytical grade perchloric acid (S.D. Fine Chemical Ltd. Mumbai, India) were added. The sample was digested without frothing and vigorous boiling using a temperature just below 200°C until the sample was reduced to 3 ml. The digested sample was then allowed to cool and made up to a volume of 25 ml using distilled water. The sample was then analyzed using the ICP/AES methodology. Data analysis was performed using the one-way analysis of variance (ANOVA) parametric test to test whether there is difference in the mean of heavy metal concentrations over the four study sites.

Results

The data obtained for heavy metal concentrations in soil are shown in table 1.

ANOVA test showed that there is no significant difference in nickel concentrations in soil between the different study sites ($F_{3, 4} = 5.62$, $P = 0.06$). There was no difference in nickel concentration in soil between the control and clear sites ($F_{3, 4} = 3.26$, $P = 0.21$). A significant difference was observed between the control and soot sites ($F_{3, 4} = 23.5$, $P < 0.04$) and the control and tar mat sites ($F_{3, 4} = 21.9$, $P < 0.04$). ANOVA test showed that there is no difference in vanadium concentrations in soil between the different study sites ($F_{3, 4} = 6.12$, P

= 0.05). No significant difference was realized in vanadium concentration between the control and clear sites ($F_{3,4}$ = 1.88, P = 0.30) and the control and soot sites ($F_{3,4}$ = 11.4, P = 0.07). A significant difference was observed between the control and tar mat sites ($F_{3,4}$ = 21.8, $P <$ 0.04).

Table 1. Mean (± s.d.) heavy metal concentrations (ppm) in soil samples

Location (N =16)	Nickel	Vanadium
Control	17.9 ± 0.64	20.3 ± 0.49
Clear	21.2 ± 2.55	22.9 ± 2.69
Soot	22.9 ± 1.34	22.8 ± 0.92
Tar mat	24.9 ± 2.05	27.9 ± 2.26

The mean nickel concentrations were highest in the tar mat soils and lowest in the control soils. Mean concentrations of vanadium were also highest in the tar mat soils but there were similar mean concentrations in the soot and clear soils. The lowest mean concentration of vanadium was in the control soils.

The data obtained for heavy metal concentrations in lizard tissues are shown in table 2.

Table 2. Mean (± s.d.) heavy metal concentrations (ppm) in total lizard tissues

Location (N =10)	Nickel	Vanadium
Control	3.1 ± 0.14	0.55 ± 0.14
Clear	4.8 ± 1.34	0.83 ± 0.04
Soot	3.6 ± 1.13	0.53 ± 0.11
Tar mat	4.5 ± 0.18	0.84 ± 0.25

ANOVA test showed that there is no significant difference in nickel concentrations in lizard tissues between the different study sites ($F_{3,4}$ = 1.60, P = 0.32). Nickel concentrations in lizard tissues did not show any difference between the control and clear sites ($F_{3,4}$ = 3.16, P = 0.21) and the control and soot sites ($F_{3,4}$ = 0.31, P = 0.63). A significant difference was evident between the control and tar mat sites ($F_{3,4}$ = 73.8, $P <$ 0.01). ANOVA test showed that there is no difference in vanadium concentrations in lizard tissues between the different study sites ($F_{3,4}$ = 2.36, P = 0.21). No significant difference was observed in vanadium concentration of lizard tissues between the control and clear sites ($F_{3,4}$ = 7.1, P = 0.21), the control and soot sites ($F_{3,4}$ = 0.04, P = 0.86) and the control and tar mat sites ($F_{3,4}$ = 1.86, P = 0.31). The mean nickel concentrations in lizard whole body tissue were highest at both the clear and tar mat sites and values were lowest at the control sites. Mean concentrations of vanadium in lizard tissues were similar at the tar mat and clear sites but were low at both soot and control sites.

CONCLUSION

The statistical analysis of the results of the present study showed some variation in nickel content between the sites in both soil and lizard tissues. Variation in vanadium content was only evident in soil but did not differ in lizard tissues between the different study sites.

Nickel concentrations in the liver and carcass of the lizard *A. stellio stellio* were found to be 3.60 $\pm$ 3.11 and 33.83 $\pm$ 22.12 (ppm) respectively (Loumbourdis, 1997). In the present study, the highest mean nickel concentration of the sand lizard *A. scutellatus* was found to be 4.80 $\pm$ 1.34 (ppm) for the whole body tissues. Thus the total concentration in the carcass of nickel in *A. stellio stellio* was much higher than that of the present study in *A. scutellatus*. Feeding habits may have a great effect on the exposure of organisms to metals as well as local concentrations of the metal. The overall few reptile species studied in the literature regarding nickel and vanadium do not facilitate easy conclusions concerning dynamics of incorporation of these elements.

Analytical chemistry is almost always required at some point to assess or predict toxic hazards of heavy metals to wildlife. Chemical analysis often serves to identify those contaminants present, and their respective concentrations and may often provide for some diagnosis of an observed wildlife problem. In many instances, however, chemical residue data obtained from field samples will prompt further laboratory and field investigations to determine if any undetected biological effects are occurring.

This chapter provided some information about the presence of nickel and vanadium in the tissues of *A. scutellatus* and although the values of these metals are relatively modest, they might well have severe effects on lizard's survival and health.

REFERENCES

Adriano, D.C., 2001. Trace Elements in Terrestrial Environments: *Biogeochemistry, Bioavailability, and Risks of Metals.* New York, Springer.

Albers, P.H., Sileo, L. and Mulhern, B.M., 1986. Effects of environmental contaminants on snapping turtles of a tidal wetland. *Archive of Environmental Contamamination and Toxicology*, 15, 39-49.

Al-Hashem, M.A., 2006. *Effects of Oil Pollution on the Ecology Behaviour and Physiology of the Sand Lizard* (Acanthodactylus scutellatus) in Kuwait. PhD thesis, Wales: University of Wales Swansea.

Al-Hashem, M., Brain, P. and Omar, S., 2007, Effects of Oil Pollution at Kuwait's Greater Al-Burgan Oil Field on Polycyclic Aromatic Hydrocarbon Concentrations in the Tissues of Desert Lizard Acanthodactylus scutellatus and their Ant Prey. *Ecotoxicology*, 16, 551-555.

Al-Hassan, J.M., 1992. *Iraqi Invasion of Kuwait: An Environmental Catastrophe*. Kuwait: Fahad Al-Marzouk Publishers.

Al-Sayegh, A., 1993. The environmental pollution in Kuwait after the Gulf War of 1991 http://alsayegh.Homepage.com/links/kwpollution.html.

Beck, A.B., 1956. The copper content of the liver and blood of some vertebrates. *Australian Journal of Zoology*, 4, 1-18.

Burger, J., Cooper, K., Saliva, J., Gochfield, D., Lipsky, D. and Gochfield, M., 1992. Mercury bioaccumulation in organisms from three Puerto Rican estuaries. *Environmental Monitoring Assessment,* 22, 181-197.

Campbell, P.G.C., 1995. In: Tessier, A., and Turner, D.R. (Eds) *Metal Speciation and Bioavailability in Aquatic Systems.* New York: John Wiley and Sons.

Golden, M.H.N. and B.E. Golden, 1981. Trace elements : Potential importance in human nutrition with particular reference to zinc and vanadium. *British Medical Bulletin*, 37, 31-36.

Herkovits, J., C.S. Pérez-Coll and F. D. Herkovits, 2000. Evaluation of nickel-zinc interactions by means of bioassays with amphibian embryos. *Ecotoxicology and Environmental Safety,* 45, 266-273.

Kaur, S., 1988. Lead in the scales of cobras and wall lizards from rural and urban areas of Punjab, India. *Science of Total Environment*, 77, 289-290.

Lambert, M.R.K., 1993. Effects of DDT ground-spraying against tsetse flies on lizards in NW Zimbabwe. *Environmental Pollution*, 82, 231-237.

Lance, V.A., T. Cort, J. Masuoka, R. Lawson and P. Saltman 1995. Unusually high zinc concentrations in snake plasma, with observations on plasma zinc concentrations in lizards, turtles and alligators. *Journal of Zoology of London*, 235, 577-585.

Loumbourdis, N.S., 1997. Heavy metal contamination in a lizard, Agama stellio stellio, compared in urban, high altitude and agricultural, low altitude areas of north Greece. *Bullettin of Environmental Contamination and Toxicology*, 58, 945-952.

Nielsen, F.H. (1984) Ultratrace elements in nutrition. *Annual Review of Nutrition*, 4, 21-41.

Omar, S. A., Y. Al-Mutawa and S. Zaman, 2000. *Vegetation of Kuwait : A Comprehensive Illustrative Guide to the Flora and Ecology of the Desert of Kuwait.* Kuwait: Kuwait Institute for Scientific Research.

Pais, I., J. Benton and Jr. Jones, 1997. *The Handbook of Trace Elements.* Boca Raton, Florida: CRC.

Saeki, K., M. Nakajima, K. Noda, T. R. Loughlin, N. Baba, M. Kiyota, R. Tatsukawa and D. G. Calkins, 1999. Vanadium accumulation in pinnipeds. *Archive of Environmental Contamination and Toxicology*, 36, 81-86.

Schmidt, J., 1984. Heavy metal analyses in Hemidactylus mabouia (Geckonidae) as a method to classify urban environmental quality. *Amazoniana,* 9, 35-42.

Sperling, K.R., R. Bahr and J. Ott, 2000. Vanadium in marine mussels and algae. *Fresenius Journal of Analytical Chemistry*, 366, 132-136.

Tochimoto, H., T. Maki, M. Afzal and S. Tanabe, 2003. Accumulation of trace metals in aquatic insect Stenopsyche marmorata Navas transferred in streams. *Ecotoxicology and Environmental Safety*, 56, 256-264.

Reviewed by: ***Dr Peter Chevins***
Department of Life Sciences
Keele University,Keele
Staffordshire, UK
p.f.d.chevins@biol.keele.ac.uk

INDEX

A

B

C

D

E

F

G

I

J

K

L

M

N

O

P

Q

R

S

T

U

V

W

X

Y

Z